Synchrotron Radiation in Materials Research

Synchrotron Radiation in Materials Research

Symposium held November 28-30, 1988, Boston,
Massachusetts, U.S.A.

EDITORS:

Roy Clarke
University of Michigan, Ann Arbor, Michigan, U.S.A.

John Gland
University of Michigan, Ann Arbor, Michigan, U.S.A.

John H. Weaver
University of Minnesota, Minneapolis, Minnesota, U.S.A.

MRS MATERIALS RESEARCH SOCIETY
Pittsburgh, Pennsylvania

CAMBRIDGE
UNIVERSITY PRESS

University Printing House, Cambridge CB2 8BS, United Kingdom

One Liberty Plaza, 20th Floor, New York, NY 10006, USA

477 Williamstown Road, Port Melbourne, VIC 3207, Australia

314-321, 3rd Floor, Plot 3, Splendor Forum, Jasola District Centre, New Delhi - 110025, India

79 Anson Road, #06-04/06, Singapore 079906

Cambridge University Press is part of the University of Cambridge.

It furthers the University's mission by disseminating knowledge in the pursuit of education, learning and research at the highest international levels of excellence.

www.cambridge.org
Information on this title: www.cambridge.org/9781558990166

Materials Research Society
506 Keystone Drive, Warrendale, PA 15086
http://www.mrs.org

© Materials Research Society 1989

First published 1989
First paperback edition 2012

Single article reprints from this publication are available through
University Microfilms Inc., 300 North Zeeb Road, Ann Arbor, MI 48106

CODEN: MRSPDH

A catalogue record for this publication is available from the British Library

ISBN 978-1-558-99016-6 Hardback
ISBN 978-1-107-41077-0 Paperback

Cambridge University Press has no responsibility for the persistence or
accuracy of URLs for external or third-party internet websites referred to in
this publication, and does not guarantee that any content on such websites is,
or will remain, accurate or appropriate.

Contents

INTRODUCTION

PART I: STRUCTURE OF SURFACES, INTERFACES AND MULTILAYERS

PART II: ABSORPTION SPECTROSCOPY AND ELECTRONIC STRUCTURE

*Invited Paper

*Invited Paper

PART VI: ABSTRACTS OF UNPUBLISHED PAPERS

*Invited Paper

Preface

This volume is a collection of invited and contributed papers presented at the Symposium on Synchrotron Radiation at the Materials Research Society Fall Meeting held in Boston, Massachusetts on November 28-30, 1988. We are now seeing the beginning of a new era in synchrotron radiation research, one in which very high brightness sources lead to experiments that were hitherto inaccessible. The construction of advanced, insertion-device-based sources such as Argonne's Advanced Photon Source, Japan's 6 GeV Source, and the European Synchrotron Radiation Facility in Grenoble, and at lower energy, the Berkeley Advanced Light Source, promises exciting opportunities for the next decade. Indeed, many of the presentations in this symposium referred to the challenges associated with the effective utilization of these new ultrahigh-brilliance sources. The broad range of topics covered during the symposium, including crystallography, electronic structure, x-ray microscopy, and lithography, ensures that the diverse materials community will be heavy users of synchrotron sources.

March, 1989
Roy Clarke
John Gland
John Weaver

Acknowledgments

It is our pleasure to acknowledge financial support of this Symposium from the following sources: The Advanced Photon Source Users Organization, Exxon Research and Engineering Company, Mobil Research and Development Corporation, and the International Centre for Diffraction Data.

We wish to thank all the Symposium contributors and the session chairs for helping to make the meeting a successful and stimulating forum for new ideas. Our sincere thanks go also to those who found time to review the manuscripts for the conference proceedings.

Finally, we express our appreciation to Rohn McNamer and Lauren Myers for their invaluable help in the organization of the Symposium and in the preparation of the proceedings.

MATERIALS RESEARCH SOCIETY SYMPOSIUM PROCEEDINGS

ISSN 0272 - 9172

Volume 26—Scientific Basis for Nuclear Waste Management VII, G. L. McVay, 1984, ISBN 0-444-00906-X

Volume 27—Ion Implantation and Ion Beam Processing of Materials, G. K. Hubler, O. W. Holland, C. R. Clayton, C. W. White, 1984, ISBN 0-444-00869-1

Volume 28—Rapidly Solidified Metastable Materials, B. H. Kear, B. C. Giessen, 1984, ISBN 0-444-00935-3

Volume 29—Laser-Controlled Chemical Processing of Surfaces, A. W. Johnson, D. J. Ehrlich, H. R. Schlossberg, 1984, ISBN 0-444-00894-2

Volume 30—Plasma Processing and Synthesis of Materials, J. Szekely, D. Apelian, 1984, ISBN 0-444-00895-0

Volume 31—Electron Microscopy of Materials, W. Krakow, D. A. Smith, L. W. Hobbs, 1984, ISBN 0-444-00898-7

Volume 32—Better Ceramics Through Chemistry, C. J. Brinker, D. E. Clark, D. R. Ulrich, 1984, ISBN 0-444-00898-5

Volume 33—Comparison of Thin Film Transistor and SOI Technologies, H. W. Lam, M. J. Thompson, 1984, ISBN 0-444-00899-3

Volume 34—Physical Metallurgy of Cast Iron, H. Fredriksson, M. Hillerts, 1985, ISBN 0-444-00938-8

Volume 35—Energy Beam-Solid Interactions and Transient Thermal Processing/1984, D. K. Biegelsen, G. A. Rozgonyi, C. V. Shank, 1985, ISBN 0-931837-00-6

Volume 36—Impurity Diffusion and Gettering in Silicon, R. B. Fair, C. W. Pearce, J. Washburn, 1985, ISBN 0-931837-01-4

Volume 37—Layered Structures, Epitaxy, and Interfaces, J. M. Gibson, L. R. Dawson, 1985, ISBN 0-931837-02-2

Volume 38—Plasma Synthesis and Etching of Electronic Materials, R. P. H. Chang, B. Abeles, 1985, ISBN 0-931837-03-0

Volume 39—High-Temperature Ordered Intermetallic Alloys, C. C. Koch, C. T. Liu, N. S. Stoloff, 1985, ISBN 0-931837-04-9

Volume 40—Electronic Packaging Materials Science, E. A. Giess, K.-N. Tu, D. R. Uhlmann, 1985, ISBN 0-931837-05-7

Volume 41—Advanced Photon and Particle Techniques for the Characterization of Defects in Solids, J. B. Roberto, R. W. Carpenter, M. C. Wittels, 1985, ISBN 0-931837-06-5

Volume 42—Very High Strength Cement-Based Materials, J. F. Young, 1985, ISBN 0-931837-07-3

Volume 43—Fly Ash and Coal Conversion By-Products: Characterization, Utilization, and Disposal I, G. J. McCarthy, R. J. Lauf, 1985, ISBN 0-931837-08-1

Volume 44—Scientific Basis for Nuclear Waste Management VIII, C. M. Jantzen, J. A. Stone, R. C. Ewing, 1985, ISBN 0-931837-09-X

Volume 45—Ion Beam Processes in Advanced Electronic Materials and Device Technology, B. R. Appleton, F. H. Eisen, T. W. Sigmon, 1985, ISBN 0-931837-10-3

Volume 46—Microscopic Identification of Electronic Defects in Semiconductors, N. M. Johnson, S. G. Bishop, G. D. Watkins, 1985, ISBN 0-931837-11-1

Volume 47—Thin Films: The Relationship of Structure to Properties, C. R. Aita, K. S. SreeHarsha, 1985, ISBN 0-931837-12-X

Volume 48—Applied Materials Characterization, W. Katz, P. Williams, 1985, ISBN 0-931837-13-8

Volume 49—Materials Issues in Applications of Amorphous Silicon Technology, D. Adler, A. Madan, M. J. Thompson, 1985, ISBN 0-931837-14-6

Volume 99—High-Temperature Superconductors, M. B. Brodsky, R. C. Dynes, K. Kitazawa, H. L. Tuller, 1988, ISBN 0-931837-67-7

Volume 100—Fundamentals of Beam-Solid Interactions and Transient Thermal Processing, M. J. Aziz, L. E. Rehn, B. Stritzker, 1988, ISBN 0-931837-68-5

Volume 101—Laser and Particle-Beam Chemical Processing for Microelectronics, D.J. Ehrlich, G.S. Higashi, M.M. Oprysko, 1988, ISBN 0-931837-69-3

Volume 102—Epitaxy of Semiconductor Layered Structures, R. T. Tung, L. R. Dawson, R. L. Gunshor, 1988, ISBN 0-931837-70-7

Volume 103—Multilayers: Synthesis, Properties, and Nonelectronic Applications, T. W. Barbee Jr., F. Spaepen, L. Greer, 1988, ISBN 0-931837-71-5

Volume 104—Defects in Electronic Materials, M. Stavola, S. J. Pearton, G. Davies, 1988, ISBN 0-931837-72-3

Volume 105—SiO_2 and Its Interfaces, G. Lucovsky, S. T. Pantelides, 1988, ISBN 0-931837-73-1

Volume 106—Polysilicon Films and Interfaces, C.Y. Wong, C.V. Thompson, K-N. Tu, 1988, ISBN 0-931837-74-X

Volume 107—Silicon-on-Insulator and Buried Metals in Semiconductors, J. C. Sturm, C. K. Chen, L. Pfeiffer, P. L. F. Hemment, 1988, ISBN 0-931837-75-8

Volume 108—Electronic Packaging Materials Science II, R. C. Sundahl, R. Jaccodine, K. A. Jackson, 1988, ISBN 0-931837-76-6

Volume 109—Nonlinear Optical Properties of Polymers, A. J. Heeger, J. Orenstein, D. R. Ulrich, 1988, ISBN 0-931837-77-4

Volume 110—Biomedical Materials and Devices, J. S. Hanker, B. L. Giammara, 1988, ISBN 0-931837-78-2

Volume 111—Microstructure and Properties of Catalysts, M. M. J. Treacy, J. M. Thomas, J. M. White, 1988, ISBN 0-931837-79-0

Volume 112—Scientific Basis for Nuclear Waste Management XI, M. J. Apted, R. E. Westerman, 1988, ISBN 0-931837-80-4

Volume 113—Fly Ash and Coal Conversion By-Products: Characterization, Utilization, and Disposal IV, G. J. McCarthy, D. M. Roy, F. P. Glasser, R. T. Hemmings, 1988, ISBN 0-931837-81-2

Volume 114—Bonding in Cementitious Composites, S. Mindess, S. P. Shah, 1988, ISBN 0-931837-82-0

Volume 115—Specimen Preparation for Transmission Electron Microscopy of Materials, J. C. Bravman, R. Anderson, M. L. McDonald, 1988, ISBN 0-931837-83-9

Volume 116—Heteroepitaxy on Silicon: Fundamentals, Structures,and Devices, H.K. Choi, H. Ishiwara, R. Hull, R.J. Nemanich, 1988, ISBN: 0-931837-86-3

Volume 117—Process Diagnostics: Materials, Combustion, Fusion, K. Hays, A.C. Eckbreth, G.A. Campbell, 1988, ISBN: 0-931837-87-1

Volume 118—Amorphous Silicon Technology, A. Madan, M.J. Thompson, P.C. Taylor, P.G. LeComber, Y. Hamakawa, 1988, ISBN: 0-931837-88-X

Volume 119—Adhesion in Solids, D.M. Mattox, C. Batich, J.E.E. Baglin, R.J. Gottschall, 1988, ISBN: 0-931837-89-8

Volume 120—High-Temperature/High-Performance Composites, F.D. Lemkey, A.G. Evans, S.G. Fishman, J.R. Strife, 1988, ISBN: 0-931837-90-1

Volume 121—Better Ceramics Through Chemistry III, C.J. Brinker, D.E. Clark, D.R. Ulrich, 1988, ISBN: 0-931837-91-X

Volume 122—Interfacial Structure, Properties, and Design, M.H. Yoo, W.A.T. Clark, C.L. Briant, 1988, ISBN: 0-931837-92-8

MATERIALS RESEARCH SOCIETY SYMPOSIUM PROCEEDINGS

Tungsten and Other Refractory Metals for VLSI Applications, R. S. Blewer, 1986; ISSN 0886-7860; ISBN 0-931837-32-4

Tungsten and Other Refractory Metals for VLSI Applications II, E.K. Broadbent, 1987; ISSN 0886-7860; ISBN 0-931837-66-9

Ternary and Multinary Compounds, S. Deb, A. Zunger, 1987; ISBN 0-931837-57-x

Tungsten and Other Refractory Metals for VLSI Applications III, Victor A. Wells, 1988; ISSN 0886-7860; ISBN 0-931837-84-7

Atomic and Molecular Processing of Electronic and Ceramic Materials: Preparation, Characterization and Properties, Ilhan A. Aksay, Gary L. McVay, Thomas G. Stoebe, 1988; ISBN 0-931837-85-5

Materials Futures: Strategies and Opportunities, R. Byron Pipes, U.S. Organizing Committee, Rune Lagneborg, Swedish Organizing Committee, 1988; ISBN 0-55899-000-3

Tungsten and Other Refractory Metals for VLSI Applications IV, Robert S. Blewer, Carol M. McConica, 1989; ISSN: 0886-7860; ISBN: 0-931837-98-7

Introduction

THE ARGONNE ADVANCED PHOTON SOURCE

DAVID E. MONCTON
Argonne National Laboratory, Argonne, IL 60439 and Exxon Research and Engineering Co., Annandale, NJ 0880

Argonne National Laboratory is preparing to build a new synchrotron radiation source, the 7-GeV Advanced Photon Source (APS), that will provide the world's most brilliant x-ray beams for research. The APS will produce x-rays for materials research, condensed-matter physics, chemistry, and biological and medical studies by researchers from industry, universities, and national laboratories.

Currently the APS project has entered the detailed design phase necessary to begin construction in October 1990 and operations in 1995. As shown in Figure 1, this synchrotron x-ray facility will include a 40-period, 7-GeV positron storage ring containing 34 five-meter insertion device locations and a similar number of bending magnet ports. The ring emittance is designed to be 7 nm-rad, more than ten times smaller than the NSLS (National Synchrotron Light Source) X-ray ring. At the operating energy of 7 GeV, a single undulator magnet of 3.3-cm magnetic period will provide radiation with a brilliance of approximately 10^{19} photons/sec-mm^2 0.1% bw. As shown in Figure 2, the first harmonic is tunable from 5 to 15 keV and the third harmonic is tunable from 15 to 45 keV. Thus, this device will produce an optimal beam for almost all hard x-ray science. An overview of the technical design is provided in Reference [1]. See Figure 2 for a comparison of the spectral output from APS undulators with the spectra from other synchrotron radiation sources.

An important aspect of the facility is the provision for substantial space for the beamline facilities, as shown in Figure 1. The experimental hall will be surrounded by eight laboratory/office modules providing a total of 256 offices and 64 600-ft^2 laboratories for the experimental and support staff. It is expected that the majority of users (say 75%) will come from outside institutions. A team of collaborating researchers will construct and operate each sector of the facility; a sector consists of an insertion device (wiggler or undulator) beamline and a companion bending magnet beamline. Policy guidelines are now being developed to govern user access, and it is expected that proposals will be sought in early 1990. Reference [2] below contains technical information for prospective users.

REFERENCES

[1] G. K. Shenoy and D. E. Moncton, Nucl. Instrum. Methods, A266, 38 (1988).

[2] G. K. Shenoy, P. J. Viccaro, and D. M. Mills, Argonne National Laboratory Report ANL-88-9, 1988.

4

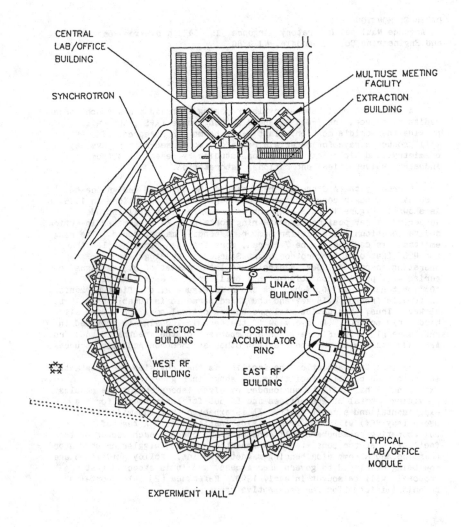

CENTRAL
LAB/OFFICE
BUILDING

MULTIUSE MEETING
FACILITY

EXTRACTION
BUILDING

SYNCHROTRON

LINAC
BUILDING

INJECTOR
BUILDING

POSITRON
ACCUMULATOR
RING

WEST RF
BUILDING

EAST RF
BUILDING

TYPICAL
LAB/OFFICE
MODULE

EXPERIMENT HALL

Figure 1.   A plan view of the conceptual design for the Argonne Advanced
Photon Source.

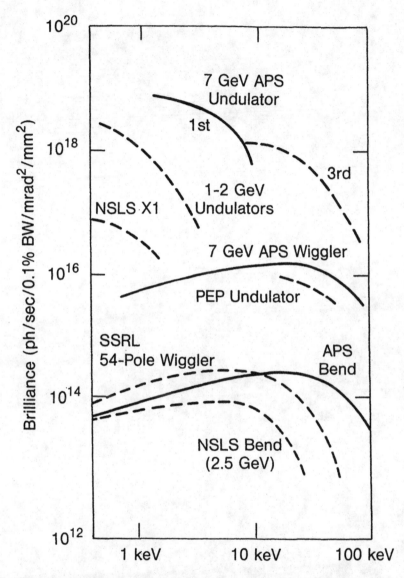

Figure 2.   Brightness of various sources as a function of energy. The
7-GeV APS undulator is the λ_o=3.3 cm device discussed in the
text; the curves represent the envelopes of the first and
third harmonics, which are tunable by gap variation. Similar
curves are also shown for the undulators on the 1-2 GeV
Synchrotron Radiation Source at Berkeley, the X-1 undulator at
the National Synchrotron Light Source (NSLS), and the
λ_o=7.7 cm undulator on the PEP ring (14.5 GeV) at Stanford.

PART I

Structure of Surfaces, Interfaces and Multilayers

FLUCTUATING INTERFACES: A SERIES OF SYNCHROTRON X-RAY SCATTERING STUDIES OF INTERACTING STACKED MEMBRANES

C.R. SAFINYA
Exxon Research and Engineering Co. Annandale, N.J. 08801, U.S.A.

ABSTRACT

In this paper we concentrate on fluctuation phenomena encountered in interacting multilayered fluid membranes using synchrotron x-ray scattering as the primary tool. These systems consisting of surfactant or lipidic surfaces, are prototype models for understanding the statistical behavior of fluctuating surfaces embedded in three-dimensions. In elucidating the nature of fluid membranes we stress a unique intrinsic property: in contrast to usual surfaces with finite shear moduli, where surface tension plays a central role, the free energy of these essentially tensionless surfaces is governed by their geometrical shape and its fluctuations. We present data in a new regime of stability for very dilute and flexible membranes with interlayer separations of order hundreds of Angstroms. This is in contrast to most rigid membranes where the interlayer interactions are dominated by detailed microscopic interactions such as hydration and van der Waals. We show that the stability of these dilute lamellar phases is associated with violent out-of-plane fluctuations of the membranes giving rise to an effectively large long-range repulsive interaction theoretically elucidated by Helfrich. <u>Because of its entropic origin, this interaction is universal</u> and we present data for two different surfactant systems. Finally, we show that this new regime is distinct from the classical regime in which largely separated membrane sheets are stabilized because of their mutual electrostatic repulsion.

INTRODUCTION

The understanding of the surface properties of both fluid and more ordered membranes has recently attracted much experimental and theoretical attention [1-10]. This focus of attention stems primarily from the inherent interest in elucidating the statistical physics of two-dimensional random surfaces. In some sense the physical properties of fluid membranes are unique: in contrast to usual surfaces with finite shear moduli with surface tension playing a central role, the free energy of these essentially tensionless surfaces is governed by their geometrical shape and its fluctuations. The rigidity k_C associated with the restoring force to layer bending is then the important modulus which in many cases will determine the physical state of the membrane. From a biophysical view point the physical nature of a fluid membrane surface may in some cases have a profound influence on the precise mechanism of membrane-membrane interactions; which influence processes such as cell-cell contact.

In usual multilayer systems consisting of stacks of alternating membranes (composed of a bimolecular sheet of a single type of lipid molecules) and water the bending modulus has been measured to be about 20 k_BT in the fluid L_α phase [7]. Presumably these interfaces are flat and posses macroscopic persistence lengths of order microns; futhermore, short wavelength thermal fluctuations are not expected to play an important role [8]. The assumption of flat interfaces is also born out by numerous experiments. For example, in studies [3,10] designed to measure the interaction of two such membranes the free energy is consistent with that between two infinitely flat charge neutral membranes interacting via electrodynamic (van der Waals) and hydration forces (classical regime). In these flat multimembrane systems the attractive van der Waals forces prevent the layers from separating more than typically 20 Å to 30 Å.

On the other hand, in the L_α phase of some quaternary and ternary systems [1,2,6] very large dilutions with intermembrane distances of several hundred ($\simeq$ 500 Å) Angstroms is possible. In our recent synchrotron studies [1,2] we demonstrated that the thermodynamic stability of these phases was due to an effectively long range repulsive interaction (which we refer to as undulation forces) arising from the mutual hinderance of fluctuating membranes with a very small bending rigidity $k_C \simeq k_BT$. Since this interaction [5] scales as k_C^{-1}, it completely overwhelms the attractive van der Waals interaction and stabilizes the membranes at large separations. Furthermore, because the undulation forces are entropically driven, this interaction is universal with its dependence determined entirely by geometric and elastic parameters such as the layer spacing and the layer rigidity. We have also shown that this new regime is distinct from the classical regime in which largely separated layered membranes are stabilized because of their mutual electrostatic repulsion. Our work is consistent with the theoretical work of Helfrich [5] which had suggested that violent out-of-plane fluctuations of membranes, if sufficiently strong, should give rise to an overwhelmingly large long range repulsive interaction leading to the stability of very dilute membranes.

EXPERIMENTAL RESULTS AND DISCUSSION

In this paper we present synchrotron x-ray scattering data of three multilayered membrane systems in the fluid L_α phase. In the first system the neutral membrane consists of water layers coated with a mixture of the surfactant sodium-dodecyl-sulfate (SDS) and cosurfactant (pentanol). The layers are separated by dodecane (i.e. an oil dilution study). In the second and third systems, the negatively charged membranes consist of a mixture of (SDS) and pentanol, while the solvent separating the membranes is either pure water or brine ($\simeq$ 0.5 mole ℓ^{-1} of NaCl).

The x-ray spectrometer used monochromator and analyzer elements which consisted of a double bounce Si(111) and a triple bounce Si(111) channel cut crystal set at 8 KeV in the non-dispersive configuration yielding a very sharp in-plane Gaussian resolution function with very weak tail scattering with half-width at half-maximum (HWHM) 8×10^{-5} Å $^{-1}$. A sharp Gaussian out-of-plane resolution function (HWHM = 10^{-3} Å) was achieved by use of extremely narrow slits. The samples were contained in sealed quartz capillaries with diameters of 1 and 2 mm, which yielded randomly oriented lamellar domains.

The first system studied was the lamellar L_α phase of the quaternary mixture of sodium dodecyl sulfate (SDS, the surfactant), pentanol (cosurfactant), water, and dodecane as a function of dodecane dilution [1]. Figure 1 shows a cut of the phase diagram mapped out by Roux and Bellocq [11] represented on a standard trianglular phase diagram in the plane with a constant water/SDS weight ratio equal to 1.55. This cut contains five one-phase regions: the microemulsion phases $\mu\epsilon_1$ and $\mu\epsilon_2$ and the liquid crystalline hexagonal (E), rectangular (R), and lamellar (L_α) phases.

The L_α phase consists of water layers (inverted membrane) embedded in oil, shown schematically in Fig. 1 with the fluid surfactant at the interface. This affords distinct advantages for studies of intermembrane

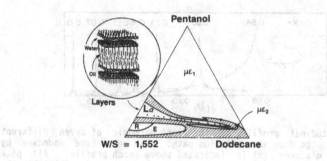

FIG. 1 A cut of the phase diagram of the quaternary mixture of SDS, pentanol, water, and dodecane shown in the plane with a constant water/SDS weight ratio=1.55 (Ref. 11). The dots in the lamellar phase labeled L_α correspond to the mixtures studied.

interactions. First, the water layers are charge neutral so that there are no intermembrane electrostatic interactions. Second, and very significantly for this plane of the phase diagram, we are able to dilute over an unusually large oil range between 0 and 80 wt.%, which corresponds to intermembrane separations between ~2.0 and larger than 20.0 nm. Third, previous work [12] on the curvature elasticity in similar dilute lyotropic L_α phases had indicated unusually small values of $k_C \sim k_BT$. These unique properties of these L_α phases thus allow for a comprehensive study of the long-range van der Waals and fluctuation induced steric interactions.

As was first pointed out by Peierls and Landau [13], three dimensional structures whose densities are periodic in only one direction such as the lyotropic L_α phase, are marginally stable to thermal fluctuations which destroy the long range order and replace the δ-function bilayer stacking structure factor peaks at $(0,0,q_m=mq_0=m2\pi/d)$ (d=interlayer spacing and $m=1,2,...$ is the harmonic number) by weaker algebraic singularities. Caillé [14] has derived the scattering cross section for a smectic-A liquid-crystal phase, which has the identical elastic free energy as that for the L_α phase:

$$F/V = \{B(\partial u/\partial z)^2 + K[(\partial^2 u/\partial x^2) + (\partial^2 u/\partial y^2)^2]\}/2 \qquad (1)$$

Here, $u(r)$ is the layer displacement in the z direction normal to the layers, and B and K are the bulk moduli for layer compression (erg/cm³) and layer curvature (erg/cm). K is related to the bending rigidity for a single bilayer: $K=k_C/d$.

The asymptotic form of the structure factor is described by power laws: $I(0,0,q_z) \sim |q_z - q_m|^{-2+\eta_m}$ and $I(q_\rho,0,q_m) \sim q_\rho^{-4+2\eta_m}$. q_ρ and q_z are components of the wave vector parallel and normal to the layers and:

$$\eta_m = m^2 q_0^2 k_BT/(8\pi(BK)^{0.5}), \qquad (2)$$

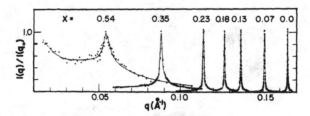

FIG. 2. Longitudinal profiles of the first harmonic of seven different mixtures along the dodecane dilution path. The percentage dodecane by weight of the mixtures (x) is indicated above each profile. All peak intensities are normalized to unity. The solid lines are fits by the Caillé power-law line shape Eq. (3).

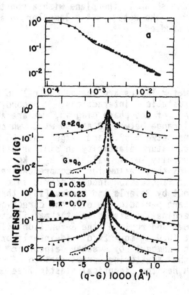

FIG. 3 (a) Profile of the first harmonic ($G = q_0$) for the mixture x=0.23 on a log-log scale which shows finite size rounding at small q-G followed by power-law behavior at larger q-G. (b) Profile of the first and second harmonics for the mixture x=0.07 on a logarithmic intensity scale. (c) Profile of the first harmonic ($G=q_0$) for three mixtures on a logarithmic intensity scale. All peak intensities are normalized to unity. The solid lines are fits by the Caillé power-law line shape [Eq. (3)].

is the exponent which describes the algebraic decay of layer correlations.

In the dodecane dilution series, we carried out detailed studies for 10 distinct mixtures: x=0, 0.07, 0.13, 0.18, 0.29, 0.35, 0.47, 0.54, and 0.62. Here x is the percent dodecane by weight of the mixtures. We show in Fig. 2 typical scattering profiles for longitudinal scans through the first harmonic for x between 0 and 0.54 in the L_α phase where the total layer spacing $d = 2\pi/q_0$ increases from 3.82 nm to 11.5 nm. In the mixtures studied, the dilution corresponds to a path where water layers with approximately constant thickness $\delta \sim 18$ Å are pushed apart with d_0/δ varying from ~1 to ~7. ($d_0 = d - \delta$ is the oil thickness between water layers.) A striking feature of the profiles in Fig. 2 is the tail scattering which becomes dramatically more pronounced as d increases. This effect is further elucidated in Fig.3(c), where we plot on a logarithmic intensity scale versus $q-q_0$ the scattering for three mixtures, x=0.07, 0.23, and 0.35. The significant difference in the profiles over the entire dilution range is now immediately clear. It is qualitatively clear that $\eta_1(d)$, which characterizes the asymptotic scattering profile and which is a measure of the ratio of the tail to peak intensity scattering, is increasing as d increases.

While the asymptotic forms for I(q) are simple, to quantitatively analyze the data we fit to the exact expression for the structure factor (also derived by Caillé [14]):

$$I(q) \sim \int dz \int d\rho \, s(z,\rho) e^{-R^2\pi/L^2} [(\sin qR)/qR] e^{iq_m z}. \qquad (3)$$

Here, $S(z,\rho) \sim (1/\rho)^{2\eta_m} \cdot \exp(-\eta_m[2\gamma + E1(\rho^2/4\lambda z)])$ is the layer-layer correlation function, [14] where $R^2 = z^2 + \rho^2$, γ is Euler's constant, E1(x) is the exponential integral function, and $\lambda = (K/B)^{0.5}$. The precise steps leading to Eq. (3) have been discussed elsewhere; [1,2] here, we summarize the essential points. First, I(q) incoporates a finite size effect because of the observed finite lamellar domain sizes typically between ~5000 and ~10000 Ångstroms (L^3 is the domain volume). Second, the structure factor has been powder averaged over all solid angles in reciprocal space because our samples consist of randomly oriented domains. The powder averaging results in the asymptotic power-law behavior for $I(q) \sim |q - q_m|^{-p}$ where, the exponent p is approximately equal to $1 - \eta$ [1,2].

The analysis consists of simultaneous fits to either two or three harmonics (depending on whether the third harmonic is observable) where the scaling of η_m ($=m^2 \cdot \eta_1$) is incorporated. The solid lines shown in Figs. 2 and 3 are the results of typical fits which give a satisfactory description of the scattering. From the fits we obtain an accurate measurement of $\eta_1(d)$.

We plot in Fig. 3(a) on a log-log scale the intensity versus $q-q_0$ for x=0.23, where the solid line is a result of the fit yielding $\eta_1 = 0.25$ and L=8640 Å. Two features in the scattering profile and the theoretical cross section are immediately apparent. While at large $q-q_0 \geq 2\pi/L$ the scattering exhibits power-law behavior $S(q) \sim |q - q_0|^{-p}$ with $P-1-\eta$, at $q \approx q_0$, the finite size effects round off the observed profile with characteristic width ~1/L. The data also confirm the scaling of η_m with m^2. We show in Fig. 3(b) the profiles and fits for the first and second harmonics for x=0.07 on a normalized logarithmic intensity scale. We find $\eta_1(q_0) = 0.14 \pm 0.02$, $\lambda = 8.59 \pm 1$ for the first harmonic and $\eta_2(2q_0) = 0.575 \pm 0.02$, $\lambda = 8.13 \pm 2$ for the second harmonic.

The value for k_c ($=K \cdot d$) is obtained from the measured values of λ ($=\sqrt{[K/B]}$) and η ($\sim 1/\sqrt{KB}$). k_c varies smoothly between 0.5 and 2 kBT. This value is significantly lower than that measured for most lipidic membrane systems and is precisely why fluctuations on short $\lambda < d$ and long $\lambda >> d$ length scales become important for these dilute membranes. In fact, for these very flexible surfaces, Helfrich has proposed that large, thermally induced, out-of plane layer fluctuations give rise to a repulsive interaction between membranes because of steric hindrance in multilayer systems. More generally, steric repulsions are known to be the dominant interactions associated with wandering walls of incommensurate phases [15].

The entropy difference between a "free" and a "bound" undulating membrane in a multilamellar phase has been calculated by Helfrich [5] with use of the Landau-deGennes elastic free energy per unit volume given by equation (1). This undulation induced interaction per unit surface is given by:

$$F_{und}/A = 0.33[(k_BT)^2/k_c(d-\delta)^2].$$ (4)

Thus, the Helfrich interaction is repulsive and very importantly long range. Moreover, it is inversely proportional to the elastic bending modulus for a single membrane k_c.

From the free energy per unit surface, we readily calculate the layer compressional modulus B and η as a function of d. Indeed, B is related to the second derivative of the free energy per unit volume:

$$B = d^2 \left| \frac{\partial^2 \frac{F}{V}}{\partial d^2} \right|_n$$ (5)

where n is the number of layers per unit length and F is the free energy.

We perform the double derivative at constant number of layers assuming that the thickness of surfactant layers δ remain constant. In this manner we obtain an expression for B as a function of d. The value of η_1 is then obtained using equation (2) together with $K = k_c/d$:

$$\eta_1^{und} = 1.33 \cdot (1-\delta/d)^2$$ (6)

We plot in Fig. 4 $\eta_1(d)$ resulting from fits to the profile at the first harmonic as a function of d. The solid line, which agrees well with the experimental data, is a plot of the predicted value for $\eta_1(d)$ (Eq. (6)) derived from the Helfrich theory. Here, we have taken the effective water thickness $\delta \approx 29$ Å to include the known excluded volume effects [11] of the surfactant tails in the oil. This then provides compelling evidence that in this SDS multimembrane system swollen by dodecane the intermembrane interactions are dominated by the Helfrich mecahnism of entropically driven undulation forces.

To further elucidate the nature of interactions in the fluid multi-membranes, we carried out a comprehensive X-ray study of competing electrostatic and undulation forces in two other multimembrane systems in

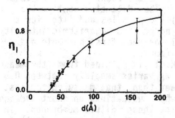

FIG. 4 Power-law exponent η_1 as a function of the intermembrane distance for mixtures along the dilution path. The solid line is the prediction of the model of Helfrich of entropically driven steric interactions.

the lamellar L_α phase as a function of the intermembrane distance. In each case, the negatively charged membrane is composed of a mixture of Sodium Dodecyl Sulfate (SDS) and pentanol, while the solvent separating the membranes is either pure water or brine (≈ 0.5 mole.l^{-1} of NaCl). The results clearly show that when diluting with pure water the interactions are dominated by long range electrostatic forces. In the brine dilution system, the addition of free ions (NaCl) to the solvent yields a small Debye length ($\lambda_D/d \ll 1$). In this case, with short range electrostatic interactions, undulation forces are restored with $\eta(d) = 1.33(1 - \delta/d)^2$ approaching a constant value for large d, as predicted by Helfrich [5].

The information we have on the thermodynamics of the system are contained in the shape of the peaks. The value of η is directly related to the interactions between the layers. Figure 5 shows the variation of η_1 as a function of d for the water dilution and the brine dilution. In both cases η_1 increases as a function of d but in a different way. For the water dilution η_1 saturates rapidly at a value around 0.3, while in the brine dilution η_1 continues to increase and saturates at a value larger than 1. This behavior can be qualitatively interpreted as resulting from the damping of layer undulations in the water dilution compared to the brine dilution system. This is due to the existence of long range repulsive electrostatic interactions for the SDS-pentanol series which consists of negatively charged bilayer membranes separated by water layers. Consequently, the power-law exponent η_1 is dominated by electrostatic forces and gives a direct measurement of k_c. In this case the interlayer electrostatic free energy can be calculated exactly from the one dimensional Poisson-Boltzmann equation and one readily derives η_1 [2]:

$$\eta_1^{elec}(d) = (k_BTL_e/2gk_c)^{0.5}[(1 - \delta/d)^{1.5}/d^{0.5}] \quad , \qquad (7)$$

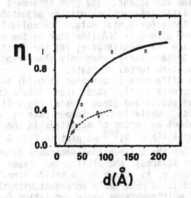

FIG. 5 Variation of the exponent η_1 as a function of the repeat distance d for the water dilution (open circles) and for the brine dilution (open square). The solid line corresponds to the prediction of the pure undulation interaction (Eq. 6), the dashed line is the solution of Posson-Boltzman equation (Eq. 7). In each case, the values of all parameters are determined experimentally. $\delta = 20$Å for both cases, for the water dilution: k_c ranges from 2 kT to 0.07 kT and a range from 80 to 190 Å^2).

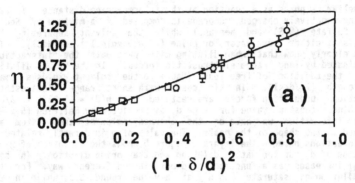

FIG. 6 Power-law exponent η_1 as a function of $(1-\delta/d)^2$ for two dilution systems. The open circles and squares are for the SDS-pentanol membranes along brine and oil dilution lines respectively. The solid line is the prediction of the Helfrich theory as discussed in the text.

where, $g = [1 - 3(D/d_W) + 6(D/d_W)^2 + ..]$. Here, $d_W = d - \delta$, $L_e = \pi e^2/\epsilon k_B T \sim 22$ Å, and $D = a/L_e$ (a is the surface area per (charged) polar head: one negative charge per SDS molecule, the alcohols are neutral, and all surface charges are asumed to be dissociated).

Figure 5 shows the results of the comparison between the experimental and theoretical values of η_1 for the water dilution with long range electrostatic forces (open circles for the experimental data and dashed curve for the theoretical prediction). k_C varies smoothly from 2 kT to 0.7 $k_B T$, and a and δ correspond to the value calculated from the known composition of alcohol and surfactant. The good agreement indicates that a simple model of electrostatic interactions with no adjustable parameters is sufficient to interpret the experimental data. The solid line in figure 5 is the prediction of η_1^{und} with δ = 20Å (the same as the water dilution). The data (open squares) are the results of η_1 for the first harmonic in the brine dilution series. Quite clearly, when only short range electrostatic forces are present, undulation forces dominate.

We stress the large difference in behavior in $\eta_1(d)$ over the dilution range between the electrostatically stabilized dilute membranes (open circle; Fig. 5) and those stabilized (open square; Fig.; 5) by entropically induced undulation forces: while in the former case η_1 changes by less than 30%, it varies by about an order of magnitude in the latter systems. To demonstrate the universality of η_1^{und} we plot in Fig. 6 the results for η_1^{und} versus $(1 - \delta/d)^2$ for the SDS-pentanol membranes diluted respectively with (i) dodecane (open squares) and (ii) brine (open circles). The theoretical prediction for η_1 is drawn as a solid line. The universal behavior is now clear and is in remarkable agreement with the prediction of the Helfrich theory[5] for multimembranes where undulation forces dominate.

CONCLUSION

In our synchrotron work we have studied the precise nature of competing forces in both neutral and charged layered surfactant systems.

The basic stabilizing force between rigid membrane sheets is known to originate in usual cases from a balance between van der Waals (electrodynamic), electrostatic and hydration forces. We have discovered a novel universal regime for floppy membranes with small rigidty moduli $k_c \simeq k_B T$, where the forces originate from the entropic undulations of layers and depend only on geometric and elatic parameters such as the layer spacing and the layer rigidity. Our work is quantitatively consistent with the theoretical work of Helfrich on the interlayer interactions associated with violent out-of-plane fluctuations of membranes.

We point out that these novel floppy membranes for which large dilutions are possible almost always contains cosurfactant in addition to surfactant molecules. In our most recent work, we have found that the primary effect of replacing surfactants with shorter chain cosurfactants in a mixed system is the thinning of the membrane which in turn leads to a reduction of the rigidity modulus k_C [16]. In this case then repulsive undulation forces completely overwhelm the van der Waals attraction and one crosses over from the classical microscopic regime for "bound membranes" to the floppy "almost unbound" regime for fluctuating dilute membranes.

ACKNOWLEDGEMENTS

The work described here has been carried out in close collaboration with D. Roux, G.S. Smith, and N.A. Clark. The author has benefited form numberous discussions with, S.A. Safran, S.K. Sinha, E.B. Sirota, and S. Milner. The National Synchrotron Light Source, Brookhaven National Laboratory, and the Stanford Synchrotron Radiation Laboratory are supported by the U.S. Department of Energy. A part of this work was supported by a joint Industry/University NSF Grant. No. DMR-8307157.

REFERENCES

1. C.R. Safinya, D. Roux, G.S. Smith, S.K. Sinha, P. Dimon, N.A. Clark, and A.M. Bellocq, Phys. Rev. Lett. 57 2718 (1986).
2. D. Roux and C.R. Safinya, J. Phys. (Paris) 49, 307 (1988).
3. G.S. Smith, E.B. Sirota, C.R. Safinya, and N.A. Clark, Phys. Rev. Lett. 60, 813 (1988); G.S. Smith, C.R. Safinya, D. Roux, and N.A. Clark, Mol. Cryst. Liq. Cryst. 144, 235 (1987).
4. D.R. Nelson and L. Peliti, J. Phys. (Paris) 48, 1085 (1987); R. Lipowsky and S. Leibler, Phys. Rev. Lett. 56, 2541 (1986); J.A. Aronovitz and T.C. Lubensky, Phys. Rev. Lett. 60, 2634 (1988).
5. W. Helfrich, Z. Naturforsch. 33a, 305 (1978).
6. F. Larche, J. Appell, G. Porte, P. Bassereau, and J. Marignan, Phys. Rev. Lett. 56, 1200 (1986).
7. M.B. Schneider, J.T. Jenkins and W.W. Webb, J. Phys. (Paris) 45, 1457 (1984).
8. W. Helfrich, J. Phys. (Paris) 46, 1263 (1985); L. Peliti and S. Leibler, Phys. Rev. Lett. 54, 1960 (1985).
9. P.G. de Gennes and C. Taupin, J. Phys. Chem. 86, 2294 (1982); S.A. Safran, D. Roux, M.E. Cates, D. Andelman, Phys. REv. Lett. 57, 491, 1986.
10. A. Parsegian, N. Fuller and R.P. Rand, Proc. Natl. Acad. Sci. 76, 2750 (1979).

11. D. Roux and A.M. Bellocq, Physics of Amphiphiles, edited by V. DeGiorgio and M. Corti (North-Holland, Amsterdam, 1985).
12. J.M. diMeglio, M. Dvolaitsky, and C. Taupin, J. Phys. Chem. $\underline{89}$, 871 (1985).
13. L.D. Landau, in Collected Papers of L.S. Landau, edited by D. Ter Haar (Gordon and Breach, New York, 1965), p. 209; R.E. Peierls, Helv. Phys. Acta. $\underline{7}$, Suppl., 81 (1934).
14. A. Caillé, C.R. Acad. Sci. Ser. $\underline{B274}$, 891 (1972).
15. S.G.J. Mochrie, A.R. Kortan, R.J. Birgeneau, and P.M. Horn, Z. Phys. B $\underline{62}$, 79 (1985).
16. C.R. Safinya, E.B. Sirota, D. Roux, G.S. Smith, Submitted to Phys. Rev. Lett., Oct. 1988.

NEAR SURFACE STRUCTURE OF ION IMPLANTED Si STUDIED BY GRAZING INCIDENCE X-RAY SCATTERING

G.Wallner, E.Burkel,H.Metzger,J.Peisl,S. Rugel.

Sektion Physik, Ludwig-Maximilians-Universität München, D-8000 MÜNCHEN 22, FRG

ABSTRACT

X-rays incident on a surface under grazing angle may undergo total external reflection and excite an interior wave field damped exponentially into the bulk. These evanescent waves are a sensitive probe for the study the real structure in the near surface region. We report results on the influence of implantation defects on Bragg diffracted and on diffuse intensities. By detailed comparison of Bragg intensities with predictions of dynamical scattering theory we detect the presence of amorphous layers and determine their thickness. For the first time defect induced diffuse scattering under conditions of grazing incidence and exit is observed and compared to recent theoretical results. Strength and symmetry of implantation induced defects can be determined as well as their depth distribution which is compared to results of a TRIM simulation: the defect distribution is found to agree with that of the deposited collisional energy.

INTRODUCTION

In recent years X-ray diffraction has been successfully adapted to the study of the near-surface structure of solids (e.g.[1-4]). Under conditions of grazing incidence and exit the distribution of scattered intensity at or near bulk Bragg-reflections is measured. The theoretical background for the analysis of experiments is supplied by kinematical [5,6] and dynamical [7,8] theory of scattering.

We study the near-surface structure of high perfection semiconductor samples after induced structural changes, e.g. ion implantation, or controlled growth of overlayers. By measuring scattered intensities as a function of exit angle we - at the cost of a weak scattering signal - achieve high resolution in reciprocal space and dispose of an additional parameter to control the scattering depth. In the present communication we focus on our experiments with Si-samples, implanted with 80 keV As ions to various doses, in order to give a consistent impression of the possibilities of the method.

Mat. Res. Soc. Symp. Proc. Vol. 143. ©1989 Materials Research Society

EXPERIMENTAL DETAILS

The samples were perfect Si single crystals with surface normal in [100]-direction. They were polished and etched according to the usual procedures, and implanted with 80 keV As ions at ambient temperature to various doses. The measurements were conducted at the beamlines D4 and W1 at the synchrotron source DORIS II at Hasylab/DESY. The experimental set-up is shown schematically in figure 1.

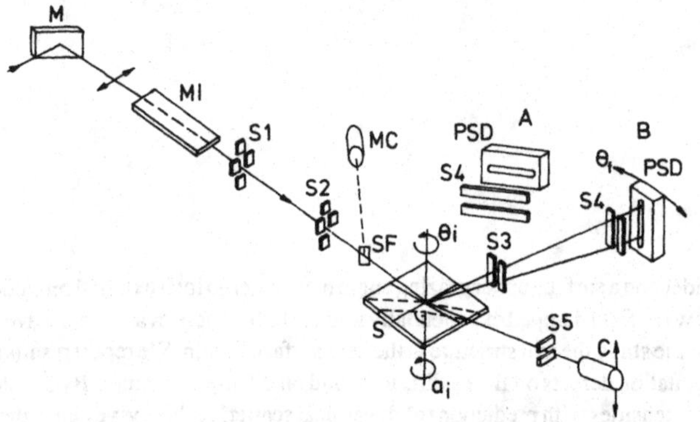

Figure 1:

Experimental set-up. M: Monochromator, MI: Mirror, S1..S5: Slits, MC: Monitor counter, SF: Scattering foil, S: Sample, PSD: Position senstive detector, C: Scintillation counter. The PSD is in position B to record intensity distributions $I(\alpha_f)$, and in position A to record intensity distributions integrated over α_f.

Intensity distributions $I(\alpha_f)$ as a function of exit angle α_f were recorded at q=0, where the Bragg condition in the plane of the surface (x-y-plane) is fulfilled for the (220)-reflection, or at different q≠0 (q is the projection of the distance from the (220) reciprocal lattice point onto the x-y-plane). Incident angles α_i were varied in the range $0 \leq \alpha_i \leq 3\alpha_c$ (the critical angle α_c is 3.43 mrad for Si at the wavelength of 1.38 Å used). Resolutions in α_i and α_f were of the order of 0.1 mrad, resolutions in the scattering plane (of Θ_i and Θ_f)were about 2 mrad.

EXPERIMENTAL RESULTS AND DISCUSSION

Figure 2 shows distributions $I(\alpha_f)$ at the (220) Bragg point for a sample implanted with $1*10^{15}$ As/cm^2, for different angles of incidence α_i. For comparison a measurement on an unimplanted reference sample (dotted curve) is also shown. High dose implantation markedly changes the intensity distribution: at α_i and α_f smaller than α_c no Bragg diffracted intensity can be observed.

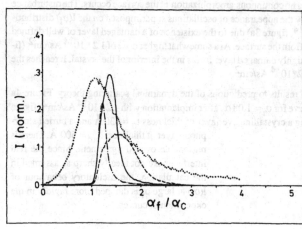

Figure 2:

Experimental intensity distributions $I(\alpha_f)$, normalized to the incident intensity for Si, implanted with $1*10^{15}$ As/cm². $\alpha_i/\alpha_c=1.06$ (dashed), $=1.24$ (solid), $=1.57$ (dash-dotted).
The dotted curve gives $I(\alpha_f)$ for an unimplanted sample at $\alpha_i/\alpha_c=1.01$

That is due to the formation of an amorphous layer at the surface of the sample of thickness t_{am} considerably larger than the minimum penetration depth of about 100 Å. A detailed comparison of experiment with calculations on the basis of dynamical diffraction theory [4] yields a mean thickness $t_{am}=(1445\pm50)$ Å, which varies by about 10% laterally (in x- and y-directions) across the sample. As in the case of the unimplanted sample [8] excellent agreement of theory and experiment is achieved.

One example for a problem that might be studied in a similar way is the analysis of structure and thickness of an amorphous or crystalline layer on a perfect crystalline substrate, e.g. in a MOS structure, and the determination of interface roughness.

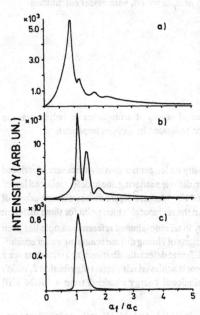

Figure 3:

Intensity distributions $I(\alpha_f)$, for Si implanted to different doses: a) $8*10^{13}$ As/cm², b) $1.2*10^{14}$As/cm², c) $2*10^{14}$ As/cm², at $\alpha_i/\alpha_c=1.08$.

At lower implantation doses no continuous amorphization at the surface occurs. The disturbance of the lattice is evidenced by the appearance of oscillations superimposed on the $I(\alpha_f)$ distributions at a low dose (0.8 $*10^{14}$, figure 3a) due to the existence of a damaged layer of well defined thickness at some distance from the surface. At a somewhat higher dose of $1.2 * 10^{14}$ As/cm^2 (figure 3c) an amorphous or highly damaged layer forms in the interior of the crystal. It reaches the surface at a dose of about $2*10^{14}$ As/cm^2.

We compare experimental results to predictions of the dynamical scattering theory. Figure 4a shows the experimental curve for $\alpha_i = 1.08\ \alpha_c$ after implantation with $1.2* 10^{14}$ As/cm^2, figure 4b the calculation assuming a crystalline overlayer of thickness $t_o = 330$ Å and a buried amorphous layer of thickness $t_{am} = 500$ Å. There is reasonable overall agreement. Since the real interfaces are not ideally sharp as assumed in the calculation, the oscillatory behaviour of $I(\alpha_f)$ at large α_f is damped more rapidly in the experimental curves.

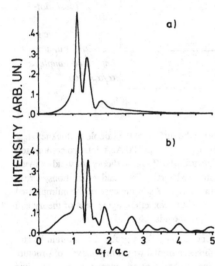

Figure 4:

Comparison of experimental distribution for Si, implanted with $1.2 *10^{14}$ As/cm^2, at $\alpha_i/\alpha c=1.08$, with model calculation

In an analoguous way implantation induced buried layers of amorphous or polycrystalline structure may be studied, e.g. a buried SiO$_2$ layer generated by oxygen implantation.

At low implantation doses Bragg diffracted intensity no longer is a sensitive measure of induced damage. At the wings of the Bragg peaks, however, diffuse scattering due to defect induced lattice distortions appears. For an implantation dose of $6*10^{13}$ As/cm^2 figure 5 shows distributions of diffuse intensity at a distance $q=2*10^{-2}$ Å^{-1} from the reciprocal lattice point for three different angles of incidence α_i. The weak diffuse intensity in an unimplanted reference sample has been subtracted. The lines correspond to model calculations of Huang diffuse scattering under conditions of grazing incidence and exit [9], assuming different defect distributions $\rho(z)$ with distance z from the surface (compare figure 6). Best agreement is achieved with defect distribution 2, which corresponds to the distribution of deposited collisional energy obtained from a TRIM [10] simulation.

Thus diffuse scattering under conditions of grazing incidence and exit offers the possibility to study a small concentration of near surface defects, determine their characteristics and depth distribution, and to observe the change of these parameters during an annealing treatment.

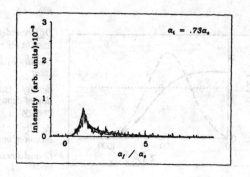

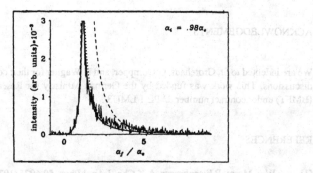

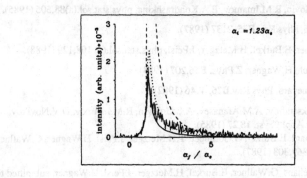

Figure 5:

*Distributions of diffuse intensity for Si,
implanted with $6*10^{13}$ As/cm^2, at different α_i. The lines correspond to model
calculations with different defect densities $\rho(z)$, compare figure 6. (after [9])*

24

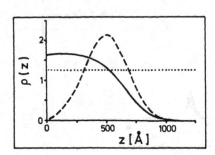

Figure 6:

Different defect densities $\rho(z)$ used in the model calculations.

Density 1 (dotted curve): constant

density 2 (solid line) : distribution of deposited collisional energy,

density 3 (dashed line): distribution of implanted As atoms

(2 and 3 results from TRIM simulation)

ACKNOWLEDGEMENTS

We are indebted to St. Grotehans, G.Gompper, and H.Wagner for their continuous interest and discussions. This work was funded by the German Ministry for Research and Technology (BMFT) under contract number 03 PE 1 LMU3 .

REFERENCES

[1]   W.C.Marra, P.Eisenberger, A.Y.Cho, J.Appl.Phys. **50**,6927 (1979).

[2]   A.L.Golovin, R.M.Imamov, E.A.Kondrashkina, phys.stat.sol.(a)**88**,505 (1985).

[3]   H.Dosch, Phys.Rev.**B35**, 2137 (1987).

[4]   G.Wallner, E.Burkel, H.Metzger, J.Peisl, phys.stat.sol.(a) **108**,129 (1988).

[5]   S.Dietrich, H.Wagner, Z.Phys. **B56**,207 (1984).

[6]   G.H. Vineyard, Phys.Rev.**B26**, 4146 (1982).

[7]   P.A.Aleksandrov, A.M.Afanasev, A.L.Golovin, R.M.Imamov, D.V.Novikov, S.A.Stepanov, J.Appl.Cryst. **18**,27 (1985).

[8]   N.Bernhard, E.Burkel, G.Gompper, T.H.Metzger, J.Peisl, H.Wagner, G.Wallner, Z.Phys.**B69**,303 (1987).

[9]   St. Grotehans, G.Wallner, E.Burkel, H.Metzger, J.Peisl, H.Wagner, submitted to Phys.Rev.B (1988).

[10]   J.P.Biersack, L.G.Haggmark, Nucl.InstrMeth **174**,257 (1980).

STUDIES OF AN InAs/GaAs HETEROJUNCTION BY TOTAL-ELECTRON-YIELD

A. KROL*, C.J. SHER*, D.R. STORCH*, S.C. WORONICK*, L. KREBS*, Y.H. KAO*, AND L.L. CHANG**
*Department of Physics, State University of New York at Stony Brook, Stony Brook, NY 11794
**IBM Thomas J. Watson Research Center, P.O. Box 218, Yorktown Heights, NY 10598

ABSTRACT

Total electron yield (TEY) of an InAs/GaAs heterojunction due to soft x-ray excitation has been studied. This heterojunction was prepared by an overgrowth of a 600 Å InAs layer on a GaAs substrate using molecular beam epitaxy. Experimental data are compared with theoretical analysis based on a modified Fresnel formulation to calculate the wave field distribution in stratified media with interfacial roughness. The TEY angular profiles obtained at a given x-ray energy reveal information on the interfacial roughness, secondary electron escape length, attenuation length of elastically scattered photo-electrons, and optical constants of the epilayer in the x-ray regime.

Measurements of total-electron-yield due to grazing angle x-ray radiation at a constant energy of incoming x-ray photons E_i were performed. The sample was mounted on an insulating plate and negatively biased. An electrometer was used in order to measure the neutralizing current provided by a biasing battery. This current was equal to the total number of electrons emitted by the sample per unit time. The actual electrons emitted by the sample are also counted simultaneously by means of a positively biased collector. The total number of photoelectrons created at a depth z in a small volume dV within the sample is proportional to the total number of x-ray photons absorbed in this volume. The resulting primary core vacancies can be filled in radiative (fluorescence emission) or non-radiative (Auger electron yield) transitions from higher shells [1]. However, elastically emitted Auger and photoelectrons give rise to only a small portion of all emitted electrons [2,3]. Most of the emitted secondary electrons have a kinetic energy below 10 eV. The radiant energy flow in a heterostructure is strongly dependent on the boundary conditions. For this reason material parameters such as the surface and interfacial roughness, optical constants and layer thickness can be deduced from the angular profile of the TEY, which is a function of the dynamical absorption. Information about the secondary electron escape length and the attenuation length of elastically scattered photoelectrons can be found from the fraction of electrons that reach the surface from a depth z as the absorption length of the oncoming x-ray increases with increasing incidence angle [4].

Dynamical absorption in heterostructures can be calculated using the optical electromagnetic wave solution of the Fresnel equations on each interface, as described in our previous publication [4]. The number of primary core holes of type q created per unit time in the infinitesimally thin layer dz at a depth z, normalized to incoming flux, is

$$\frac{dN_q(z)}{F_0} = - \quad Cn_q(z)\sigma_q \frac{\sin\phi_1(z_0)}{\sin\phi_0(z_0)} \frac{dP_1(z,\theta)}{P_0} \tag{1}$$

where C is a constant, θ the grazing incidence angle, $n_q(z)$ the concentration of atoms giving rise to q-type core holes at a depth z, σ_q the partial photo-ionization cross section, F_0 incoming flux, $P_1(z,\theta)$ and P_0 are the Poynting vector below and above the surface of the sample, and ϕ_0 and ϕ_1 the directions of Poynting vectors incident and at a depth z.
The TEY normalized to incoming flux can be written in the following form:

$$i_{TEY}(0)/F_0 = \sum_q \{[i_p^q + i_{sp}^q] + \sum_{rp} [i_a^{qrp} + i_{sa}^{qrp}]\} \tag{2}$$

i_p^q is the current due to elastically scattered photoelectrons with kinetic energy $E = E_i - E_q$ excited from shell q,

$$i_p^q = \int_0^{2\pi} \int_0^{z_f} \frac{dN_q(z)}{F_0 \cos\phi} \exp(- \frac{z}{a_p(E)\cos\phi}) dzd\phi \tag{3}$$

$a_p(E)$ is the attenuation length of elastically scattered photoelectrons with energy E, ϕ is the angle between the direction of emission and the normal to surface.
The i_{sp}^q is the secondary electron current created by photoelectrons excited from shell q. We employ the model of secondary electron emission proposed by Erbil et al. [3] which assumes that the TEY current consists mainly of secondary electrons excited by photo- and Auger electrons, which are created uniformly and isotropically in the volume of a sphere of radius equal to the range of the primary electrons R(E). Each primary electron excites N secondary electrons, where $N = E/\epsilon$, ϵ being the average energy of secondary electrons.

$$i_s^{pq} = \int_0^{2\pi} \int_0^{R_p} \frac{3N\,N_q}{8F_0\,R_p(E)} [1 - (\frac{z}{R_p(E)})^2 + 2\frac{z\cos\phi}{\alpha R_p(E)^2} - \frac{2\cos^2\phi}{(\alpha R_p(E))^2}] dzd\phi$$

$$+ \int_0^{2\pi} \int_{R_p}^{z_f} \frac{3N\,N_q}{8F_0\,R_p(E)} [2\frac{\cos\phi}{\alpha R_p(E)} - \frac{2\cos^2\phi}{(\alpha R_p(E))^2}] \exp[-\frac{\alpha(z-R_p(E))}{\cos\phi}] dzd\phi \tag{4}$$

where $R_p(E)$ is the range of photoelectrons with energy E and $1/\alpha$

is the secondary electron escape length.

The term arising from Auger electrons may be written in a similar manner, taking into account the different ranges $R_A(E)$ associated with the different energies of the primary Auger electrons, and with the number of secondary electrons $N=a_q E_{grp}/\epsilon$, where a_q is the Auger yield of q-type subshell.

In our experiment we studied InAs/GaAs heterojunctions in the soft x-ray energy range. The pertinent Auger electron energies are listed in Table I. In our calculations we have neglected non-radiative decays of holes in N and higher shells, due to the lower energy, and thus smaller range of this type of Auger electron. The ranges of the primary electrons were calculated using data published by Seah and Dench [5], and are also presented in Table I.

Table I. Auger electron penetration ranges calculated using data published by Seah and Dench [5].

Transition	E(eV)	Range (Å) InAs	GaAs
In			
$M_V N_{23} N_{45}$	340	41	
$M_{IV} N_{23} N_{45}$	348	41	
$M_V N_{45} N_{45}$	403	44	
$M_{IV} N_{45} N_{45}$	411	45	
$M_{III} N_{23} N_{45}$	561	52	
$M_{III} N_{45} N_{45}$	624	52	
Ga			
$M_{III} M_{45} M_{45}$	56		17
$M_I M_{45} M_{45}$	111		18
$M_{II} M_{45} M_{45}$	60		22
As			
$M_{III} M_{45} M_{45}$	43	18	17
$M_{II} M_{45} M_{45}$	49	18	17
$M_{III} M_{45} N_{23+2}$	87	21	20
$M_I M_{45} M_{45}$	106	23	22

The influence of surface and interfacial roughness on radiative energy flow in stratified media was discussed in reference [4]. The influence of surface roughness is most pronounced in the pre-critical angle region, while the influence of interfacial roughness is exhibited by extinguishing interference oscillations in the post-critical angle region. We would like to emphasize that although the majority of the secondary electrons are emitted from a relatively thin surface layer (of order 50Å), the information depth is much larger, since the number of primary core holes is proportional to the flux, which in turn is defined by the boundary conditions. In this way even a remote interface can affect the shape of the angular TEY. A detailed analysis of the contributions of different factors to the angular TEY will be given in a future publication.

Experimental data were taken at the U-15 beamline of the National Synchrotron Light Source at Brookhaven National Laboratory. An example of data points vs. theoretical fit is shown in Fig. 1.

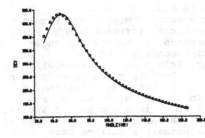

Fig. 1. Total-electron-yield with respect to angle in an InAs(600Å)/GaAs hetero-structure. The energy of the incident photons is 600 eV.

However, we have concluded that better accuracy in the analysis of the experimental TEY angular profile may be obtained by fitting the angular derivative of TEY, since small features, such as interference oscillations, are enhanced. In Fig. 2 three experimental derivatives are shown.

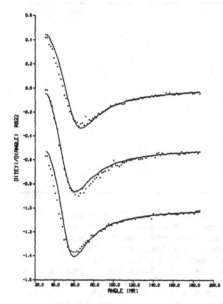

Fig. 2. The derivative of total-electron-yield with respect to angle in an InAs(600Å)/GaAs hetero-structure. The energy of the incident photons is 600 eV for the upper curve, 665 eV for the middle curve and 700 eV for the lower curve.

The data are fitted to theory based on calculated values of the ranges and attenuation lengths of Auger electrons. The optical constants, surface and interface rms roughness, secondary electron escape length and primary photoelectron attenuation length were obtained from our fits. The best fit parameters are listed in Table II. The values obtained for surface and interfacial rms roughness are in reasonable agreement with our reflectivity studies. The attenuation length for primary electrons and the secondary electrons escape length are similar to those predicted by Seah and Dench [5].

Table II. Material parameters obtained from best fits of experimental data. Attenuation length refers to the average length of the most energetic photoelectrons produced at given energy of incident photons. = 8 eV.

top roughness 10±5 Å
interface roughness 25±5 Å
secondary electron escape length 10±5 Å

Energy E_i(eV)	InAs delta(10^{-3})	beta(10^{-4})	GaAs delta(10^{-3})	beta(10^{-4})	Attenuation length (Å)
600	1.47±0.1	8.86±0.1	1.66±0.1	3.89±0.1	31±5
665	1.18±0.1	7.52±0.1	1.35±0.1	2.80±0.1	34±5
700	1.24±0.1	6.49±0.1	1.22±0.1	2.32±0.1	28±5
827	0.85±0.1	3.65±0.1	0.87±0.1	1.46±0.1	45±5

In summary, we have measured grazing angle total-electron-yield of a InAs/GaAs heterostructure in the soft x-ray range. The angular profiles were fitted to the proposed theory of secondary electron production in layered structures. Some important material parameters (optical constants, rms roughness, secondary electron escape length and primary photoelectron attenuation length) are derived. These measurements showed that the information length of TEY is much larger than the characteristic escape lengths of both primary and secondary electrons.

ACKNOWLEDGEMENTS

This research is supported by ONR under grant No. N0001483K0675, and by DOE under grant No. DE- FG02-87ER45283.

References

1. W. Bambynek, B. Crasemann, R.W. Fink, H.U. Freund, H. Mark, C.D. Swift, R.E. Price, and P. Venugopala Rao, Rev. Mod. Phys. 44, 716 (1972).
2. J. Stohr, C. Noguera, and T. Kendelewicz, Phys. Rev. B 30, 5571 (1984).
3. A. Erbil, G.S. Cargill III, R. Frahm, and R.F. Boehme, Phys. Rev. B 37, 2450 (1988).
4. A. Krol, C.J. Sher, and Y.H. Kao, Phys. Rev. B 38, 8579 (1988).
5. M.P. Seah and W.A. Dench, Surface and Interface Analysis 1, 2 (1979).

X–RAY DIFFRACTOMETER FOR THE STUDY OF A MONOMOLECULAR FILM SPREAD ON A LIQUID SUBSTRATE

KARL M. ROBINSON and J. ADIN MANN JR.
Case Western Reserve University, Dept. of Chemical Engineering, Cleveland, OH 44106

ABSTRACT

The study of ultrathin films spread on liquid substrates has been limited to macroscopic observations of general film behavior. Synchrotron radiation has provided the means of observing the two–dimensional crystalline properties of monolayers. Our recent experiments performed on NSLS X–23B beam line was the trial run of a new liquid surface diffractometer. Major emphasis has been placed on vibrational and environmental isolation of the film, in addition to 0.1 micron resolution in the control of beam positioning. The film is examined in a closed, ultraclean environment consisting of a fused silica trough, purified N_2 atmosphere, 0.1°C temperature control and chromatography grade or better solvents and surfactants. Surface properties (surface tension, and visco–elastic moduli) are monitored by surface laser light scattering spectroscopy. The x–ray scattering pattern from the liquid surface contains the expected specular reflection as well as scattered peaks at small wave numbers. We believe these to be caused by high frequency surface ripple fields.

INTRODUCTION

The range of applications of Langmuir–Blodgett films provides motivation to seek the molecular structure of spread monolayers. Although x–ray and electron diffraction techniques have been used on the deposited film [1,2], the structure information explains little concerning the film properties before and during deposition. The high intensity of synchrotron radiation provided the first opportunity to determine the molecular structure directly as opposed to the indirect methods based on isotherm techniques. Surface density profiles have been measured by Pershan et.al.[3,4,5] through studies of the variation of the reflected beam intensity in the plane of incidence. Attempts at studying in plane and out of plane scattering have shown the extreme difficulty in handling the surface properly to achieve a high signal to noise ratio[6,7].

The liquid surface diffractometer described below is similar to the Chapman, Bloch fluorescence experiment[8] but with the capability of scanning around the reflected beam for in and out of plane scattering. The first runs of the new diffractometer were done with several well characterized surfactants. The first film was n–pentadecanoic acid which has been thoroughly studied by Pallas et.al [9,10]. The second film was a diacetylene compound suspected of island formation [2]. The third film was a new phthalocyanine compound[12]. In addition to each of these films, several films deposited on a silicon substrate by the Langmuir–Blodgett technique were placed in the diffractometer. The diffractometer was positioned on X–23B of the National Synchrotron Light Source at Brookhaven National Laboratory.

LIQUID SURFACE DIFFRACTOMETER

The design of the diffractometer can be broken into two major systems, the beam control system and the surface control system. Beam control consists of the equipment necessary to deflect the beam toward the sample, monitor intensity and monitor the scattered intensity and position. Surface control consists of the equipment necessary to maintain the film integrity, i.e. surface pressure, temperature and humidity, and the purity of the surface film. These two components will be discussed separately.

Beam Control

The diffractometer has three main components, the mirror to deflect the synchrotron beam downward to the sample, the trough housing within which the sample is contained and the detectors which monitor the beam position. The X–23B beamline provides the advantage that the beam position does not change with energy selection from the monochromator. This allows a continual selection of various wavelengths with which to probe the surface without realignment of the diffractometer. The beam control is shown in figure 1 with the geometry shown in figure 2.

The beam entering the experimental hutch is defined by entrance slits S1 and the ion chamber IC. The beam is then deflected downward by a quartz based Ni coated mirror (Diamond Electro–Optics, Inc., Waltham, Ma.), M, and further defined by slits S2 and monitored by scintillation detector SC1. The maximum angle of incidence obtained is limited by its low (<2°) critical angle but it does maintain the horizontal polarization of the beam. The incident angle is remote controlled by the user and is known to ±.001°. The signals from IC and SC1 are stored in the source compensation channel of the controller for the scattered beam detector.

The deflected beam enters the trough housing through a kapton entrance window. The position of the water surface is controlled by two manual x–y positioners and a computer controlled z stepper motor. The water surface height is adjusted for the particular incident angle such that the incident beam strikes the center of the surface. The center is further monitored by SC2 to detect the maximum signal from the incident

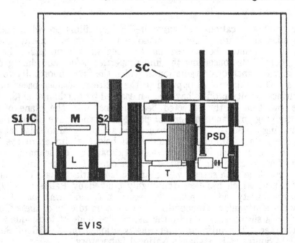

Figure 1. The liquid surface diffractometer. The beam enters from the left through slits S1 and ion chamber IC. A portion of the beam is reflected downward by mirror M through slits S2. I_0 is monitored by the scintillation detector SC after the S2. The liquid surface is contained within the trough housing, T and beam position is monitored by a scintillation detector, SC. Scattered radiation is detected by the position–sensitive detector, PSD as it rotates about the trough. The laser, L, provides the light for the capillary ripple spectroscopy.

beam on the water surface. The front of the trough is composed of a large kapton window through which the reflected beam and the scattered radiation exit the housing.

The scattered radiation is monitored by an EG&G PARC 1412XR silicon photo–diode array detector, PSD. The array consists of 1024 pixels in 27.05 x 2 mmsq. area held in the vertical, z, direction. The detector array is held at a fixed distance (20.5cm) from the center of the trough and is rotated about that point by a stepper motor driven rotary table. The detector is evacuated and cooled to −40ºC to reduce dark current noise. The detection of low scattering irradiance is enhanced by electing to use a long exposure time and a large number of independent scans to average out background noise. The PSD, IC and SC1 signals are stored in the PSD controller with the source compensation channel set as an integrated value over the PSD exposure time. This protocol eliminates the need for complex gating of detectors. The data is retrieved from the detector controller by the computer which controls all the beam positioning equipment.

Surface Control

The signal from the spread monolayer is highly susceptible to vibration and impurities in both the water and atmosphere [11,12]. Each of these difficulties have been addressed in the design of the diffractometer since without careful control here the signal, if any, will mean very little. The ideal set–up for the diffractometer would be to mount it on a massive isolation table in a Class 10 clean room facility. However, since this is not practical on the light source floor, several modifications have been incorporated in the design to provide a clean and still liquid surface.

The water surface is extremely sensitive to any vibration inside the housing. In addition, the effect of the Debye–Waller factor in the scattering scattering will eliminate any interference pattern when the surface vibration is larger than the normal thermal capillary waves [11]. Even then q_z must be small enough for diffraction to be observed.

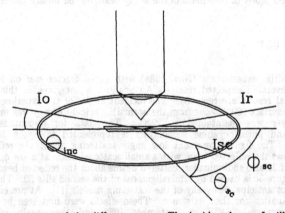

Figure 2. The geometry of the diffractometer. The incident beam, I_0, illuminates the surface at an incident angle Θ_{inc}. Depending on the particular incident angle, a portion of I_0 is reflected, I_r, or scattered by the surface, I_{sc}. For I_r, $\Theta_{inc} = \Theta_{sc}$, otherwise the scattered beam is defined by ϕ_{sc} and Θ_{sc} (which is monitored by the different pixels of the PSD). The point depicts the center of rotation for the PSD.

The diffractometer design eliminates the large amplitude low frequency (0.1 − 1000 Hz) floor oscillations with a Newport EVIS active feed back isolation table. The table measures only 3 x 2 sqft but has an effective mass of 10^6 tons in this frequency range. A self supporting dust cover encloses the entire diffractometer effectively isolating any air currents in the hutch from the trough housing. These two components together adequately isolate the diffractometer from the surrounding low level noise. High frequency noise can still penetrate the isolation table but are usually of such a small amplitude that the massiveness of the diffractometer minimizes their effect. Water oscillations induced by the diffractometer itself during stepper motor motion are allowed to die out by providing a time delay after the motor moves the PSD and the beginning of the detector exposure period.

The removal of impurities from the water surface is an extremely important issue in the design. Careful studies by Pallas et.al. [9] have shown the degree to which cleanliness must be taken to assure a proper spread monolayer. Two troughs are used; a teflon coated brass trough and a solid fused silica quartz trough. The teflon trough is used for aligning the system while the quartz trough is used with the clean monolayer. The Pallas technique is used for maintaining the high purity of the quartz trough and of all the chemicals which come in contact with the quartz trough. The troughs measure 25cm in diameter and 1cm deep. The water surface is maintained at a constant height by thermodynamic equilibrium with the atmosphere inside the housing, which is ultra−pure N_2 at 100% humidity and held at a constant temperature by the same brass heating/cooling mantle that maintains the trough temperature.

The circular troughs eliminate curvature effects caused by the corners in standard square Langmuir troughs. A major compromise in the use of the circular trough is that there is no barrier with which to compress the film. Surface pressure is increased by a continual addition method, which limits the maximum available compression to the equilibrium spreading pressure, but also gives a very stable film in thermal and mechanical equilibrium. The surface pressure is monitored by a non−invasive laser light scattering system. A class 3 Argon−Ion laser is used to illuminate a small patch of the surface along the normal to the surface. The light scattering system monitors the 3−4 Å amplitude capillary ripples (q= 200 1/cm) without introducing impurities from a Whilhelmy plate or surface barrier. The autocorrelation of the capillary wave light scattering signal contains information about the surface tension as well as the visco−elastic response function of the monolayer. Monitoring of the x−ray patch and the non−illuminated areas of the monolayer make possible a direct study of the effects of the x−ray beam on the surface properties.

INITIAL RESULTS

The initial experiments (Nov. 1988) with the diffractometer on NSLS X−23B resulted in several unexpected results. An analysis is not given in this manuscript, however initial results are reported. The incident angle for the experiment was 0.180 deg off of the surface (89.820 deg from the normal). Incident energy was set at 7600EV. The monolayer was maintained at 22ºC. The data for n−pentadecanoic acid, diacetylene and phthalocyanine monolayers is represented in figure 3a,3b and 3c respectively. The figures represent low angle scattering about the reflected beam. Figure 3a shows the reflected beam with a small scattered beam at a low ϕ_{sc}. Similarly, figures 3b and 3c also show smaller scattered beams about the reflected beam. The large doublet on figure 3c is caused by a misalignment of the second slits S2. These scattered peaks were not anticipated by any of the scattering models [11]. At present there is no definite explanation for their existence. These effects were first seen by us during a preliminary run on X−23B in September 1988 and then again in November. Investigators conducting similar experiments [3−8] have not reported similar results however they were not probing the low wavenumber region represented in figure 3.

We have tested for the obvious artifacts caused by stray scattering. The peaks we show are not due to stray beams either from the mirror that deflects the x−ray beam

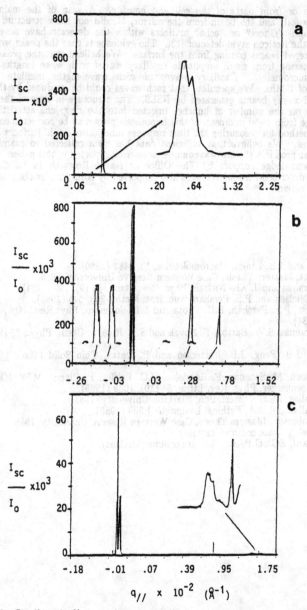

Figure 3. Small angle ($\phi_{sc} < 1^\circ$) scattering for a. n—pentadecanoic acid, b. diacetlyene and c. phthalocyanine. In each experiment, E = 7600EV, Θ_{inc} = .180° and trough temperature is 22.0°C. The scattered peaks are magnified to show their internal structure. Little peak structure is seen in the n—pentadecanoic acid while appreciably more is seen in both the diacetlyene and phthalocyanine.

to the surface or from parts of the cell and housing. A scan of the main beam (attenuated by foil) and the beam from the mirror, M, did not show structure due to stray scattering. "Ghost" or "echo" artifacts within the detector have never been reported with the Reticon style detector [13]. The evidence is that the peaks we report represent scattered beams coming from the surface. We believe that the peaks could represent scattering from small amplitude capillary waves with wave lengths of the order of 1 micrometer. Capillary waves of such wavelengths oscillate in the neighborhood of 1 Mhz. We speculate that such waves could be produced by the very intense, pulsed x–ray beams generated by NSLS. The structure in time of the x–ray beam depends on the number of bunches injected into the ring; characteristic time constants range from 1–50 microsec. If this hypothesis proves to be correct then a new and exciting method for measuring the time response function of liquid surfaces at very high frequencies. We believe that sufficient data has been collected to examine the mechanism that produces the low wavenumber scattering reported in this paper.

We acknowledge support by The Office of Naval Research in building the diffractometer described herein. We were also supported by Amp Inc. for the monolayer work reported herein.

REFERENCES

1. D. Day and J.B. Lando, Macromolecules, 13, 1483 (1980).
2. J. Schutt, Masters Thesis, Case Western Reserve University, 1986.
3. P.S. Pershan and J. Als–Nielsen, Phys. Rev. Lttrs., 52 (9), 759 (1984).
4. J. Als–Nielsen and P.S. Pershan, Nuc. Inst. Meth., 208, 545 (1983).
5. J. Collett, P.S. Pershan, E.B. Sirota and L.B. Sorensen, Phys.Rev.Lttrs., 52 (5), 356 (1984).
6. B.N. Thomas, S.W. Barton, F. Novak and S.A. Rice, J. Chem. Phys., 86 (2) 1036 (1987).
7. B. Lin, J.B. Peng, J.B. Ketterson and P. Dutta, Thin Solid Films, 159, 111 (1988).
8. J.M. Bloch, M. Sansone, F. Rondolez, D.G. Peiffer, P. Pincus, M.W. Kim and P.M. Eisenberger, Phys. Rev. Lttrs., 54 (10), 1039 (1985).
9. N.R. Pallas, Ph.D. Dissertation, Clarkson University, 1981.
10. N.R. Pallas and B.A. Pethica, Langmuir, 1, 509 (1985).
11. K.M. Robinson, Masters Thesis, Case Western Reserve University, 1986.
12. D. Batzel (private communications).
13. R. Howard, EG&G PARC (private communications).

IN SITU X-RAY SCATTERING OF MONOLAYERS ADSORBED AT ELECTROCHEMICAL INTERFACES.

Michael F. Toney and Owen. R Melroy
IBM Research Division, Almaden Research Center, 650 Harry Road, San Jose, CA 95120-6099

ABSTRACT

Surface x-ray scattering has been used to study *in-situ* the structure of Pb monolayers electrochemically adsorbed on Ag (111) electrodes. Pb forms an incommensurate, hexagonal two-dimensional (2D) solid, which is rotated approximately 4.5° from the substrate symmetry directions and compressed relative to bulk Pb. Between monolayer formation and bulk deposition, the Pb-Pb near neighbor distance decreases linearly with applied potential. Due to the chemical equilibrium between the Pb monolayer and the Pb in solution, the isothermal compressibility of the monolayer can be measured and is in good agreement with that calculated for a 2D non-interacting free electron gas model of the monolayer. It is observed that the intensity of surface diffraction from the Ag substrate (the Ag crystal truncation rod) decreases when the Pb monolayer is adsorbed, although the cause of this is not known.

INTRODUCTION

Understanding the structure of electrochemical surfaces and adsorbed layers is fundamental both to understanding their chemical reactivity and the growth of any material that is subsequently deposited. Despite this, our knowledge of the metal/solution interface is far less detailed than for the solid/vacuum interface. This is so because the powerful surface science techniques (developed for ultrahigh vacuum (UHV)) based on electron or ion scattering are unsuitable for use in condensed phases, since the mean free path is far too short. Although these UHV techniques have been applied *ex-situ* and have provided significant insight into the nature of adsorbed species [1], the question of possible changes occurring during transfer from the electrolyte to vacuum remains unanswered. In addition, all information concerning the structure of the solvent is lost. Optical methods such as Raman spectroscopy [2], second harmonic generation [3], and surface infrared spectroscopy [4] have been successfully used to study the solid/liquid interface in-situ, but yield only indirect information about the geometry of the surface. Hard x-rays, however, are well suited to study the electrode/electrolyte interface, since they have a large penetration depth in aqueous solution and provide direct information on crystallographic structure. The availability of synchrotron radiation, with its high brightness, high collimation, and wide spectral range, has recently made direct, *in-situ* structural characterization of this interface possible using surface x-ray diffraction. In this paper we describe the use *in-situ* surface x-ray scattering to study metal monolayers adsorbed at a metal electrode/electrolyte interface.

Electrochemical deposition of metals onto a foreign metal substrate frequently occurs in distinct stages [5]. The initial peaks in the current/potential profile correspond to the formation of different ad-layers on the electrode surface and occur at electrode potentials positive of the reversible thermodynamic (Nerst) potential for bulk deposition. They are, thus, termed underpotential deposition (UPD) peaks. On single crystals, these initial peaks presumably correspond to the formation of well defined, ordered monolayers [6]. For Pb on Ag(111) one large, sharp UPD peak is observed and a typical current response to a linear sweep of potential and the corresponding adsorption isotherm (integral of the current) are shown in Fig. 1. The first peak at approximately -350mV (vs. Ag/AgCl) corresponds to the deposition of a single monolayer of Pb [7,8]. Between this potential and the onset of bulk deposition (-550mV), a single monolayer of Pb is adsorbed on the Ag surface and the monolayer coexists in chemical and thermal equilibrium with the Pb^{2+} ions in solution.

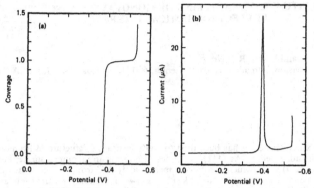

Fig. 1. (a) Voltammogram for the deposition of Pb on Ag(111). (b) Corresponding adsorption isotherm. Potentials are measured relative to Ag/AgCl. The scan rate is 20 mV/s.

EXPERIMENTAL APPROACH

Grazing incidence x-ray scattering (GIXS) is becoming an established technique for the structural determination of surfaces and adsorbed layers [9]. In this geometry the incident beam impinges on the sample surface at an angle close to the critical angle for total external reflection, which results in a significant increase in the x-ray scattering intensity from the surface and also results in a decrease in background scattering. One of the key obstacles to overcome before applying GIXS to electrochemical systems is the development of a suitable cell. The requirements for x-ray scattering and good thin layer electrochemistry are difficult to meet simultaneously. The x-ray scattering must be measured with only a thin layer of electrolyte covering the electrode to minimize the diffuse background scattering. This, however, accentuates cell resistance problems (e.g. causes a large ir drop). As a compromise, the electrochemical cell (see Fig. 2) was designed so that the metal monolayer could be deposited with a relatively thick layer of electrolyte covering the electrode (and characteristic voltammograms obtained) and then reconfigured to a thin layer cell in which the x-ray scattering was measured [8].

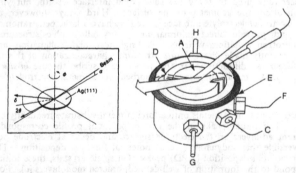

Fig. 2. Electrochemical Cell: A) Silver (111) electrode, B) Ag/AgCl reference electrode, C) Platinum counter electrode, D) Polypropylene window, E) O-ring holding polypropylene to cell, F) External electrical connection to the working electrode, G) Solution inlet, and H) Solution outlet. Inset: GIXS geometry showing the incident angle α, the exit angle δ, the scattering angle 2θ, and the azimuthal angle ϕ. I_0 is the incident beam, I_s the scattered beam, and I_r is the specular reflection. Note that for clarity the cell and GIXS geometry are shown horizontally; in the experiments they are vertical.

The Pb monolayer was deposited at -400 mV from a 0.1M sodium acetate, 0.1M acetic acid, and 5 x 10^{-3}M lead acetate electrolyte and reconfigured to the thin layer geometry at this potential. Experiments at different potential were conducted by changing the potential after the cell was in the thin layer configuration. For studies of the initial stages of bulk deposition, a predetermined amount of Pb was deposited, and then solution removed from the cell, leaving only the thin layer covering the surface. This prevented additional deposition during the x-ray measurements. To prevent the layer of electrolyte from drying and to avoid complications from the slow oxidation of the Pb due to diffusing oxygen, the Pb was electrochemically removed and redeposited between diffraction scans.

Fig. 2 shows the GIXS geometry used in these experiments. The sample was held in the vertical plane of the diffractometer with the incident angle of x-rays α equal to the exit angle δ at which the scattered beam was collected (e.g. symmetric ω = 0 mode). The incident x-ray beam size was restricted by slits in the vertical and horizontal directions to limit diffuse scattering by the electrolyte. The diffracted direction was determined with either two sets of slits or Soller slits and the intensity measured by a scintillation detector. The data were collected at the Stanford Synchrotron Radiation Laboratory (SSRL) on either the focussed 54-pole wiggler beam line (6-2) [8] or the focussed 7-pole wiggler beam line (7-2) [10]. The incident x-ray energy was chosen to be either 12350 eV (1.003 Å) or 8080 eV (1.534 Å) for the experiments on beam lines (6-2) and (7-2), respectively.

RESULTS AND DISCUSSION

Monolayer Structure

Fig. 3a and b shows, respectively, azimuthal and radial scans from a monolayer of Pb adsorbed on Ag(111) at a potential slightly greater than for monolayer completion [8]. The relatively large background is primarily due to scattering from the thin electrolyte layer. The diffraction was confirmed to result from the Pb monolayer by repeating the radial scan with the electrode held at 0 V (vs. Ag/AgCl). At this potential, Pb is oxidized to Pb^{2+}, dissolves in the electrolyte, and the monolayer and its diffraction peaks are not observed. Identical diffraction peaks are observed at 60° intervals consistent with the expected six fold symmetry for a hexagonal layer and the (11) and (20) reflections from the adlayer are also observed. These data directly show the Pb monolayer forms an incommensurate hexagonal layer.

Fig. 4 shows the structure of one domain of the Pb monolayer deduced from these data. This structure agrees with that suggested by electrochemical data where the charge measured for the UPD is consistent with the formation of a closed packed monolayer of Pb atoms, assuming two electrons are transferred in the deposition of each Pb^{2+} ion. The monolayer Pb-Pb near-neighbor (nn) distance of a_{nn} = 3.45Å is about 1% smaller than the bulk Pb nn

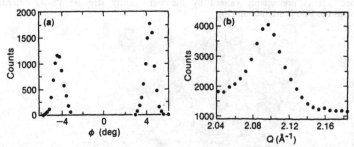

Fig. 3. The (10) reflection from UPD Pb/Ag(111) for a potential slightly greater than for monolayer completion. The Ag ($\bar{2}$11) direction is defined as ϕ = 0. (a) Azimuthal scan at Q = 2.10Å^{-1} with the background subtracted. The +/-symmetry with respect to the substrate results from diffraction from different Pb domains. (b) Radial scan at ϕ = 4.5°. The 0.037Å^{-1} width of the peak indicates a domain size of about 150Å.

spacing of 3.50Å. This contraction is probably caused by the stronger Pb-Ag bonds compared to the Pb-Pb bonds, which is reflected in the 0.3 eV excess adsorption energy [7]. It is interesting to note that the structure shown in Fig. 4 is essentially the same as observed for vapor deposited Pb on Ag (111), although the contraction of the Pb lattice observed in these experiments was less than observed here [7,11]. This similarity is surprising considering the significant difference between the two environments: in solution the monolayer is covered by a large concentration of water and other ions, whose adsorption may have an important effect, while in vacuum there is essentially nothing above the monolayer.

Potential Dependence

As the potential is made more cathodic, the Pb-Pb nn spacing decreases from $a_{nn} = 3.45$Å at -400mV to 3.40Å at -550mV. This compression of the monolayer with applied potential is interpreted as a measure of the 2D compressibility of the Pb monolayer [10]. Since the adsorbed Pb monolayer is in chemical equilibrium with the ions in solution, varying the potential in the region between monolayer formation and bulk deposition changes the chemical potential and is thus analogous to varying the vapor pressure of a gas in equilibrium with its physisorbed monolayer. Equilibrium thermodynamics can thus be used to calculate the 2D isothermal compressibility κ_{2D} and for Pb on Ag(111) we find $\kappa_{2D} = 0.98$Å^2/eV [10]. Recall that for most bulk metals, the compressibility is dominated by the electron compressibility [12], and hence, a similar domination is expected for metal monolayers. A simple model for a metal monolayer is a 2D free electron gas [12], in which the electrons are treated as non-interacting spin-1/2 particles. Using this model, it is a classic graduate-level problem to calculate κ_{2D} and for Pb we estimate $\kappa_{2D} = 0.3$Å^2/eV. This is in good agreement with the measured value, which is probably a result of the free-electron nature of Pb. While this agreement is gratifying, a more realistic calculation is desirable and would involve a 2D band structure calculation that included effects of the Ag substrate.

The interpretation given above is predicated on several conditions. The first is that the monolayer is in equilibrium. This is very reasonable, since no kinetic effects are observed for Pb/Ag(111) in the potential range of the experiments described above [13] and no systematical changes in the diffraction peakshape or linewidth are observed with varying potential [10]. The second is that the Ag substrate is rigid and does not participate in the adlayer compression. The validity of this is not obvious and remains to be tested.

The voltammogram in Fig. 1 and other x-ray diffraction data [10] show that a second Pb layer does not form atop the first monolayer, but rather bulk Pb is deposited (Stranski-Kranstonov growth). This is apparently due to the large mismatch between the bulk and monolayer lattice constants, since at the potential for bulk deposition, the monolayer is compressed 2.8% compared to bulk. This lattice mismatch creates a large strain energy that quenches the epitaxy of further layers and results in the growth of bulk crystallites. It is presumably caused by the significantly stronger Pb-Ag interaction compared to the Pb-Pb interaction [10].

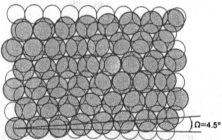

Fig. 4. Schematic representation of one domain of monolayer Pb on Ag(111). Open circles represent the Ag atoms of the (111) surface and shaded circles represent the Pb atoms. The rotational epitaxy angle between the Ag and Pb lattices is $\Omega = 4.5°$.

Fig. 4 shows that the monolayer is rotated 4.5° from the Ag ($\overline{2}11$) direction, which is not unexpected for an incommensurate monolayer, since such an alignment, referred to as rotational epitaxy, has been observed for incommensurate physisorbed gases and chemisorbed alkali atoms [14]. A model for rotational epitaxy was first developed by Novaco and McTague [15] and later generalized by Shiba [16]. The periodic substrate-adsorbate interaction potential (the potential energy of an adatom as it moves over the surface) makes it energetically favorable for the adlayer to assume a rotation that is not along a high symmetry direction of the substrate. Using the Novaco and McTague model [15] a rotational epitaxy angle of about 5-6° is predicted. Although this is in reasonable agreement with the measured value of 4.5°, we did not observe any changes in the rotational epitaxy angle as the lattice spacing in monolayer decreased from 3.45 to 3.40Å. This is unexpected since either the Novaco-McTague or Shiba models predict changes of 0.5° or more, which would easily be observable. The reasons for this are uncertain, however, very recent data for Tl electrochemically adsorbed on Ag(111) suggest that trace impurities can affect the rotational epitaxy angle [17].

Effect of Pb Monolayer on Ag Substrate

It is important to determine what effect the deposition of the Pb monolayer has on the underlying Ag substrate, since this provides information on the atomic interaction between the Pb overlayer and the Ag substrate and may relate to the compressibility interpretation given above. Some information about this can be obtained by measuring the Ag crystal truncation rods (CTRs). These are diffuse rods of scattered intensity that result from the truncation of a crystal at a sharp boundary, e.g. a surface or interface. They connect Bragg points, are spread across reciprocal space, and can provide important information on surface and interface topography [18].

Fig. 5 shows radial scans of the Ag $1/3(\overline{4}22)$ CTR at -500mV, where the Pb monolayer is adsorbed, and at -300mV, where no Pb is adsorbed. The incidence angle is slightly larger than the critical angle. These data indicate that the lineshape of the CTR does not change when the Pb adsorbs, but the intensity drops significantly. Since the attenuation by the Pb monolayer is negligible, this shows the adlayer does affect the Ag substrate, although these data do not unambiguously show the nature of the effect. One possible explanation is that the adlayer causes the substrate surface to become rougher, although other interpretations are also possible. To better understand this effect, it is necessary to measure the intensity of the Ag CTR as a function of incidence angle both with and without the adlayer. These experiments are in progress.

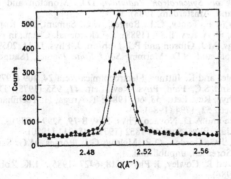

Fig. 5. Radial scan of Ag $1/3$ ($\overline{4}22$) crystal truncation rod at $Q_z \simeq 0$. The solid circles and solid line is for a potential of -300mV, with the no Pb on the surface. The triangles and dashed lines are for -500mV, where a full Pb monolayer has formed.

SUMMARY

Surface x-ray scattering has been used to study the structure of Pb monolayers electrochemically adsorbed on Ag (111). The Pb monolayer forms an incommensurate, hexagonal two-dimensional (2D) solid that is rotated approximately 4.5° from the Ag (211) direction. As the potential is varied between monolayer formation and bulk deposition, the Pb-Pb near neighbor distance decreases linearly and from this the 2D isothermal compressibility is measured. It is in good agreement with that calculated using a 2D non-interacting free electron gas model for the Pb monolayer. At the potential for bulk deposition the monolayer is compressed 2.8% compared to bulk Pb and this lattice mismatch apparently results in the Stranski-Kranstonov growth mode. The intensity of the Ag crystal truncation rod decreases when the Pb monolayer is adsorbed, which might be caused by a roughening of the substrate surface.

ACKNOWLEDGMENTS

The work described here was partially done at SSRL, which is funded by the Department of Energy.

REFERENCES

1. A. Hubbard, Acc. Chem. Res. *13*, 177 (1980); E. Yeagar, J. Electroanal. Chem. *128*, 1600 (1981); P.N. Ross, Surf. Sci. *102*, 463 (1981).

2. M. Fleischmann, P.J. Hendra, and A.J. McQuillan, Chem. Phys. Lett. *26*, 173 (1974); D.J. Jeanmaire and R.P. Van Duyne, J. Electroanal. Chem. *84*, 1 (1977).

3. G.L. Richmond, J.M. Robinson, and V.L. Shannon, Prog. Surf. Sci. *28*, 1 (1988).

4. S. Pons, J. Electroanal. Chem. *150*, 495 (1983); A. Bewick, J. Electroanal. Chem. *150*, 481 (1983).

5. D.M. Kolb in *Advances in Electrochemistry and Electrochemical Engineering*, Vol 11, H. Gerischer and C. Tobias, Eds., (Wiley, New York, 1978) p. 125.

6. D.M. Kolb, J. Vac. Sci. Technol. *A4*, 1294 (1986) and references therein.

7. K. Takayanagi, D.M. Kolb, K. Kambe and G. Lehmpfuhl, Surf. Sci. *100*, 407 (1980); K. Takayanagi, Surf. Sci. *104*, 527 (1981).

8. M.G. Samant, M.F. Toney, G.L. Borges, L. Blum and O.R. Melroy, Surf. Sci. *193*, L29 (1988); J. Phys. Chem. *92*, 220 (1988).

9. J. Als-Nielsen in *Structure and Dynamics of Surfaces*, Vol. II, W. Schommers and P. von Blanckenhagen, eds. (Springer-Verlag, Berlin, 1987) pp.181; I.K. Robinson in *Handbook on Synchrotron Radiation*, D.E. Moncton and G.S. Brown, eds. (North Holland, Amsterdam, 1988).

10. O.R. Melroy,M.F. Toney, G.L. Borges, M.G. Samant, J.B. Kortright, P.N. Ross, and L. Blum, Phys. Rev. B *38*, (1988); J. Eelctroanal. Chem., in press.

11. K.J. Rawlings, M.J. Gibson and P.J. Dobson, J. Phys. D *11*, 2059 (1978).

12. N.W. Ashcroft and N.D. Mermin, *Solid State Physics*, (Saunders, Philadelphia, 1976).

13. H. Siegenthaler and K. Juttner, Electrochimica Acta *24*, 109 (1979).

14. C.G. Shaw and S.C. Fain, Phys. Rev. Lett. *41*, 955 (1978); D. Doering and S. Semancik, Phys. Rev. Lett. *53*, 66 (1984); T. Aruga, H. Tochihara, and Y. Murata Phys. Rev. Lett. *52*, 1794 (1984).

15. J.P. McTague and A.D. Novaco, Phys. Rev. B *19*, 5299 (1979).

16. H. Shiba, J. Jap. Phys. Soc. *46*, 1852 (1979); *48*, 211 (1980).

17. M.F. Toney, J.G. Gordon, O.R. Melroy, G.L. Borges, M.G. Samant, G. Kau, D. Yee, and L. Sorensen, unpublished.

18. S. Andrews and R. Cowley, J. Phys. C18, 6427 (1985). I.K. Robinson, Phys. Rev. B33, 3830 (1986).

PART II

Absorption Spectroscopy and Electronic Structure

SYNCHROTRON RADIATION STUDIES OF MAGNETIC MATERIALS

J.L.ERSKINE, C.A.BALLENTINE, JOSE ARAYA-POCHET, AND RICHARD FINK
University of Texas, Department of Physics, Austin, Texas 78712

ABSTRACT

New opportunities for research on magnetic materials are emerging as a result of quiet revolutions in several areas including: materials synthesis techniques, surface characterization capabilities, new magnetic sensitive detectors and spectroscopic techniques, improved synchrotron radiation instrumentation, and predictive modeling based on first principals calculations. This paper describes some of the more recent advances and assesses some of the new opportunities that are emerging in the field of magnetic materials research.

INTRODUCTION

Magnetism is one of the oldest subfields of solid state physics. In spite of decades of work in the field of magnetism and magnetic materials, renewed interest in the field is being generated by several factors that now form a basis for unprecedented opportunities for probing magnetism especially of novel magnetic materials. The new opportunities are not based on a single breakthrough (as for example, the recent discovery of new high Tc superconductors), but are emerging as a result of a number of technological advances in several related areas. There are at least five important areas in which new scientific capabilities are driving interest in magnetic materials research. These areas include: 1) the advent of new materials synthesis techniques, 2) refinements in surface structure characterization techniques, 3) development of new compact spin detectors and spin sensitive techniques for probing magnetism, 4) the successful operation of undulators at storage ring facilities, and 5) remarkable achievements in predictive capabilities of first-principles calculations. In addition, the tremendous technological importance of magnetic materials in data storage applications, electromechanical devices and power generation and distribution has continued to stimulate broad interest in the field.

MATERIALS SYNTHESIS AND CHARACTERIZATION

Perhaps the most important factor responsible for new interest in magnetic materials is the evolution of techniques for synthesizing and characterizing thin film magnetic structures. The advent of molecular beam epitaxy (MBE) and broad advances in surface and thin film characterization capabilities now makes possible routine growth of epitaxial metal films having precisely determined structure. The recent introduction of scanning tunneling microscopy [1] and surface structure analysis [2] based on x-ray diffraction using synchrotron radiation, represent important advances in structure sensitive probes. More importantly, MBE permits stabilization of novel phases of matter that do not occur in nature, (i.e., bcc Ni and fcc Fe), and provides a mechanism for deliberate modifications in the atomic-level structure of thin films due to lattice mismatches and template crystal structure'. By deliberately modifying the lattice constant, the local symmetry or the coordination of an epitaxial film, systematic trends in magnetic properties as influenced by these microscopic changes can be investigated.

FIRST-PRINCIPLES PREDICTIVE MODELING; THEORY

A second very important factor stimulating interest in magnetic materials is the outstanding achievements in predictive capabilities based on the local density approximation. These advances have been made by a number of groups, but here only a few of the results most relevant to the present discussion are mentioned. Confidence in the predictive capabilities of large-scale first principles calculations has been established methodically by addressing, in order, the electronic and magnetic properties of bulk materials, intrinsic surface properties (i.e., dead layers, etc.) [3] and thin epitaxial films. Novel material phases of bulk materials and thin films of normally non-ferromagnetic materials (Cr, V) have also been studied.[4,5]

We now address some specific issues and predictions that are candidates for detailed tests using new experimental probes of magnetic properties including spin-polarized photoemission and synchrotron radiation. In 1985, Fu et al. [5] reported predictions of enhanced magnetic moments in epitaxial transition metal films on noble metal surfaces. Reduced symmetry, lower coordination of neighboring atoms, and electronic structure effects (surface and interface states) were factors that influenced the enhanced moment formation. Related work by Tersoff and Falicov [6] addressed the role of sp-d hybridization in affecting the magnetic properties of epitaxial monolayers of Ni grown on Cu(100) and Cu(111) surfaces. A striking prediction of the work by Tersoff and Falicov was the quenching of the magnetic moment of a monolayer Ni film on Cu(111) but not on Cu(100). An important calculation by Gay and Richter [7] treated spin-orbit effects, and arrived at the conclusion that in very thin films (monolayers) spin-orbit effects introduce a term in the magnetic anisotropy that is of sufficient strength to govern the spin direction in certain cases. In particular, for a monolayer of Fe on Ag(100), the calculations predict the preferred spin direction is perpendicular to the surface. This result is in contrast with intuition for thin film behavior. The (volume dependent) shape anisotropy term normally forces spins in a thin film to lie in the film plane.

There are other related issues that can be addressed simultaneously in experiments dealing with the magnetic properties of ultra-thin films. One example relevant to results presented in this paper is the temperature dependence of saturated magnetization of thin films and its relationship to two-dimensional models of phase transformations.

The remainder of this paper deals with recent experimental progress towards probing thin film magnetism. Some of the new results that have been obtained in two specific thin film systems are described. These new results are used as a basis for judging future activity and opportunities in the field, particularly in regard to experiments based on synchrotron radiation.

MACROSCOPIC MAGNETIC PROPERTIES OF ULTRA-THIN FILMS

Advances in synchrotron radiation instrumentation and spin detection (described in the following section) offer the opportunity to probe in great detail the atomic-level factors that govern thin film magnetic behavior. However, spin-dependent detection of the energy and angle resolved distribution of photoemitted electrons from a thin epitaxial film is a technically difficult and costly experiment. Detailed knowledge of macroscopic magnetic properties such as coercive forces, anisotropies, remanent magnetism, and conditions (temperature, sample thickness, surface roughness) that affect these properties are prerequisites for any synchrotron radiation based photoemission studies. These properties can be effectively studied using magneto-optical techniques, and the results provide an essential basis for planning more difficult electronic structure measurements based on polarized electron spectroscopy.

Figure 1 displays layer dependent hysteresis curves for p(1x1) epitaxial Ni layers grown by MBE on Ag(111). The epitaxy of this system is excellent judged from low energy electron diffraction (LEED), and contamination levels of the substrates and films resulting from residual gas adsorption prior to and during measurements is below 1% of a monolayer as judged by Auger analysis. The hysteresis loops were generated by using the surface magneto-optic Kerr effect (SMOKE) as described previously.[8] The measurements corresponding to data displayed in Figure 1 were conducted with the applied magnetic field H parallel to the film plane. Corresponding measurements conducted with H perpendicular to the plane failed to detect Kerr effect signals showing that the spin orientation is in the plane.

In other measurements [8] carried out on p(1x1) Fe layers on Ag(100), hysteresis loops for H perpendicular to the surface were observed for a film thickness of $n = 1$ ML, but not for $n \geq 2$ ML. For H parallel to the film plane, the same films yielded no Kerr effect signals for $n = 1$ ML, and square loops for $n \geq 2$ ML. These measurements and related studies using other methods such as polarized electron emission, [9] Mossbauer measurements,[10] and ferromagnetic resonance [11] have established the existence of spin-orbit induced surface spin anisotropy. The spin orientation of a 1 ML p(1x1) Fe film on Ag(100) at $T = 100K$ is perpendicular to the surface.

Returning to Figure 1, it is profitable to examine in greater detail the information represented by the hysteresis loops, especially in relation to more sophisticated experiments such as spin-polarized photoemission. First, it is clear that the coercive force Hc is small (about 100 Oe) and weakly dependent on thickness. (For bcc Fe on Ag(100), the coercive force is

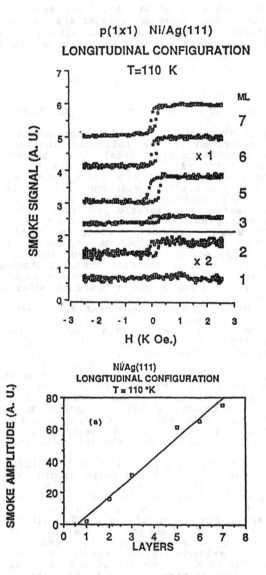

Figure 1 Upper panel: Surface magneto-optic Kerr effect (SMOKE) derived hysteresis loops for p(1x1) Ni on Ag(111) as a function of film thickness in monolayers (ML). Lower panel, SMOKE amplitude vs thickness.

much smaller, ~ 10 Oe for H parallel to the plane). These properties, along with the high degree of remanent magnetization apparent from the nearly rectangular loop, are fortuitous for spin-polarized studies. A simple wire loop near the film energized by a capacitor is sufficient to establish and reverse the magnetization direction. Also, because the remanent magnetization is essentially the same direction and magnitude as the saturation magnetization, the film exhibits a single domain (for H = 0) at least over the area probed by the light beam (1 mm^2) and probably over the entire film. This is also a useful property of the films for polarized electron experiments.

Referring to Figure 1, it is clear that for n = 1 ML, the magnetic moment is significantly reduced compared to n > 2 ML. This apparent quenching of the magnetism at n = 1 ML is consistent with predicted effects due to sp-d hybridization.[6] Corresponding Kerr effect studies [8] of Ni films on Cu(100) do not exhibit magnetic quenching. Well-defined hysteresis curves which yield linear behavior as a function of n down to single ML coverage are observed showing that ferromagnetic behavior persists at n = 1 ML. This result (crystal surface dependent quenching of magnetism) is very interesting and suggests that a surface related electronic structure effect plays a critical role in surface and interface magnetism. Angle and spin-resolved photoemission experiments should provide additional insight into this mechanism affecting magnetism of thin films.

Before describing progress toward the capability to routinely conduct spin-polarized angle resolved photoemission experiments using synchrotron radiation, we briefly illustrate some interesting new results also obtained using Kerr effect measurements. Figure 2 displays the layer and temperature dependent magnetic behavior of p(1x1) Ni films on Cu(111).

These data exhibit the striking departure of physical properties of ultra-thin epitaxial magnetic films from those of bulk materials. Note that the Curie temperatures of the films vary rapidly with film thickness, and that the temperature dependence of the saturated magnetization becomes a linear function for n = 1 ML. This particular system (Ni on Cu(111)) appears to exhibit excellent epitaxy, and is stable (no apparent diffusion at the interface) over a temperature range extending to ~ 200 C. These properties render the system suitable for detailed studies of magnetic two-dimensional phase transformation.

SPIN-POLARIZED PHOTOEMISSION EXPERIMENTS

Spin-polarized angle resolved photoemission spectroscopy using synchrotron radiation is a source limited technique. It is not feasible to conveniently incorporate multichannel detection, therefore the source and spin detector offer the only alternatives for improving sensitivity. As an approximate bench mark for judging the problems facing polarized electron spectroscopy, it is only necessary to consider the efficiency of current spin detectors (10^{-4}) and the typical counting rates of non spin-resolved experiments. A good bending magnet beamline (accepting 25 m rad) with monochromator and energy analyzer

NI/Cu(111)
LONGITUDINAL

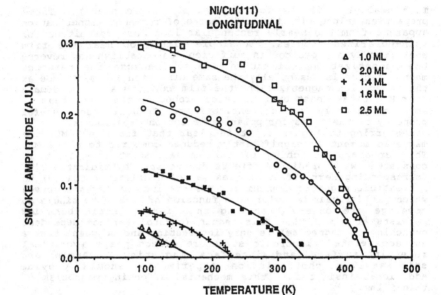

Figure 2 Layer and temperature dependence of magnetism of ultra
-thin films of Ni on Cu(111). Note the dramatic difference in
the Curie temperature vs film thickness.

operating at 100 meV energy resolution and 1° angle resolution
will achieve counting rates of the order of 10^6/sec under
favorable conditions. Under the same conditions, a
spin-resolved measurement will achieve a counting rate of
10^2/sec.

New compact low energy spin detectors [12-14] are small
enough to be easily incorporated into movable angle-resolving
energy analyzers. The figure of merit for these detectors is
comparable to the large Mott scattering detectors (10^{-4}).
Although a number of refinements have recently been made in
compact spin detectors, it is clear that significant (factors of
10) additional improvements are unlikely. The best prospects of
improving sensitivity in spin-polarized photoemission
experiments is by improving the source intensity.

Figure 3 displays the spectrum of a new undulator being
constructed for the U5 insertion device beamline at the National
Synchrotron Light Source. Undulator radiation exhibits
properties that are qualitatively different from the radiation
emitted from bending magnets. Figure 3 illustrates some of the
principle features that render undulator radiation useful for
spin-polarized photoemission experiments. The most significant
property is the increase in useable flux from an insertion
device. The radiation from a magnetic lattice is produced in
harmonics at particular energies. The strength of a particular

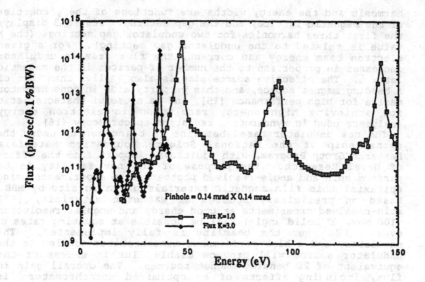

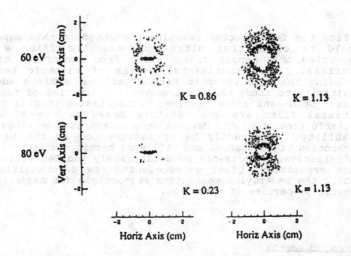

Figure 3 Upper panel: Computed flux from a 33 pole undulator (P. J. Viccaro, G. K. Shenoy, S. H. Kim, and S. D. Bader, Rev. Sci. Instru.(in press)); lower panel, radiation patterns at various photon energes at a point 1000 cm from center of an undulator for two gap settings. (D. Anacker and J. L. Erskine, ref. 15).

harmonic and the energy widths are functions of the properties of the magnetic lattice, and the gap setting. Figure 3 displays the first three harmonics for two undulator gap settings (the K value is related to the undulator gap setting). For a given electron beam energy and current, the flux from an undulator increases in proportion to the number of periods in the magnetic lattice. The effective source size is also smaller than that of a bending magnet source, and this property aids in monochromator design for high performance [15], and is a useful characteristic for achieving high energy resolution in electron energy analyzers used in synchrotron radiation spectroscopy [16].

The new undulator based beamline is being set up under the sponsorship of the National Science Foundation Materials Research Group Program, with additional support from the Office of Naval Research. The purpose of the new facility is to perform spin and angle-resolved photoemission experiments using epitaxial thin film magnetic materials grown in situ by MBE. Based on preliminary studies of magnetic materials, [17] spin-resolved experiments at good energy and angular resolution (100 meV, 2° solid angle) will be possible at counting rates of over 1 kHz when the beamline is fully implemented. The undulator, and a high performance monochromator coupled to the undulator source will produce usable flux in excess of the equivalent of 20 bending magnet sources. The overall gain in flux, including effects of an optimized monochromator is sufficient that routine spin and angle resolved photoemission studies of magnetic materials will be possible. These capabilities coupled with the novel materials synthesis capabilities of MBE will provide important new opportunities for fundamental studies of magnetism and magnetic materials.

OUTLOOK

From the few selected examples presented in this paper, it should be clear that ultra-thin magnetic films exhibit properties that depart dramatically from properties of bulk materials. New experimental probes of magnetic behavior including the magneto-optic Kerr effect can achieve adequate sensitivity to study the novel magnetic properties of monolayer films. Spin- and angle-resolved photoemission studies of thin epitaxial films are now feasible based on the high flux available from undulator devices on electron storage rings. The capability is currently being implemented at the National Synchrotron Light Source, and will soon permit routine spin- and angle-resolved experiments on deliberately modifed structures. These experiments offer unprecedented new opportunities for probing the underlying mechanisms responsible for magnetism and megnetic properties of materials.

Acknowledgements

This work was sponsored by the NSF DMR-86-03304 (MRG Program), DMR-87-02848, and by the Joint Services Electronic Program AFOSR-F49620-86-C-0045.

References

1. Recent developments in STM are described in *J. Vac. Sci. Technol.* **6** (1988). (Entire issue devoted to STM).
2. S. Brennan, P.H. Fuoss, P. Eisenberger, *Phys. Rev.* **B33**, 3678 (1986).
3. An extensive but not complete reference to electronic structure calculations for bulk ferromagnetic materials and their surfaces can be found in: A.M. Turner, A.W. Donoho, and J.L. Erskine, *Phys. Rev.* **B29**, 2986 (1984); and A.M. Turner and J.L. Erskine, *Phys. Rev.* **B30**, 6675 (1984).
4. S. Blugel, M. Weinert, and P.H. Dederichs, *Phys. Rev. Lett.* **60**, 1077 (1988).
5. C.L. Fu, A.J. Freeman, and T. Oguchi, *Phys. Rev. Lett.* **54**, 2700 (1985); A.J. Freeman and C.L. Fu, *J. Appl. Phys.* **61**, 3356 (1987).
6. J. Tersoff and L.M. Falicov, *Phys. Rev.* **B26**, 6186 (1982).
7. J.G. Gay and R. Richter, *Phys. Rev. Lett.* **56**, 2728 (1986); *J. Appl. Phys.* **61**, 3362 (1987).
8. Jose Araya-Pochet, C.A. Ballentine, and J.L. Erskine, *Phys. Rev.* **B38**, xxx (1988).
9. M. Stampanoni, A. Vaterlaus, M. Aeschlimann, and F. Meier, *Phys. Rev. Lett.* **59**, 2483 (1987).
10. N.C. Koon, B.T. Jonker, F.A. Volkening, J.J. Krebs, and G.A. Prinz, *Phys. Rev. Lett.* **59**, 2463 (1987).
11. B. Heinrich, K.B. Urquhart, A.S. Arrott, J.F. Cochran, K. Myrtle, and S.T. Purcell, *Phys. Rev. Lett.* **59**, 1756 (1987).
12. L.A. Hodge, T.J. Moravec, F.B. Dunning, and G.K. Walters, *Rev. Sci. Instrum.* **50**, 5 (1979); F.B. Dunning, L.G. Gray, J.M. Ratliff, F.-C. Tang, X. Zhang, and G.K. Walters, *Rev. Sci. Instrum.* **58**, 1706 (1987).
13. J. Krischner and R. Reder, *Phys. Rev. Lett.* **47**, 1008 (1979).
14. J. Unguris, D.T. Pierce, and R.J. Celotta, *Rev. Sci. Instrum.* **57**, 1314 (1986).
15. D. Anacker and J. L. Erskine, *Nucl. Instrum. Methods* **A266**, 336 (1988).
16. G. K. Ovrebo and J. L. Erskine, *J. Electron Spectros. Rel. Phenom.* **24**, 189 (1981); H. A. Stevens, A. M. Turner, A. W. Donoho, and J. L. Erskine, *J. Electron Spectros. Rel. Phenom.* **32**, 327 (1983).
17. P. D. Johnson, A Clark, N. B. Brooks, S. L. Hulbert, B. Sinkovic and N. V. Smith, *Phys. Rev. Lett.* **61**, 2257 (1988).

SYNCHROTRON PHOTOEMISSION STUDIES OF SURFACES AND OVERLAYERS

T.-C. CHIANG
Department of Physics and Materials Research Laboratory, University of
Illinois at Urbana-Champaign, 1110 W. Green Street, Urbana, IL 61801

ABSTRACT

High-resolution core-level photoemission spectroscopy allows the
distinction of atoms in different layers and in inequivalent sites by their
binding energy shifts. By comparison with model structures and reference
samples, the number of atoms in each distinct chemical configuration can be
determined. The chemical shifts induced by adsorption can be correlated
with the electronegativity difference between the substrate and the
adsorbate atoms. These observations provide a quantitative description of
the interaction and reaction between adsorbates and surfaces, and important
information about the atomic structure and the electronic properties can be
deduced. Results from several representative systems including the
adsorption of In, Ag, and Sn on Si(100) will be discussed.

INTRODUCTION

In this paper recent developments in the application of photoemission
spectroscopy to the studies of surfaces and adsorbates will be described.
The atomic structure and electronic properties of surfaces have attracted
considerable interest in recent years. Core-level spectroscopy is a
particularly powerful tool for such investigations.[1] The core-level
binding energies of the surface atoms of a clean surface often show small
shifts relative to the bulk values. If the surface reconstructs to exhibit
inequivalent sites, then even the site-dependent core-level shifts can often
be detected. This information can be used to examine the various structural
models for the reconstruction. With the adsorption of atoms on the surface,
the core levels of the surface atoms involved in the adsorption bonding will
generally show chemical shifts. By examining the evolution of the various
surface core-level components as a function of adsorbate coverage, the
interaction and reaction between the adsorbate and the substrate surface
atoms can be studied in detail.

While the above concepts are simple and straightforward, most of the
work in core-level spectroscopy has been limited to qualitative or semi-
quantitative studies.[1] There are several reasons for this limitation: (1)
The site-dependent shifts and adsorbate-induced chemical shifts are usually
quite small and difficult to resolve. To deduce the relative populations of
different species, it is necessary to obtain accurate measurements of the
relative intensities. (2) The surfaces under study may exhibit a mixture of
differently reconstructed domains, causing considerable difficulty in the
identification and counting of surface atoms in inequivalent sites. This is
in fact a fairly common problem for semiconductor systems. (3) The surfaces
may have defects which should be accounted for in such studies. (4)
Accurate predictions of core-level shifts are often unavailable.

Problem (1) mentioned above is now largely resolved with the use of
synchrotron-radiation light sources and state-of-the-art monochromators
which provide an intense and highly monochromatic beam of photons. Problems
(2) and (3) can be resolved by combing information obtained from scanning
tunneling microscopy (STM) and photoemission. It is important to prepare
the samples for both the STM and photoemission experiments under the same
conditions; thus, the mutually complementary information obtained from these
two techniques can be directly related. A solution to problem (4) is to
perform experimental investigation of the chemical trend to identify the

material parameters relevant to the core-level shifts. In the following, an empirical correlation will be established between the chemical shifts and the adsorbate-substrate electronegativity difference and the bonding coordination number.

To illustrate these ideas, we will present the results for the adsorption of In, Ag, and Sn on Si(100)-(2x1).[2-4] The In- and Ag-Si(100) systems were prepared by deposition with the substrate at nearly room temperature; therefore, the main effect of adsorption should be the saturation of the dangling bonds on the surface, and the substrate structure is expected to be nearly unaffected. The Sn-Si system was prepared at higher temperatures, and more complicated atomic rearrangement on the surface is likely to occur.

Si(100)-(2x1)

Figure 1 shows a photoemission spectrum for the Si 2p core level taken from Si(100)-(2x1); for comparison, a spectrum for Si(111)-(7x7) is also included.[1,5] The photoelectron escape depth in the sample is rather short for the photon energy used; therefore, the spectra contain substantial emission from the surface layer. By comparison with spectra taken under more bulk-sensitive conditions (lower photon energies) and with the use of a computer fitting procedure described in detail elsewhere, the spectra can be decomposed into the surface and bulk contributions.[1,5] The results of the fit and the decomposition into the bulk (B) and surface (S, S1, and S2) contributions are indicated by the various curves. The ability to detect and identify surface emission is essential for quantitative chemisorption studies.

To count the number of surface atoms, it is necessary first to determine the relation between the measured photoemission intensity and the number of atoms contributing to the emission. The STM results of Hamers, Tromp, and Demuth indicate that the (2x1) reconstruction of Si(100) is a result of dimer formation on the surface to reduce the number of dangling bonds from 2 to 1 for each surface atom.[6] Figure 2 presents a top view of the (2x1) reconstruction. They also find that both buckled (asymmetric) and nonbuckled (symmetric) dimers are present with about equal population, and about 10% of the surface area consists of "missing-dimer defects"

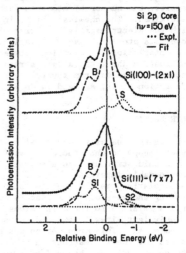

Fig. 1   Si 2p core-level spectra for Si(100)-(2x1) and Si(111)-(7x7). The results of the fit and the decomposition into the various components are indicated. The binding energy is referred to the bulk Si $2p_{3/2}$ component.

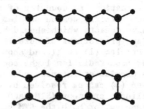

Fig. 2   Top view of the dimerized Si(100)-(2x1) structure.

(that is, the dangling bonds are not seen by STM). The buckled dimers are mostly located near the defects. It has been speculated that the apparently nonbuckled dimers are in fact dynamically-buckled dimers, and the defects tend to stabilize nearby dimers into the stationary buckled configuration. Since there is always the question about the reproducibility and quality of the sample in any surface experiment, we have also performed STM studies of Si(100) prepared under the same conditions as those used for our photoemission measurements. The results are similar to those described above except that we saw somewhat more nonbuckled dimmers than the buckled dimmers. The upper panel of Fig. 3 shows a STM image over a large area. The bright periodic stripes are dimer rows, and there is a domain boundary near the middle of the picture (the very bright and extended appearance of the boundary has to do with the way the picture was taken). An examination of an enlarged version of the same picture shows the presence of both buckled and nonbuckled dimers.

Since the core-level spectra (see Fig. 1) show only one surface component, it is not clear which of the three kinds of dimer atoms (two for the buckled dimer and one for the nonbuckled dimer) contribute to the surface shift. An independent absolute intensity calibration in needed; this is done by comparing the results of Si(100) with those of Si(111).[5]

The two spectra in Fig. 1 were taken under identical experimental conditions (same photon energy, same analyzer, same experimental geometry, etc.); therefore, the intensities can be directly compared. The S2 component for the Si(111)-(7x7) surface has been associated with the 12 surface "adatoms" in a (7x7) unit cell.[5,7] Our STM results, in agreement with previous studies done by other groups, show that the (7x7) surface prepared under the usual conditions has few defects, and all unit cells are equivalent (a STM image is shown in the lower panel of Fig. 3). Therefore, the S2 component is a good intensity reference. From a series of such measurement involving 20 pairs of (100) and (111) samples, we have deduced that the S component for Si(100)-(2x1) corresponds to emission from 0.92 ± 0.07 monolayers (ML).[5] Since this number is fairly close to the ideal value of 1 ML, we conclude that all three kinds of dimer atoms contribute to the S emission, and the departure from the ideal value of 1 ML is due to the defects. The surface core-

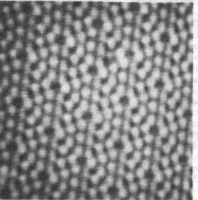

Fig. 3 STM images of Si(100)-(2x1) (upper panel) and Si(111)-(7x7) (lower panel). The distance between neighboring dimer rows on Si(100) is 7.68 Å, and the distance between two neighboring "corner holes" on Si(111) is 26.9 Å.

level shift for clean Si(100)-(2x1) is due to the difference in atomic environment; the surface atoms are 3-fold coordinated with a dangling bond, while the bulk atoms are 4-fold coordinated. There is a higher average electronic charge at the surface atomic site due to the dangling bond. This results in a higher average electrostatic potential energy, so the surface core-level binding energy is reduced relative to the bulk value.

The Ge(100) surface exhibits a very similar (2x1) reconstruction. It also shows a surface core-level shift towards smaller binding energies. We have determined the number of surface atoms contributing to the surface emission in this case, too.[8] For intensity reference, we used a sample of Si(100) covered by 0.25 ML of Ge. The Ge coverage in this case was calibrated in an absolute manner based on the intensity oscillation in high-energy electron diffraction during the epitaxial growth of Ge on Ge(100). Again, core-level spectra were taken from the Ge(100) and the Ge-on-Si(100) systems under the same conditions; from the intensity comparison, we have deduced that the Ge(100) surface emission corresponds to 0.87 ± 0.09 ML. This number is in close agreement with that for the Si(100) case, thus providing a strong support for the above interpretation.

Several earlier theoretical calculations indicated the possibility of a significant charge transfer between the two atoms in a dimer when buckling occurs. The charge transfer should lead to a corresponding core-level shift due to the electrostatic potential change. Since our results indicate that all three kinds of dimer atoms have the same binding energy, the charge transfer associated with buckling must be small, estimated to be less than about 0.1 e (e is the electronic charge). The relation between charge transfer (or bond ionicity) and the core-level binding-energy shift will be examined below.

In-Si(100)

Having characterized the clean Si(100)-(2x1) surface, we can now examine the physics of adsorption of foreign atoms. The first system to be discussed is In on Si(100)-(2x1).[2] The bulk solubility of In in Si is very low at room temperature, and there is no known bulk compounds formed between In and Si. Therefore, for deposition at nearly room temperature, it is expected that the In will simply saturate the Si dangling bonds with little disturbance to the Si substrate structure. From HEED studies, it is known that for In coverages θ < 0.1 ML the In adatoms are dispersed.[9] Here the coverage unit ML is defined with respect to the Si(100) substrate. At higher coverages the In forms (2x2) islands, and by about 1/2-ML coverage the (2x2) structure becomes fully developed.[9]

Figure 4 shows the Si 2p core-level spectra for increasing coverages of In. Clearly, the surface-shifted (S) component is gradually converted to become

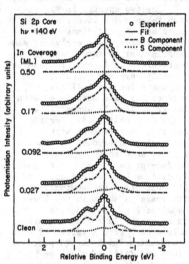

Fig. 4   Si 2p core-level spectra taken with a photon energy of 140 eV for Si(100) covered with various amounts of In. The results of the fit and the decomposition into the surface (S) and bulk (B) components are indicated. The binding energy is referred to the bulk Si $2p_{3/2}$ component.

bulklike, and for In coverages beyond about 0.4-0.5 ML the shifted component is no longer observed. The simplest interpretation is that the initially 3-fold coordinated Si surface atom becomes 4-fold coordinated (and hence bulklike) after the dangling bond is converted to an In-Si chemisorption bond, so the surface shift is suppressed. For θ = 0.4-0.5 ML where the (2x2) structure is fully developed, all of the surface dangling bonds are saturated. Since the clean Si(100) surface has about 0.9 ML of surface atoms with a dangling bond, this result yields a bonding coordination number of 2 for each In adatom at this coverage. In other words, each In adatom is attached to two Si surface atoms, so two Si dangling bonds are saturated.

Figure 5(a) shows the simplest possible structural model for the In-induced (2x2) structure, where the defects have been ignored. This model was originally proposed by Knall, Sundgren, Hansson, and Greene based on electron-diffraction, Auger, and scanning-electron-microscopy studies.[9] The model explains nicely the present core-level results, namely, each In adatom saturates two Si dangling

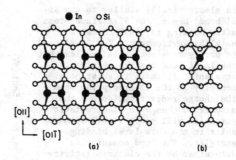

Fig. 5   Structural models for (a) Si(100)-In(2x2) (In coverage 1/2 ML) and (b) an isolated In adatom on Si(100)-(2x1).

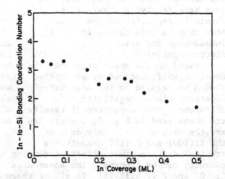

Fig. 6   The average number of surface Si atoms affected by an In adatom (In-to-Si bonding coordination number) as a function of In coverage.

bonds. In this model, the In atoms form dimers; this is necessary to account for the chemical valence of In, which is three. The readers can easily verify that the In coverage is 1/2 ML, and the structure is (2x2).

Since the measured intensity of the surface-shifted Si 2p core-level component in Fig. 4 reflects the number of remaining dangling bonds, the In-to-Si bonding coordination number can be determined as a function of In coverage. The results are shown in Fig. 6. This number is about 3 at low coverages where the In adatoms are dispersed, and decreases to 2 at about 1/2-ML coverage as discussed above. The low-coverage result can be easily understood, as the bonding coordination number is just the chemical valence of In. Figure 5(b) shows schematically the structural model for an isolated In adatom. To satisfy the chemical valence of In, each adatom must saturate 3 Si dangling bonds. For intermediate coverages between 0.1 and 1/2 ML, the system is likely to be a mixture of dispersed adatoms and (2x2) islands; therefore, the bonding coordination number falls between 2 and 3.

CHEMICAL SHIFT AND ELECTRONEGATIVITY DIFFERENCE

In the above analysis, we have implicitly assumed that the In-Si bond

is electronically similar to the Si-
Si bond; hence, the Si surface atoms
after bonding to the In "see" a
bulklike environment. The Si-Si bond
is purely covalent; this implies that
the In-Si bond must be mainly
covalent. If there is substantial
ionicity in the chemisorption bond,
the charge redistribution and the
associated change in electrostatic
potential will cause a corresponding
shift in the core-level binding
energies. The bond ionicity is
determined by the electronegativity
difference between the two atoms
involved in the bonding. Here we
will adopt the electronegativity
table given by Sanderson,[10] which
was derived from a rather extensive
data base involving a variety of
chemical properties of the bond.
According to this table, In and Si
indeed have the same
electronegativity.

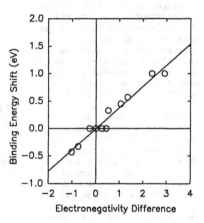

Fig. 7   Adsorbate-induced surface
chemical shifts (relative to the
bulk atoms) for Si(100) and Ge(100).
Only data points corresponding to a
single chemisorption bond per
surface atom are included. The
abscissa is the difference in
electronegativity between the
adsorbate and the substrate atoms.

Based on the above argument, the
chemical shift induced by adsorption
should be related to the adsorbate-
substrate electronegativity
difference. The experimental results
are summarized in Fig. 7; the circles
are the data points. Only data for
the Si(100) and Ge(100) substrates
are included, so other effects including the differences in chemisorption
geometry can be ignored. The adsorbates include Ag, In, Sn, Ga, Sb, As, S,
Cl, O, and F.[2-4,11-15] In all of these cases, care was taken to include
only data points for which only one of the four bonds of the substrate atom
is a chemisorption bond. From previous studies of O and F on Si, it is
known that the chemical shift for multiple chemisorption bonds is just the
number of chemisorption bonds times the shift for a single chemisorption
bond.[11,12] Thus, the chemical shift appears to be additive. The data for
Si(111) and Ge(111) were not included in Fig. 7, because the surface
structures of clean Si(111) and Ge(111) are rather complex and the
chemisorption geometries are generally unknown.

The results in Fig. 7 show an approximately linear relation. For
systems with small (unresolved) chemical shifts which cannot be deduced
reliably, the shifts are usually taken to be zero; this accounts for the
behavior of the data points near the origin. The ionicity of the
chemisorption bond, or the Coulomb effect, is thus the dominant factor in
determining the chemical shift. The straight line in the figure is a linear
fit to the data; the slope is 0.38 eV. If we assume (or define) the charge
transfer between the most electronegative atom F (electronegativity 5.75)
and the most electropositive atom Cs (electronegativity 0.28) to be the
electronic charge e, the ratio R between the chemical shift and the charge
transfer can be obtained:

$$R = 2.1 \text{ eV/e.}$$

This conversion factor was used to estimate the upper bound of charge
transfer for the Si(100) buckled dimer discussed above.

Ag-Si(100)

Ag and Si do not form silicides, and the bulk solubility of Ag in Si at room temperature is very small. Thus, for deposition at nearly room temperature, the main effect is expected to be the saturation of Si dangling bonds by Ag as in the case of In on Si(100). The electronegativity difference between Ag and Si is very small, so the chemical shift should be negligible. Indeed, the Si 2p core-level spectra displayed in Fig. 8 show that the surface shift is simply suppressed by Ag adsorption. The overall shift of the peaks at higher coverages is due to a change in band bending induced by Ag.[1]

Electron-diffraction measurements and photoemission studies of the Ag 4d-band formation as a function of coverage indicate that the Ag adatoms at low coverages are dispersed, and Ag(111) clusters form at higher coverages. The growth mode is three dimensional.[3] The Ag-to-Si bonding coordination number is shown in Fig. 9 as a function of coverage; this number is 2 at low coverages. Based on these results, the simplest structural model for an isolated Ag adatom

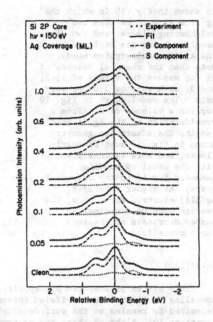

Fig. 8   Si 2p core-level spectra for Si(100)-(2x1) covered with various amounts of Ag. The results of the fit and the decomposition into the surface (S) and bulk (B) components are indicated. The abscissa shows the binding energy referred to the Fermi level offset by a constant to bring the bulk Si $2p_{3/2}$ component for the clean surface to the origin.

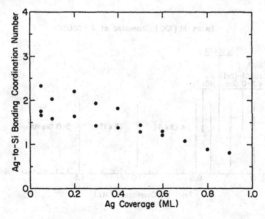

Fig. 9   The Ag-to-Si bonding coordination number as a function of Ag coverage.

is shown in Fig. 10 in which the Ag atom sits in between two neighboring dimers, and two dimer dangling bonds are replaced by the Ag-Si chemisorption bonds. Note that the chemical valence of Ag can assume the values of 1, 2, and 3, with valence 3 being very rare. The model shown in Fig. 10 implies a valence of 2. From a consideration of the Ag-Si bond length, the adsorption geometry shown in Fig. 10 must be nearly linear, and this is consistent with the usual sp bonding geometry for Ag in the valence-2 state. At higher coverages, Ag(111) clusters form; thus, the average bonding coordination number decreases as observed experimentally.

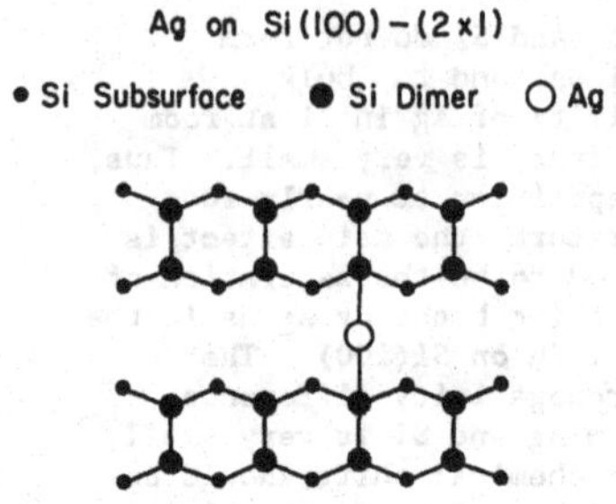

Ag on Si(100)-(2×1)

● Si Subsurface ● Si Dimer ○ Ag

Fig. 10 Top view of a model for an isolated Ag adatom on Si(100)-(2x1).

Sn-Si(100)

This system is prepared by deposition at room temperature followed by annealing at 550°C. Photoemission intensity measurements indicate that the deposited Sn remains on the surface after annealing for submonolayer coverages.[4] Although there are no compounds formed between Sn and Si and the bulk solubility is low, the annealing could result in substantial modification to the Si surface structure; therefore, the system could be much more complex than the two systems discussed above. Indeed, a number of reconstructions are observed as indicated in Fig. 11. Furthermore, photoemission spectra of the Sn 4d core level, displayed in Fig. 12, indicate the presence of two inequivalent adsorption sites (labeled S1 and S2 for the majority and minority species, respectively) whose relative populations and binding energies vary for increasing coverages. For comparison, the In 4d core level for the In-on-Si system discussed above

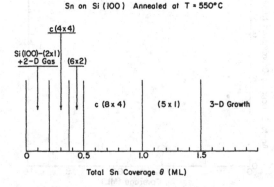

Sn on Si (100) Annealed at T = 550°C

c(4x4)

Si(100)-(2x1) +2-D Gas (6x2)

c (8 x 4)   (5 x 1)   3-D Growth

0   0.5   1.0   1.5

Total Sn Coverage θ (ML)

Fig. 11 The surface phase diagram for Sn on Si(100).

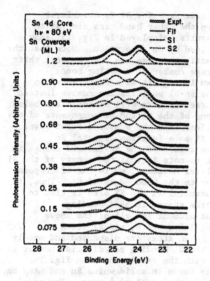

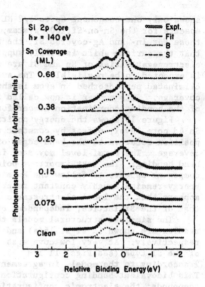

Fig. 12 Sn 4d core-level spectra
for various Sn coverages on
Si(100). The results of the fit
and the decomposition into the S1
and S2 components are indicated.
The binding-energy scale is
referred to the Fermi level.

Fig. 13 Si 2p core-level spectra
for various Sn coverages on
Si(100). The results of the fit
and the decomposition into the
surface (S) and bulk (B) components
are indicated. The binding-energy
scale is referred to the bulk Si
$2p_{3/2}$ component.

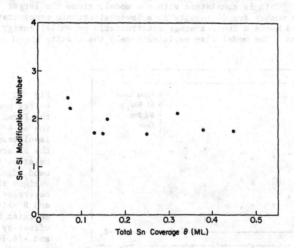

Fig. 14 The average number of Si surface atoms affected by a Sn adatom as
a function of Sn coverage.

shows only one component up to 1/2-ML coverage. The Si 2p core-level spectra for the Sn-on-Si(100) system, on the other hand, are similar to those for In- and Ag-covered Si; the results, displayed in Fig. 13, show that the surface shift is simply suppressed by Sn adsorption. Note that the electronegativities of Sn and Si are nearly the same, so the chemical shift is expected to be negligible. The average number of dangling bonds eliminated per adsorbed Sn atom is shown in Fig. 14 as a function of Sn coverage. This number is now called the Sn-Si modification number instead of the bonding coordination number for reasons that will become clear below.

Figure 15 shows the energy positions of the S1 and S2 components of the Sn 4d core level, offset by constant values, and the measured Fermi-level position relative to the band edges of Si as a function of total Sn coverage. The Fermi-level movement is a result of the Sn-induced change in band bending. The most important point to note is that the energy of the S1 component appears to follow the Fermi-level movement, while the S2-component energy remains roughly constant relative to the band edges of Si. Thus, it appears that the local electrostatic potential at the S2 site is tied to that of the Si lattice, independent of the change in band bending.

The simplest structural models that account for all of the above results are shown in Figs. 16(a) and 16(b) for the S1 and S2 sites, respectively. At very low coverages, the Sn atoms are dispersed and mostly of the S1 type (see Figs. 11 and 12). The S1 bonding coordination number is 2 according to the model, in agreement with the result shown in Fig. 14. This bicovalent bonding configuration is found in solid white Sn and many Sn compounds; the electronic configuration is the usual s^2p^2 type. For total Sn coverages larger than about 0.2 ML, the S2 population becomes significant. The model for the S2 site involves the replacement of a dimerized Si atom by a Sn atom. The Sn-Si modification number is thus 1 for this site. Since the S2-to-S1 population ratio is about 1:2 for $0.2 < \theta < 0.5$ ML, the average Sn-Si modification number becomes somewhat less than 2 (about 1.7) in this coverage range, in agreement with the result shown in Fig. 14. The Sn in the S2 site has a sp^3 bonding configuration just like that in Si; this configuration is found in solid grey tin and a large number of Sn compounds. The S2 site has a higher core-level binding energy than the S1 site. This is consistent with the model, since the larger coordination number for S2 results in a lower electronic concentration at the site, and hence a lower average electrostatic potential energy for the core electron. The model also explains nicely the results shown in Fig. 15.

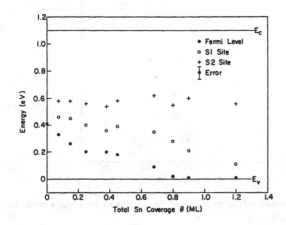

Fig. 15 The energy positions of the Fermi level and the Sn $4d_{5/2}$ core-level components of the S1 and S2 sites relative to the Si band edges for various Sn coverages. The S1- and S2-site binding energies have been offset by -24.05 and -24.90 eV, respectively.

Because the S2 site is a replacement site, the Sn atom is like an impurity atom in the Si lattice. Therefore, the local electrostatic potential for the S2 site should be tied to that of the Si lattice, independent of the band bending. The S1 adsorption site is probably located outside the surface dipole layer, so the binding energy is most likely tied to the local vacuum level, as in a typical chemisorption situation. If the local work function does not change much as a function of coverage, the S1 core-level binding energy should be roughly a constant relative to the Fermi level.

Models for all of the surface reconstructions observed (see Fig. 11) can be constructed based on combinations of the S1 and S2 structural units. One possible configuration for the c(4x4) phase at a total Sn coverage of 3/8 ML is shown in Fig. 16(c). The population ratio between S2 and S1 is 1:2 in accordance with the measured ratio. The models for other reconstructions are not shown here.

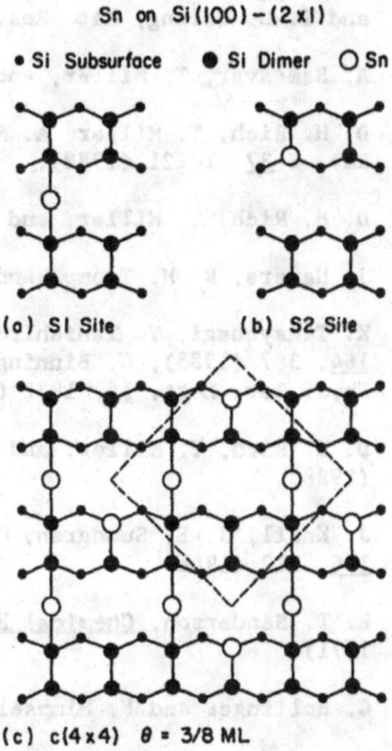

Sn on Si(100)-(2x1)

● Si Subsurface   ● Si Dimer   ○ Sn

(a) S1 Site   (b) S2 Site

(c) c(4x4) θ = 3/8 ML

Fig. 16 Structural models for Sn on Si(100).

SUMMARY AND CONCLUSIONS

The characterization of the atomic interaction and reaction on surfaces is an important area of research. High-resolution core-level spectroscopy allows the identification and counting of atoms in different layers and in inequivalent sites. The chemical shifts induced by adsorption can be related to the Coulomb effects associated with charge redistribution. These results combined with topographical information obtainable from STM studies can provide a detailed description of the electronic and structural properties of surfaces and adsorbates.

ACKNOWLEDGEMENTS

This review is based upon our recent work separately supported by the Department of Energy, Division of Materials Sciences, under Contract No. DE-AC02-76ER01198, and the National Science Foundation under Contracts No. DMR-8352083 and No. DMR-8614234. We would like to acknowledge equipment and personnel support from the E. I. du Pont de Nemours and Company (Wilmington, DE) and the IBM T. J. Watson Research Center (Yorktown Heights, NY). The Synchrotron Radiation Center of the University of Wisconsin-Madison is supported by the National Science Foundation under Contract No. DMR-8020164.

REFERENCES

1.   T.-C. Chiang, CRC Crit. Rev. Solid-State Mater. Sci. 14, 269 (1988).

2.   D. H. Rich, A. Samsavar, T. Miller, H. F. Lin, T.-C. Chiang, Phys. Rev.

Lett. $\underline{58}$, 579 (1987); D. H. Rich, A. Samsavar, T. Miller, H. F. Lin, and T.-C. Chiang, Mat. Res. Soc. Symp. Proc. $\underline{94}$, 219 (1987).

3. A. Samsavar, T. Miller, and T.-C. Chiang, Phys. Rev. (to be published).

4. D. H. Rich, T. Miller, A. Samsavar, H. F. Lin, and T.-C. Chiang, Phys. Rev. B $\underline{37}$, 10221 (1988).

5. D. H. Rich, T. Miller, and T.-C. Chiang, Phys. Rev. B $\underline{37}$, 3124 (1988).

6. J. Hamers, R. M. Tromp, and J. E. Demuth, Phys. Rev. B $\underline{34}$, 5343 (1986).

7. K. Takayanagi, Y. Tanishiro, S. Takagashi, and M. Takahashi, Surf. Sci. $\underline{164}$, 367 (1985); G. Binning, H. Rohrer, Ch. Gerber, and E. Weibel, Phys. Rev. Lett. $\underline{56}$, 1972 (1986).

8. D. H. Rich, T. Miller, and T.-C. Chiang, Phys. Rev. Lett. $\underline{60}$, 357 (1988).

9. J. Knall, J.-E. Sundgren, G. V. Hansson, and J. E. Greene, Surf. Sci. $\underline{166}$, 512 (1986).

10. R. T. Sanderson, Chemical Bonds and Bond Energy, (Academic, New York, 1971).

11. G. Hollinger and F. Himpsel, Phys. Rev. B $\underline{28}$, 3651 (1983).

12. F. R. McFeely, F. F. Monar, N. D. Shinn, G. Landgren, and F. J. Himpsel, Phys. Rev. B $\underline{30}$, 764 (1984).

13. M. A. Olmstead, R. D. Bringans, R. I. G. Uhrberg, and R. Z. Bachrach, Phys. Rev. B $\underline{34}$, 6041 (1986); R. D. Bringans, M. A. Olmstead, R. I. G. Uhrberg, and R. Z. Bachrach, Phys. Rev. B $\underline{36}$, 9569 (1987).

14. T. Weser, A. Bogen, B. Konrad, R. D. Schnell, C. A. Schug, and W. Steimann, Phys. Rev. B $\underline{35}$, 8184 (1987).

15. R. D. Schnell, F. J. Himpsel, A. Bongen, D. Rieger, and W. Steinmann, Phys. Rev. B $\underline{32}$, 8052 (1985).

BONDING IN THIN EPITAXIAL CoSi$_2$ FILMS ON Si(100)

DAVID D. CHAMBLISS,* T.N. RHODIN,* J.E. ROWE** AND S.M. YALISOVE**
* School of Applied and Engineering Physics, Cornell University,
    Ithaca NY 14853
** AT&T Bell Laboratories, Murray Hill NJ 07974

ABSTRACT

Epitaxy of CoSi$_2$ layers on Si crystal surfaces can be strongly
influenced by growing appropriate template layers. The electronic
structure of thin (~7Å) epitaxial CoSi$_2$ films on Si(100) has been
studied with angle-resolved photoemission to investigate atomic
bonding in the layers and at their boundaries. Most Co atoms in the
layers are in a CoSi$_2$-like environment, including those Co atoms near
the free surface. Cobalt atoms at the Si-CoSi$_2$ interface seem to have
fewer Si neighbors.

1. INTRODUCTION

Epitaxy of CoSi$_2$ on surfaces of silicon single crystals has attracted much scientific attention in recent years. Because CoSi$_2$ and Si have
FCC lattices with only a 1.2% lattice mismatch, epitaxial interfaces of
very high crystalline quality have been grown. These interfaces hold
great promise for electronic devices such as the metal base transistor.
But while interfaces with (111) orientation have been the easiest to
produce, conduction through an ideal Si(111)-CoSi$_2$(111) interface is
expected to be poor because there are no electronic states in CoSi$_2$
matching energy and parallel momentum with states near the Si conduction band minimum.[1] Since Si(100)-CoSi$_2$(100) interfaces should have
these matching states[1] there is great interest in achieving and
understanding high-quality epitaxy with this orientation.

Epitaxial growth of CoSi$_2$ on Si(100) has only recently been reported. It has been shown[2] that depositing thick Co layers (>200Å)
on heated Si(100) surfaces, and then annealing, yields a CoSi$_2$ layer
with grains of multiple orientations: (110), (100), (112),(221),
(551), (111), and (772). A later study has shown that CoSi$_2$ can be
grown epitaxially on Si(100) with only two dominant epitaxial
orientations: CoSi$_2$(100) on Si(100) where [100]$_{CoSi_2}$ ‖ [100]$_{Si}$ and
[011]$_{CoSi_2}$ ‖ [011]$_{Si}$, and CoSi$_2$(110) on Si(100) where [110]$_{CoSi_2}$ ‖ [100]$_{Si}$
and [1$\bar{1}$0]$_{CoSi_2}$ ‖ [011]$_{Si}$.[3] These films were grown by a combination of
room-temperature deposition and the initial growth of a very thin
"template" layer. The same study showed that films of predominantly
one orientation can be grown using a template technique if both Co and
Si were deposited at room temperature before any subsequent annealing.
The composition profile of the metal and Si as-deposited layer is a
critical parameter and can cause the resulting film to vary from
polycrystalline to monocrystalline. Growth on wafers with different
starting surface topographies (roughness, miscut, etc.) produced
varying fractions of (100) and (110) grains and varying grain size.
On a particular macroscopically rough Si(100) wafer, a film was grown
of entirely (110) orientation; the details of this dependence are not
yet completely understood.

Thin films made by deposition of Co alone on nominally oriented, very smooth wafers, followed by annealing, generally produced silicide layers which displayed poor LEED patterns and contained very small grains (with diameters ~200Å) of silicide. The same amounts of Co deposited on wafers which were macroscopically rough or intentionally miscut 4° toward the [110] direction produced silicide films with sharp LEED patterns. TEM studies show that these films contain only (100) and (110) oriented grains.

The annealed $CoSi_2$(100) surfaces display one of two character-istic LEED patterns according to whether the terminations are cobalt-rich or silicon-rich. The fourfold-symmetric Co-rich LEED pattern has been described as a $\sqrt{2}\times\sqrt{2}$ R45° pattern.[3] In addition to (1 0) pri-mary spots, it has strong $(\frac{1}{2}\,\frac{1}{2})$ spots and weaker $(\frac{1}{2}\,1)$ spots; no $(\frac{1}{2}\,0)$ spots are apparent. The LEED pattern does not change significantly as the fraction of (100) to (110) grains changes. The I–V charac-teristics of the LEED spots may show differences between (100) and (110) grains but have not been measured.

As techniques for growing epitaxial $CoSi_2$ on Si(100) are devel-oped, there arises the question of what the atomic structure of these epitaxial layers is. In particular, what bonding configurations exist at the interface and the surface, and what electronic states are asso-ciated with them? Is there a difference in electronic structure between the films grown on macroscopically rough substrates and those grown on miscut wafers? We have used angle-resolved photoemission to address these questions, as it is a sensitive probe of surface elec-tronic structure. This work probes the electronic structure of very thin $CoSi_2$ grown on both intentionally miscut wafers and macroscopic-ally rough wafers under different conditions of Co deposition and annealing.

2. EXPERIMENTAL METHODS AND SURFACE CHARACTERIZATION RESULTS

The primary samples studied in this work were thin $CoSi_2$ films grown by depositing typically 3Å Co on Si(100) surfaces and annealing to 460°C. Most samples were prepared on Si substrates cut and polished 4° or 5° off (100) orientation ("off-axis" samples) so as to tend to suppress single-layer steps. Other films were prepared on a macroscopically rough substrate. All substrates were cleaned in cycles of Ne^+ bombardment (1 keV) and annealing at 900°C measured by an infrared (IR) pyrometer. This yielded substrates with Auger electron spectroscopy (AES) peak height ratios C(KLL):Si(LVV) of less than 0.005 with no detectable AES oxygen peak (intensity ratio O(KLL):Si(LVV) <0.0003) measured using a single-pass cylindrical mirror analyzer. The substrates had sharp (2x1) LEED patterns. The patterns for off-axis substrates showed a predominance of one domain orientation and clearly resolved splitting of most spots; LEED patterns for the rough substrate had an even mixture of (2x1) and (1x2) domains and showed no splitting.

Cobalt was deposited on Si(100)-(2x1) substrates from a Co target heated by electron bombardment and surrounded by a shuttered colli-mating shroud cooled with liquid nitrogen. Temperature of the Co target was monitored with a thermocouple and kept constant to ±5°C. The deposition rate was ~0.6Å/min of Co as determined both with an *in situ* Quartz-crystal film thickness monitor and by using Rutherford backscattering on samples after photoemission experiments were completed. A 5-minute deposition caused a pressure rise of less than 3×10^{-10} torr. No significant increase in carbon or oxygen contamination was observed using AES after 3Å of Co deposition.

Annealing temperatures were determined from sample heater power; the calibration of temperature vs. power was measured with an IR pyrometer at temperatures above 600°C and extrapolated to lower temperatures. The temperature of silicide reaction, as indicated by the formation of a LEED pattern, was measured at 460°C, in agreement with other studies.[3]

After preparation and characterization by LEED and AES, the samples were transferred *in vacuo* into a separate chamber for analysis using angle-resolved photoemission. Photon energies in the range 14-80eV were used; sample orientation and emission angle were varied to investigate polarization and dispersion effects. The properties investigated here proved somewhat insensitive to contamination, showing no change after several hours' exposure at a typical pressure of 1×10^{-10} torr.

Various samples prepared by depositing Co in the range 2.0-3.0Å and annealing to 460°C reproduced well the $(\frac{1}{2}\,\frac{1}{2})$ and $(\frac{1}{2}\,1)$ spots and intensities previously reported.[3] LEED patterns from films grown on off-axis and macroscopically rough substrates appeared identical; thus the film of ~7-9Å thickness apparently removed the ordered array of double-layer steps on the clean Si(100)4° surfaces. Photoemission spectra in both cases are characteristic of a well-ordered $CoSi_2$ layer; the differences in detail indicate differences at the $CoSi_2/Si$ interface. Deviations from the growth procedure described above yielded samples with different LEED and photoemission signatures.

3. PHOTOEMISSION RESULTS AND DISCUSSION

3.1. Photoemission Studies of $CoSi_2$(100) Films Grown on Miscut Surfaces

Spectra collected at normal emission from a silicide film grown on an off-axis substrate for a range of photon energies are shown in Fig. 1. The spectra are dominated by a compound peak around E=-2eV due primarily to Co(3d) nonbonding states. The density of states (DOS) feature around E=-7eV characteristic of bulk Si is weak or absent for these spectra; this suggests that the sample is a continuous reacted layer, more or less free of pinholes. Significant intensity is also seen around E=-4eV, the energy characteristic of Co(3d)-Si(3p) bonding hybrid states in $CoSi_2$. (The bonding states are weaker in photoemission than the nonbonding states because hybridization with Si(3p) levels reduces their photoionization cross section in this photon energy range.) Many features in normal- and off-normal-emission spectra, in particular the peak labeled A in Fig. 1, agree well with bulk $CoSi_2$ band structure[1] using a free-electron-final state model that has proved successful for epitaxial $CoSi_2/Si(111)$ as well.[4,5] However, the almost dispersionless peak B at E=-1.3eV does not agree with bulk energy band structure. Since the films contain only ~3-4 layers of Co atoms, it is not surprising that the bulk-like dispersion effects would break down. State B may be a surface or interface state as it shows no dispersion with photon energy, and spectra at higher photon energies suggest it is localized at the Si–CoSi$_2$ interface. As the photon energy increases toward roughly 65eV, the escape depth of the final-state electron approaches a broad minimum on the order of 5Å. In these surface-sensitive spectra, component B is weak; the peak shape is very similar to the bulk DOS of CoSi$_2$ and to comparable spectra for a Si-rich CoSi$_2$(111) surface.[5] Thus the Co near the surface appears to be in a CoSi$_2$-like bonding environment.

With lower photon energies (~30eV) we probe deeper into the CoSi$_2$ layer, and the shifted component B grows in intensity. Changes in photoionization cross section[6] in this photon energy range are too small to account for this change; it is due to the increase in final-state escape depth. This suggests that below the surface, probably at the Si–CoSi$_2$ interface, Co atoms are bonded in a different config-uration which produces peak B. This could be due to a typical inter-face coordination of fewer than 8 Si atoms. Assuming a minimum escape depth of 5Å, the estimated relative intensities of B at photon energies 29eV and 70eV are consistent with its being localized at the interface rather than at the surface.

The observation that the outermost Co atoms are in a CoSi$_2$-like environment suggests that even the "Co-rich" surface of CoSi$_2$(100) is terminated with Si atoms. This is not surprising in light of results that eightfold coordination to Si is energetically favored at CoSi$_2$(111) surfaces[7] and at Si(111)–CoSi$_2$ interfaces.[1,8] On the other hand, it is unlikely that Co at the CoSi$_2$(100)/Si(100) interface can be bonded to more than six Si atoms, which would probably require

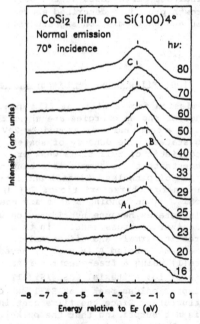

Figure 1. Angle-resolved photo-emission (ARP) spectra for differ-ent photon energies from a CoSi$_2$ film grown on Si(100) cut 4° off the [100] orientation. Energies are specified relative to the Fermi level $E_F=0$. Feature A is related to a bulk energy band of Δ_1 symmetry, and feature B is probably due to a Co(3d) state localized near the Si–CoSi$_2$ interface. We assign the peak C to a CoSi$_2$ DOS feature.

that some Si atoms at the interface have only two-fold coordination.
Preliminary results with cross-sectional TEM suggest the interface Co
atoms are sixfold coordinated.[9,10]

3.2. Photoemission Studies of $CoSi_2(100)$ Films Grown on Macroscopically Rough Substrates

In photoemission, samples grown on rough substrates (Fig. 2) are
somewhat similar to those grown on off-axis substrates (see Fig. 1).
The spectra are dominated by DOS features and band-structure features
from a $CoSi_2(100)$ layer. However, the peak B is now weak or absent,
suggesting that fewer of the interface Co atoms in this case are
undercoordinated. Surface roughness may facilitate relaxation or
reconstruction of the interface so as to satisfy the unfilled Co
states. As noted in the introduction, it seems also to facilitate
growth of (110) grains; these may have fewer underbonded Co atoms.

Some samples were annealed significantly hotter (~550°C) than the
desired annealing temperature of 460°C; the resulting samples had LEED
patterns that superposed relatively weak ($\frac{1}{2}$ $\frac{1}{2}$) spots on the substrate
(2x1) pattern. Photoemission results shown in Fig. 3 confirm the sus-
picion that these over-annealed samples had clustered islands of $CoSi_2$
and patches of Si(100) with low coverage of Co. The spectra are
relatively complicated, but can be approximately described as super-
positions of spectra from well-ordered $CoSi_2$ layers and spectra from
samples with low-coverage deposits of Co, which are probably somewhat
disordered. An important difference, however, is the apparent absence
of the state B assigned to the $Si/CoSi_2$ interface. This is expected,
for agglomeration of islands will both reduce the amount of interface
area and hide it under a thicker $CoSi_2$ layer.

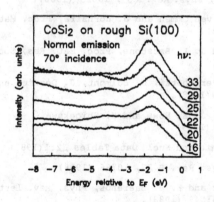

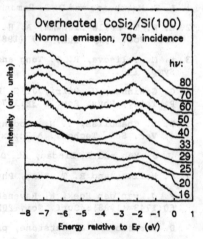

Figure 2. ARP spectra from a
$CoSi_2$ layer grown on macroscopically rough Si(100). The
intensity of component B has
decreased significantly compared to the results shown in
Fig. 1.

Figure 3. ARP spectra from an
overheated $CoSi_2$ thin film grown
on Si(100). The strong Si DOS
intensity (E=-7eV) is suggestive
of island formation, which
exposes regions of Si(100).

4. CONCLUSIONS.

Thin $CoSi_2$ layers grown on Si(100) by well-controlled cobalt deposition and annealing exhibit distinctive photoemission spectra. These results suggest that, for ~3Å Co followed by ~460°C annealing, a thin, well-ordered $CoSi_2$ layer forms. We find that the "Co-rich" surface that results from this growth method is terminated with enough Si atoms to maintain a $CoSi_2$ bonding environment for the outermost Co atoms. At the interface, Co atoms probably do not have a full eight Si neighbors, as in bulk $CoSi_2$. Layers grown on a macroscopically rough substrate, which are thought to have a greater proportion of (110) grains, show much less emission from a peak assigned to inter-face Co. The energy difference between undercoordination of Co in some interfaces formed on Si(100) and full bulk-like eightfold coordination at the (111) interface [1,8] may account for the strong dependence of film quality on small variations in Co coverage, annealing, and substrate topography.

ACKNOWLEDGMENTS

Research carried out at the National Synchrotron Light Source, Brookhaven National Laboratory, which is supported by the U. S. Department of Energy, Division of Materials Sciences and Division of Chemical Sciences (DOE contract number DE-AC02-76CH00016). We are grateful for the assistance of King Tsang and Zhangda Lin. Chambliss and Rhodin acknowledge support by the Materials Science Center, Cornell University (NSF-DMR16616-A01).

REFERENCES

1. L. F. Mattheiss and D. R. Hamann, Phys. Rev. B 37, 10623 (1988).

2. C. W. T. Bulle-Lieuwma, A. H. van Ommen and J. Hornstra, Proc. Mat. Res. Soc. Symp. 102, 377 (1988).

3. S. M. Yalisove, R. T. Tung and J. L. Batstone, Proc. Mat. Res. Soc. Symp. 160.

4. G. Gewinner, C. Pirri, J. C. Peruchetti, D. Bolmont, J. Derrien and P. Thiry, Phys. Rev. B 38:1879(1988).

5. D. D. Chambliss, T. N. Rhodin, J. E. Rowe and H. Shigekawa, J. Vac. Sci. Technol. A 7:xxx (1989).

6. J. J. Yeh and I. Lindau, At. Data and Nucl. Data Tables 32:1(1985).

7. F. Hellman and R. T. Tung, Phys. Rev. B 37, 10786 (1988).

8. P. J. van den Hoek, W. Ravenek and E. J. Baerends, Phys. Rev. Lett. 60:1743 (1988); Surf. Sci. 205:549 (1988).

9. D. Loretto and J. Batstone, private communication, 1988.

10. D. Cherns, G. R. Anstis, J. L. Hutchison and J. C. H. Spence, Phil. Mag. A46,849 (1982).

VALENCE BAND ELECTRONIC STRUCTURE
OF CLEAVED IRON OXIDE SINGLE CRYSTALS
STUDIED BY RESONANT PHOTOEMISSION

ROBERT J. LAD[*] AND VICTOR E. HENRICH
Applied Physics, Yale University, 15 Prospect Street, New Haven, CT 06520

ABSTRACT

Synchrotron radiation has been used to perform resonant photoemission measurements across the 3p→3d photoabsorption threshold from cleaved Fe_xO (x ≈ 0.945), Fe_3O_4, and α-Fe_2O_3 single crystal surfaces. The resonant enhancement of the Fe 3d photoelectrons allows the Fe 3d-derived final states in the valence band to be distinguished from the overlapping O 2p states. Using well-characterized single crystals, the distributions of Fe 3d-derived states associated with the ferrous (Fe^{2+}) and ferric (Fe^{3+}) cations have been identified. The Fe 3d-derived states are found to extend about 18 eV below the Fermi level in each oxide, which can be attributed to a significant amount of hybridization between the Fe 3d and O 2p orbitals.

INTRODUCTION

Valence band photoemission spectra from iron compounds containing ferrous (Fe^{2+}) and ferric (Fe^{3+}) cations have generally been interpreted by assuming localized 3d cation levels and correlating the photoemission features to calculated multiplet and crystal-field split final states [1-6]. However, recent calculations of the electronic structure for transition-metal compounds [7-9] have suggested that cation-ligand hybridization must also be taken into account in order to correctly describe their valence band electronic structure.

By performing resonant photoemission experiments on cleaved single crystals of Fe_xO, Fe_3O_4, and α-Fe_2O_3 at the 3p→3d photoabsorption threshold, we have been able to provide new information concerning the distribution of Fe 3d and O 2p states in the valence band of these materials. This information is extracted from highly stoichiometric single crystal samples, which contain well-defined and uniform cation environments. Our results reveal the importance of the hybridization between cation and ligand states, and they are in good agreement with recent configuration interaction cluster calculations for iron oxides performed by Fujimori et al. [8].

EXPERIMENTAL

The photoemission experiments were performed on beamline U14 at the National Synchrotron Light Source (NSLS) at Brookhaven National Laboratory. Angle-integrated valence band photoemission spectra were measured from cleaved iron

[*] Present address: Laboratory for Surface Science & Technology and Department of Physics and Astronomy, University of Maine, Orono, ME 04469.

oxide surfaces, using a double-pass cylindrical mirror analyzer and photon energies between $30 \leq h\nu \leq 100$ eV. For these photon energies, the overall energy resolution in the measured spectra ranged between 0.2 to 0.5 eV. The iron oxide single crystals were cleaved in the analysis chamber at pressures below 2×10^{-10} Torr; the cleavage planes were Fe_xO (100), Fe_3O_4 (110), and $\alpha\text{-}Fe_2O_3$ (10$\bar{1}$2). All of the data are normalized to the incident photon intensity, as determined by a flux monitor, and are referenced to the Fermi level (E_f) measured from a clean gold foil. Other experimental details can be found in Ref. 10.

RESONANT PHOTOELECTRON YIELD

In transition-metal compounds, the 3p→3d excitations remain localized on the cation site, and a quasiatomic interpretation of the resonance behavior appears to be justified [11]. Within the assumption of localized 3p excitations, the resonant process can be described in terms of atomic configurations: $3p^6 3d^n + h\nu \rightarrow [3p^5 3d^{n+1}]^* \rightarrow 3p^6 3d^{n-1} + e^-$. In iron oxides, the resonant enhancement of the Fe 3d states occurs at photon energies between $52 \leq h\nu \leq 60$ eV. Figure 1 shows the total photoelectron yield measured across the 3p photoabsorption threshold from each of the single crystals, which clearly illustrates the resonant behavior. The onset of the resonance depends on the cation valence and correlates with the Fe 3p binding energies determined by XPS in Refs. 3 and 4. The profiles contain features that can be attributed to the different multiplets of the $[3p^5 3d^{n+1}]^*$ intermediate states. In $\alpha\text{-}Fe_2O_3$, for example, the 3p excitations may occur via either the ^6P or ^6D intermediate states; likewise, structure from the ^5P and ^5D multiplets is evident in Fe_xO. These multiplet features are discussed more completely in Ref. 10.

The resonance behavior can be used to distinguish the cation 3d-derived states from the overlapping unhybridized O 2p states in the valence band. Since only the Fe 3d-derived states are involved in the resonant process, the difference

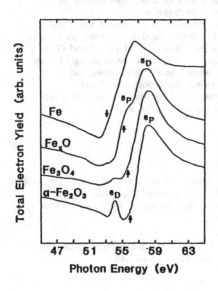

FIGURE 1. Total photoelectron yield from clean metallic Fe and from single crystal iron oxides measured across the 3p photoabsorption threshold. The arrows denote the Fe 3p binding energies in each of the compounds, as determined by XPS in Refs. 3 and 4.

between valence band spectra measured just below and just above the resonance removes the non-resonating O 2p contribution. The resulting difference spectrum contains a measure of the distribution of Fe 3d-derived states in the valence band; these may be purely Fe 3d states or Fe 3d-derived states that are hybridized with the neighboring oxygen ligands. In the next section, results of this resonance difference method are presented for each of the iron oxides.

VALENCE BAND PHOTOEMISSION

A. Fe$_x$O (100)

In the Fe$_x$O rocksalt structure, Fe^{2+} cations are located in octahedral interstices of the fcc O^{2-} anion lattice. Each Fe^{2+} cation has a high spin 3d^6 (t$_{2g}^4$e$_g^2$), ^{5}T$_{2g}$ ground state configuration. Figure 2 shows an estimate of the Fe 3d-derived states in Fe$_x$O (100) obtained by subtracting valence spectra measured at photon energies of 57 eV and 54 eV; these energies correspond to above and below the 3p→3d resonance, as shown in Fig. 1. This measurement of the Fe 3d-derived valence states exhibits maxima at 1.2, 4.1, 6.9, and 11.0 eV. The relative intensities and locations of the 3d^5 final states predicted by Bagus et al. [5] using crystal-field-theory are shown by the vertical lines below the measured spectrum. Although the general shape of the upper valence band region is predicted by calculations of the localized crystal-field-split cation states, the emission extending to 16 eV below the Fermi level is not accounted for. Recently, this additional intensity has been attributed to ligand-to-metal charge transfer effects [12]. Constant-initial-state (CIS) spectra measured across the 3p threshold [10,12] reveal that the 0 - 9 eV region is formed primarily from 3d^{6}L final states (where L represents a ligand hole), and that the 10 - 16 eV region is derived from 3d^5 final states.

B. α-Fe$_2$O$_3$ (10$\bar{1}$2)

α-Fe$_2$O$_3$ has the corundum (trigonal) crystal structure. Each Fe^{3+} cation is situated in a slightly distorted octahedron of O^{2-} anions; the ground state configuration is 3d^5 (t$_{2g}^3$e$_g^2$), ^{6}A$_{1g}$. Figure 3 shows a difference spectrum

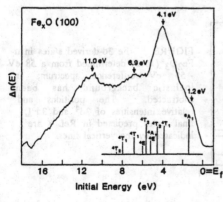

FIGURE 2. Fe 3d-derived states in Fe$_x$O (100) determined by taking the difference between valence band spectra measured above and below the 3p→3d resonance (57 eV - 54 eV); an inelastic background has been removed. The vertical lines represent the relative intensities and positions of the 3d^5 final states predicted in Ref. 5.

between valence band spectra measured at 58 eV and 55 eV, corresponding to above and below the 3p→3d resonance in α-Fe$_2$O$_3$ (see Fig. 1). A three peaked structure is visible in the Fe 3d-derived states with maxima at approximately 3.1, 6.2, and 12.9 eV. This measurement is in excellent agreement with predictions from a configuration interaction (CI) calculation of the 3d-derived final states for a (FeO$_6$)$^{9-}$ cluster, performed by Fujimori et al. [8]. The CI cluster model considers ligand-to-metal charge transfer in both the initial and final states and explains the experimental spectrum in terms of 3d^4 and 3d^{5}L final states, as shown in Fig. 3. CIS measurements at the 3p threshold substantiate these final state assignments.

C. Fe$_3$O$_4$ (110)

Fe$_3$O$_4$ has the inverse spinel structure and contains both Fe^{3+} and Fe^{2+} cations in a ratio of 2:1. Consequently, the valence band electronic structure is much more complex than that of the other iron oxides. The Fe^{2+} and half of the Fe^{3+} cations are octahedrally coordinated with O^{2-} ligands, while the remaining Fe^{3+} cations are in tetrahedrally coordinated sites. Due to the presence of both Fe valence states, the 3p→3d resonance profile shown for Fe$_3$O$_4$ in Fig. 1 is broader than for the other oxides. Figure 4 shows a 58 eV - 54 eV difference spectrum determined from valence spectra measured just above and just below the resonance, which reflects the total distribution of the Fe 3d-derived states from both valence states. By taking a 57 eV - 54 eV difference spectrum, the final states from the Fe^{2+} cations are emphasized, since most of the Fe^{2+}-derived resonance occurs within this energy range (as shown for Fe$_x$O in Fig. 1). Likewise, a 58 eV - 55 eV difference spectrum accentuates the Fe^{3+}-derived features (as shown for α-Fe$_2$O$_3$). This effect is illustrated by the bottom two spectra in Fig. 4. The Fe 3d-derived states associated with each cation environment in Fe$_3$O$_4$ is discussed more extensively in Ref. 13.

SUMMARY

The resonant photoemission phenomena has been utilized to experimentally separate the Fe 3d-derived states from the overlapping O 2p states in the valence band of Fe$_x$O, Fe$_3$O$_4$, and α-Fe$_2$O$_3$ single crystals. The resonant photoelectron yield at the 3p→3d excitation threshold was shown to be sensitive to the cation

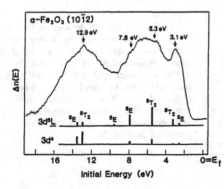

FIGURE 3. Fe 3d-derived states in α-Fe$_2$O$_3$ (10$\bar{1}$2) determined from a 58 eV - 55 eV difference spectrum; an inelastic background has been subtracted. The positions and relative intensities of 3d^4 and 3d^{5}L final states predicted in Ref. 8 are indicated by the vertical lines.

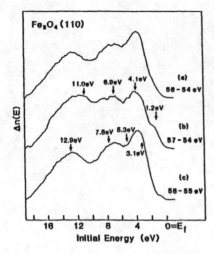

FIGURE 4. Difference spectra measured from Fe_3O_4 (110) which emphasize (a) total, (b) Fe^{2+}-related, and (c) Fe^{3+}-related 3d-derived final states; an inelastic background was removed from each spectrum. The maxima shown in Figs. 2 and 3 are indicated by the arrows.

valence state, and this effect was employed to distinguish the Fe^{2+} from the Fe^{3+} valence band features in Fe_3O_4. The large width of the Fe 3d-derived structure that was measured in each oxide has been attributed to ligand-to-metal charge transfer effects.

ACKNOWLEDGEMENTS

We wish to thank Kevin E. Smith and the NSLS staff for their assistance and helpful advice. We are also grateful to C.R. Brundle, D.J. Buttrey, J.P. Remeika, and A. Revcolevschi for providing the single crystals. This work was partially supported by NSF Solid State Chemistry Grant #DMR-87-11423 and was performed at the National Synchrotron Light Source at Brookhaven National Laboratory, which is supported by DOE Grant #DE-AC02-76CH00016.

REFERENCES

[1] D.E. Eastman and J.L. Freeouf, Phys. Rev. Lett. 34, 395 (1975).
[2] S.F. Alvarado, M. Erbudak, and P. Munz, Phys. Rev. B 14, 2740 (1976).
[3] C.R. Brundle, T.J. Chuang, and K. Wandelt, Surface Sci. 68, 459 (1977).
[4] N.S. McIntyre and D.G. Zetaruk, Anal. Chem. 49, 1521 (1977).
[5] P.S. Bagus, C.R. Brundle, T.J. Chuang, and K. Wandelt, Phys. Rev. Lett. 39, 1229 (1977).
[6] G. Grenet, Y. Jugnet, T.M. Duc, and M. Kibler, J. Chem. Phys. 72, 218 (1980); ibid. 74, 2163 (1981).
[7] A. Fujimori, F. Minami, and S. Sugano, Phys. Rev. B 29, 5225 (1984); A. Fujimori and F. Minami, ibid., 30, 957 (1984).
[8] A. Fujimori, M. Saeki, N. Kimizuka, M. Taniguchi, and S. Suga, Phys. Rev. B 34, 7318 (1986).
[9] J. Zaanen, G.A. Sawatzky, and J.W. Allen, Phys. Rev. Lett. 55, 418 (1985); J. Magn. Mat. 54-57, 607 (1986).
[10] R.J. Lad and V.E. Henrich, Phys. Rev. B (to be published).
[11] L.C. Davis, J. Appl. Phys., 59, R25 (1986).
[12] A. Fujimori, N. Kimizuka, M. Taniguchi, and S. Suga, Phys. Rev. B 36, 6691 (1987).
[13] R.J. Lad and V.E. Henrich, J. Vac. Sci. Technol. A (to be published).

PHOTOEMISSION EXAFS FROM InP(110) AND Al/InP(110)

K. M. CHOUDHARY, P. S. MANGAT, D. KILDAY* AND G. MARGARITONDO*
University of Notre Dame, Materials Science and Engineering, Notre Dame, Indiana 46556
* Synchrotron Radiation Center and Department of Physics, University of Wisconsin-Madison,
Stoughton, Wisconsin 53589

ABSTRACT

Photoemission EXAFS (PEXAFS) studies of InP(110), InP(110) + 1 Å Al and InP(110) + 3 Å Al are presented. In each case, photoemission from the P 2p core-level was monitored using a cylindrical mirror analyzer by a two-point CIS (constant-initial-state spectroscopy) method in 150-280 eV photon energy range. The data were analyzed by conventional Fourier analysis procedures using the theoretical phase function of McKale et al. plus absorber phase function of Teo and Lee. No diffraction effect other than EXAFS is observed. For the clean InP(110) surface, the PEXAFS results show relaxation (contraction) of the surface in excellent agreement with the LEED results. But a small contraction in the P-P bond length (surface unit mesh parameter, a_0) has also been measured, which was not reported in the LEED studies. For the InP(110) surface covered with 1 Å Al, metal (cluster)-induced surface structural changes in the substrate are determined which include removal of relaxation of the surface unit mesh in combination with change in the P-In bond length within the surface unit mesh. At the 3 Å Al-coverage on the InP(110) surface, the interface resembles AlP.

INTRODUCTION

A knowledge of the atomic geometries of semiconductor surfaces and interfaces is important for understanding their electronic properties and electrical behavior on a microscopic scale [1,2]. Although several tools of surface crystallography, such as low-electron electron diffraction (LEED) [3], surface extended X-ray absorption fine structure (surface EXAFS or SEXAFS) [4,5], scanning tunneling micorscopy (STM) [6], surface X-ray diffraction [7] etc., have been used   for such investigations, the atomic geometries of metal/III-V compound semiconductor interfaces are rarely known. This is because the techniques other than surface EXAFS usually require ordered surfaces whereas the metal/III-V compound semiconductor interfaces are known to be disordered, where interfacial reaction or interdiffusion occurs beyond a critical coverage of metal [8-10].

The extended X-ray absorption fine structure (EXAFS) in the photon energy dependence of the absorption coefficient is due to interference of the outgoing photoelectron waves, created in the photoabsorption process, with the backscattered waves, which are scattered by atoms surrounding the absorber atom [11]. The sinusoidal modulations of an EXAFS spectrum contain information on the short-range order of atoms in the solid. Theoretically, in order to get EXAFS in photoemission from a core-level the photoelectron signal must be averaged over a 4π solid angle [12]. Otherwise, it had been thought that photoelectron diffraction [e.g., normal photoelectron diffraction (NPD) [13] or angular resolved photoemission extended fine structure (ARPEFS)] [14] might be superposed on the EXAFS signal [12]. Although EXAFS-like oscillations in d-core-level photoemission from single crystal surfaces have been reported [15], the early PEXAFS experiments focussed on studying polycrystalline surfaces and a cylindrical mirror analyzer (CMA) was used for collecting the electrons, which averages the signal over a 2π solid angle (maximum) [16,17]. Recently, we have acquired P 2p PEXAFS from InP(110) [18], Al/InP(110) [18] and Na/InP(110) [19,20] with the use of a CMA. There is no evidence of any diffraction effect other than EXAFS. On the other hand, the technique is very surface sensitive and it is found that P 2p EXAFS from the InP(110) surface yields average interatomic bond lengths corresponding to the topmost atomic layer on the surface. Furthermore, for metal-covered surfaces changes in the EXAFS modulations are clearly observed in the spectra when the PEXAFS curves are plotted for several coverages [21]. Hence, the technique probes the short-range order of atoms in the topmost atomic layer. The very high surface sensitivity of the technique is due to the limited escape depth of elastic photoelectrons (around 5 Å) in the kinetic energy range of our PEXAFS data. The significant advantages of the PEXAFS method for studying the structure and chemistry of solid surfaces and interfaces are the following: (1) The short-range order of atoms in the substrate as well as in the adlayer can be probed; (2) The signal level is excellent and data collection time is short, which allows us to probe the structure of interfaces for several adatom coverages using the same sample.

In this paper, we descibe our recent PEXAFS studies of Al/InP(110) interfaces for 1 Å and 3 Å Al-coverages. The Al/InP(110) system has been extensively studied in the past by surface science spectroscopic techniques. It is well known that at the 1 Å Al-coverage the metal-atoms are present as clusters on the surface whereas exchange reaction occurs at the 3 Å Al-coverage [8,9]. For the InP(110) + 1 Å Al surface, the results derived from P 2p EXAFS are particularly interesting where metal-induced structural changes within the surface unit mesh has been observed. At the 3 Å Al-coverage, the well known exchange reaction [22,23] is confirmed. The results demonstrate that the PEXAFS technique is very useful for studying clean or adatom-covered reconstructed surfaces.

EXPERIMENTS

The PEXAFS experiments were performed at the Synchrotron Radiation Center (SRC) on the 1 GeV electron storage ring with the "Grasshopper" MARK II beamline. The ultrahigh vacuum (UHV) experimental system consisted of an electron spectrometer, flux monitor system, miniature molecular beam deposition (MBD) facility, sample transfer system and crystal-cleaving attachments. Single crystals of n-type InP were cleaved along the (110) planes. The aluminum metal was deposited from a home-made Knudsen cell in the MBD system. The Knudsen cell consists of a pyrolytic boron nitride (PBN) crucible, alumina parts, Mo-coil heater, multilayer radiation shielding with Ta-foils and a W5%Re-W26%Re thermocouple. The deposition rate has been calibrated against the cell temperature. The base pressure was 2×10^{-10} torr and the pressure remained in the same range during deposition, since the Knudsen cell was kept hot throughout the experiment. The photon flux was monitored by measuring the total electron yield from a high transmission (80%) nickel grid coated with LiF in situ. The grid was placed in the path of the photon beam. The total electron yield was collected by a channel electron multiplier (Galileo CEM 4821) in the analog-detection mode of data collection.

For each sample, energy distribution curves (EDC) in the region of In 4d, P 2p, Al 2p and valence band were taken. The intensity of photoemission from the P 2p core-level was monitored in the constant-initial-state (CIS) mode of photoemission spectroscopy in 150-280 eV photon energy range. The photoelectrons were analyzed by a CMA in the fixed retarding ratio mode (electron pass energy/kinetic energy = constant) [24]. We used a two-point CIS method. The P 2p CIS data for InP(110), InP(110) + 1 Å Al and InP(110) + 3 Å Al were obtained for one point of the peak and for one point of the background. The P 2p CIS parameter, $E_b + \Phi$, where E_b is the core-level binding energy and Φ is work-function, was 134 eV for the peak and 129 eV for the background. For InP(110) + 1Å Al and InP(110) + 3 Å Al, the corresponding parameters were 134.2 eV and 129.2 eV. The photon flux was measured simultaneously and the electron counts were normalized to the photon flux. The background data were shifted in energy scale by 5 eV to compensate for the difference in kinetic energy. Then the background was subtracted from the peak. This method of background subtraction removed P_{LVV} and Al_{LVV} Auger peaks from the spectra. The P 2p PEXAFS curves thus obtained are shown in Fig. 1, where the absorpotion coefficient is the (normalized) photoemission intensity. The number of CIS scans were five, six and four for InP(110), InP(110) + 1 Å Al and InP(110) + 3 Å Al, respectively.

DATA ANALYSIS AND RESULTS

The PEXAFS data were analyzed by conventional Fourier analysis procedures [4,11] using the theoretical backscattering phase function of McKale et al. [25] plus absorber phase function of Teo and Lee [26]. The photon energy values were converted to wavevectors using

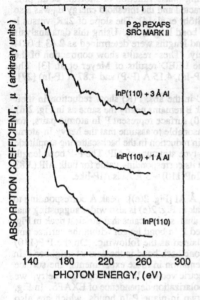

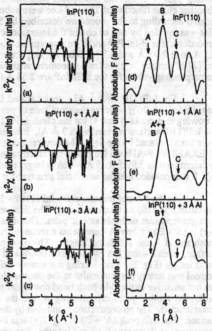

FIG. 1. P 2p PEXAFS for cleaved InP(110), InP(110)+1 Å Al and InP(110)+3 Å Al.

FIG. 2. (a) $k^2\chi(k)$ function for the InP(110) surface; (b) Fourier transform F(R) of the EXAFS function shown in (a); (c) $k^2\chi(k)$ for InP(110)+1 Å Al; (d) Fourier transform of the EXAFS function shown in (c); (e) $k^2\chi(k)$ for InP(110)+3 Å Al; (f) Fourier transform of the EXAFS function shown in (e).

the formula, $k = 0.5123 \, (h\nu - E_0)^{1/2}$, where $h\nu$ is the photon energy value and E_0 is the EXAFS threshold energy. A starting E_0 value of 132 eV was used. The EXAFS function was obtained from the equation, $\chi(k) = (\mu - \mu_0)/\mu_0$, where μ_0 is the absorption coefficient and μ_0 is the smooth background computed by a fourth-degree polynomial fit to the PEXAFS curves. The EXAFS function was Fourier transformed in a conventional manner. Figs. 2(a), 2(b) and 2(c) show the experimental EXAFS function $k^2\chi(k)$ for InP(110), InP(110) + 1 Å Al and InP(110) + 3 Å Al. The absolute values of their Fourier transform, absolute F(R), are depicted in Figs. 2(d), 2(e) and 2(f). The peaks in absolute F(R) represent near neighbor distances (modified pair distribution function) for the absorber-backscatterer atomic pairs.

For the InP(110) surface, the peaks marked as A, B and C in Fig. 2(d) represent P-In, P-P and P-In near neighbor distances. The peaks were filtered with the use of a window function, W(R), and Fourier backtransformed to obtain $k^2\chi_j(k)$, where j represents an absorber-backscatterer atomic pair, provied that the peak in F(R) corresponds to a single atomic pair. The complex $k^2\chi_j(k)$ were decomposed into amplitude and phase parts. From the phase, $2kR_j + \phi_j(k)$, the theoretical backscattering phase function of McKale et al. plus absorber phase function of Teo and Lee were subtracted and the threshold energy E_0 was also floated according to the procedure described by Stöhr et al. [4]. The slope of $2kR_j$ versus k plot was divided by two to obtain the interatomic bond length, R_j. Using this data analysis procedure, the P-In, P-P and P-In interatomic bond lengths were determined as 2.43 ± 0.04 Å, 4.06 ± 0.05 Å and 4.87 ± 0.06 Å, respectively. These results show contraction of the InP(110) surface in excellent agreement with the LEED results of Meyer et al. [3]. The corresponding values for the bulk InP are 2.54 Å (P-In), 4.15 Å (P-P) and 4.87 Å (P-In) [27].

For InP(110) + 3 Å Al, peaks A and C in the abs. F(R) show reduction in their amplitude [Fig. 2(f)] whereas the amplitude of peak B remains nearly the same as in Fig. 2(d). Knowing that peaks A and C for the clean InP(110) surface represent P-In atomic pairs, for the InP(110) surface covered with 3 Å Al, it is reasonable to assume that the heavy In-atoms have been replaced by the light Al-atoms resulting in reduction in the backscattering amplitude of EXAFS for P-Al bond lengths in the zincblende-type structure of AlP. The P-P bond length is determined as 3.99 ± 0.05 Å which is larger than its corresponding value for bulk AlP (3.83 Å). It is concluded that the interfacial structure for InP(110) + 3 Å Al is AlP-like.

In the Fourier transform for InP(110) + 1 Å Al [Fig. 2(e)], peak A corresponding to the clean InP(110) surface is missing. The first peak in Fig. 2(e) is also wider suggesting that it might represent multiple atomic pairs. Our data analysis indicates that the first peak in Fig. 2(e) is (A'+ B), where A' represents a reconstructed P-In bond length within the surface unit mesh and B is the P-P bond length. This is explained as the following. On the InP(110) surface, each P-atom has two in-plane bonds and one back bond in the nearest neighbor P-In bond length (Fig. 3). Considering the sample orientation in these experiments, the surface normal was nearly perpendicular to the photon electric vector. In this sample geometry, we are not sensitive to the P-In back bond due to the polarization dependence of EXAFS. In Fig. 2(e), the atomic pairs representing A' are the two in-plane P-In bonds, which are also reconstructed. The reconstruction is induced by the low-coverage-adatoms (Al-clusters) on the surface. The first peak (A' + B) of Fig. 2(e) was Fourier filtered and analyzed by curve-fitting procedures to determine the bond lengths. For curve-fitting, the experimental amplitude and phase functions for P-In and P-P atomic pairs were extracted from the data for the clean InP(110) surface. The floating parameters were the following: R_{P-In}, R_{P-P} (bond lengths), and f_{P-In}, f_{P-P} (fudge factors). The curve-fit is shown in Fig. 4. From this analysis, the in-plane P-In bond length was determined as 2.94 ± 0.05 Å and the P-P bond length was found as bulk-like (4.12 ± 0.05 Å).

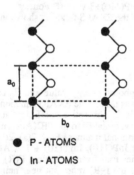

● P - ATOMS

○ In - ATOMS

FIG. 3. Top view of the InP(110) surface. The surface unit mesh in the topmost atomic layer is shown by dotted lines. Each P atom has two in-plane bonds and one back bond. The back bond is not shown in the figure.

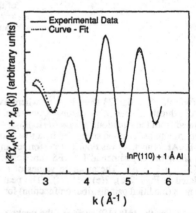

FIG. 4. Imaginary $k^2[\chi_A(k)+\chi_B(k)]$ (solid line) and curve-fit (dotted line) for the InP(110) surface covered with 1 Å Al. The curve-fitting procedure is described in the text.

DISCUSSION

For the clean InP(110) surface, our results indicate that the shortest P-P bond length, which is the surface unit mesh parameter a_0 (Fig. 3), is contracted by 0.09 ± 0.05 Å in comparison to its bulk value. It was not noticed in the LEED intensity calculations of Meyer et al. [3]. The P-In bond length is found as relaxed (contracted) in agreement with the LEED results.

For InP(110) + 1Å Al, the P 2p PEXAFS results show metal (cluster)-induced surface structural changes in the substrate. The structural changes include removal of relaxation of the surface unit mesh, since the P-P bond length is unrelaxed, in combination with reconstruction of the in-plane P-In bond length within the surface unit mesh (change in the basis). This surface structure should still give a (1x1) LEED pattern, as observed by Kahn et al. [28]. It is to be noted that the Fermi-level on the semiconductor surface is pinned at the 1 Å Al-coverage. Hence, the metal-induced reconstruction of the substrate might be influencing the Fermi-level pinning at Al/InP(110) interfaces in the low-metal-coverage regime. A similar result was recently obtained for the 1/8 monolayer Na/InP(110) interface [19], in which case the surface unit mesh becomes bulk-like when covered with the Na-atoms but the change in the basis of the surface unit mesh is different. At higher Na-coverage (≥ 0.5 monolayer), severe interfacial reaction has been observed where the Na-atoms are strongly bonded to the anions [20].

For InP(110) + 3 Å Al, the well known exchange reaction [22,23] between Al and In-atoms is observed. The P-Al bond length is not determined since curve-fit analysis is necessary whereas the experimental P-Al amplitude function is not available. However, the P-P bond length was measured as 0.16 Å larger than its value for the bulk AlP. The expansion in the P-P bond length might be due to strain induced by the substrate. These results indicate that the interface resembles AlP.

CONCLUSIONS

The results demonstrate that the PEXAFS from a core-level of atoms in the substrate probes the near neighbor distances in the topmost atomic layer. Hence, the technique can be used to determine the atomic geometries of clean surfaces and surfaces covered with adatoms for low to high-metal-coverages. For the clean InP(110) surface, the interatomic bond lengths derived from P 2p PEXAFS agree with the LEED results with the exception that the P-P bond length (a_0) of the surface unit mesh is determined as slightly contracted. For 1 Å Al-coverage on the InP(110) surface, low-coverage metal (cluster)-induced structural changes in the substrate are found. The structural changes include removal of relaxation of the surface unit mesh in combination with change in the in-plane P-P bond length within the surface unit mesh. For InP(110) + 3 Å Al, where reaction is known to occur with the release of In-atoms, we confirm the formation of an AlP-like compound.

ACKNOWLEDGEMENTS

The research was supported partially by the Jesse H. Jones Faculty Research Fund Program, University of Notre Dame, and by the Office of Naval Research. We thank the staff of the Synchrotron Radiation Center, University of Wisconsin-Madison for their support.

REFERENCES

[1]   "The Structure of Surfaces II", edited by J. F. van der Veen and M. A. Hove (Springer-Verlag, New York, 1988).
[2]   R. H. Williams in "Physics and Chemistry of III-V Compound Semiconductor Surfaces", edited by C. W. Wilmsen (Plenum, New York, 1985).

[3] R. J. Meyer, C. B. Duke, A. Paton, J. C. Tsang, J. L. Yeh, A. Kahn and P. Mark, Phys. Rev. B22, 6171 (1980).

[4] J. Stöhr, R. Jaeger and S. Brennan, Surf. Sci. 117, 503 (1982).

[5] G. Rossi, in "Semiconductor Interfaces: Formation and Properties", edited by G. Le Lay et al. (Springer-Verlag, New York, 1987), p.69.

[6] R. M. Tromp, E. J. van Loenen, R. J. Hamers and J. E. Demuth, in "The Structure of Surfaces II", edited by J. F. van der Veen and M. A. Van Hove (Springer-Verlag, New York, 1988), p. 282.

[7] I. K. Robinson, in "The Structure of Surfaces", edited by M. A. Van Hove and S. Y. Tong (Springer-Verlag, New York, 1985).

[8] (a) D. M. Hill, F. Xu, Zhangda Lin and J. H. Weaver, Phys. Rev. B38, 1893 (1988); (b) J. J. Joyce, M. Grioni, M. del Giudice, M. W. Ruckman, F. Boscherini and J. H. Weaver, J. Vac. Sci. Technol. A5, 2019 (1987); F. Xu, D. M. Hill, Zhangda Lin, S. G. Anderson and J. H. Weaver, Phys. Rev. B37, 10295 (1988).

[9] L. J. Brillson, C. F. Brucker, A. D. Katnani, N. G. Stoffel, R. Daniels and G. Margaritondo, in "Surfaces and Interfaces: Physics and Electronics", edited by R. S. Bauer (North-Holland, Amsterdam, 1983), p. 212.

[10] (a) W. E. Spicer, T. Kendelewicz, N. Newman, K. K. Chin and I. Lindau, Surf. Sci. 168, 240 (1986); (b) I. Lindau et al., Surf. Sci. 162, 591 (1985).

[11] P. A. Lee, P. H. Citrin, P. Eisenberger and B. M. Kincaid, Rev. Mod. Phys. 53, 769 (1981).

[12] P. A. Lee, Phys. Rev. B13, 5261 (1976).

[13] M. Sagurton, E. L. Bullock and C. S. Fadley, Surf. Sci. 182, 287 (1987).

[14] L. J. Terminello, X. S. Zhang, Z. Q. Huang, S. Kim, A. E. Schach von Wittenau, K. T. Leung and D. A. Shirley, Phys. Rev. B38, 3879 (1988).

[15] G. Margaritondo and N. G. Stoffel, Phys. Rev. Lett. 42, 1567 (1979).

[16] G. M. Rothberg, K. M. Choudhary, M. L. den Boer, G. P. Williams, M. H. Hecht and I. Lindau, Phys. Rev. Lett. 53, 1183 (1984).

[17] K. M. Choudhary, S. T. Kim, J. H. Lee, S. N. Shah, M. L. denBoer, G. P. Williams and G. M. Rothberg, J. Phys. (Paris) Colloq. 47, C8-203 (1986).

[18] For InP(110) and InP(110) + 3 Å Al, the PEXAFS results have been published. K. M. Choudhary, P. S. Mangat, A. E. Miller, D. Kilday, A. Flipponi and G. Margaritondo, Phys. Rev. B38, 1566 (1988).

[19] K. M. Choudhary, P. S. Mangat, H. I. Starnberg, Z. Hurych, D. Kilday and P. Soukiassian, Phys. Rev. B, accepted as a Rapid Communication.

[20] K. M. Choudhary, P. S. Mangat, H. Starnberg, Z. Hurych and P. Soukiassian, to be presented in this Meeting, Symposium D: Advanced Techniques for Characterizing Surfaces/Interfaces of Materials.

[21] This paper. Also see Ref. [20].

[22] L. J. Brillson, C. F. Brucker, A. D. Katnani, N. G. Stoffel, R. Daniels and G. Margaritondo, J. Vac. Sci. Technol. 21, 564 (1982).

[23] Te-X Zhao, R. R. Daniels, A. D. Katnani and G. Margaritondo, J. Vac. Sci. Technol. B1, 610 (1983).

[24] M. H. Hecht and I. Lindau, Nucl. Inst. Methods 195, 339 (1982).

[25] A. G. McKale, B. W. Veal, A. P. Paulikas, S. K. Chan and G. S. Knapp, J. Am. Chem. Soc. (to be published).

[26] B. K. Teo and P. A. Lee, J. Am. Chem. Soc. 101, 2815 (1979).

[27] "Crystal Structures", edited by R. W. G. Wyckoff, 2nd ed. (Wiley, New York, 1964).

[28] A. Kahn, C. R. Bonapace, C. B. Duke and A. Paton, J. Vac. Sci. Technol. B1, 613 (1983).

SURFACE LATTICE STRAIN AND THE ELECTRONIC STRUCTURE OF
THIN MERCURY OVERLAYERS

Shikha Varma*, Y.J. Kime*, P. A. Dowben* and M. Onellion[†]
*Department of Physics, Syracuse University, Syracuse, New York 13244-1130
†Department of Physics, University of Wisconsin, Madison, Wisconsin 53706

ABSTRACT

The stress in the surface of a thin (0-10 monolayer) film of mercury can be relieved by either changing the substrate and consequently changing adlayer lattice constants or by growing thicker films. An empirical relationship between strain and the valence electronic structure of mercury overlayers can be demonstrated, though we present evidence that this relationship can be altered and destroyed by crystallography changes.

INTRODUCTION

Recently [1,2], a strong correlation between the lattice constant and the electronic structure of one and two monolayer thin films of Hg has been demonstrated. Theoretical studies have also suggested that the electronic structure and work function of Hg monolayers depends upon lattice constant [3,4]. The interaction between Hg adatoms is believed to be a contributing factor in the formation of a band structure from the Hg 5d levels [1, 5-7], commonly regarded as the shallow core levels.

Hg overlayers on $Cu_3Au(100)$ [1], Ag(100) [5,6], Fe(100) [8] and W(100) [9,10] adopt the substrate crystallography forming p(1x1) overlayers, i.e. grow pseudomorphically. In such growth, misfit dislocations are absent and the substrate lattice constant a is equal to the overlayer lattice spacing c.

We shall first consider the homogeneous strain energy for pseudomorphic growth, where the total overlayer strain energy is given by van der Merwe, Markov, and Stoyanov [11-13]. If (a_x, a_y) and (b_x, b_y) are the natural equilibrium lattice constants of the substrate and overlayer respectively then we can define the misfit $f_i = \dfrac{b_i - a_i}{a_i}$ (i = x,y). Then the homogeneous strain energy E_2^e is given by [2]

$$E_2^e = \frac{\mu \Omega}{1-\nu} \left[\frac{a_x^2}{b_x^2} f_x^2 + \frac{a_y^2}{b_y^2} f_y^2 + 2\nu \frac{a_x}{b_x} \frac{a_y}{b_y} f_x f_y \right] \qquad (1)$$

where $\Omega = b_x b_y b_z$ is the natural overgrowth volume per interfacial atom and $\mu = \dfrac{2 \mu_0 \mu_s}{\mu_0 + \mu_s}$. The μ_0 is the shear modulus of the overlayer and μ_s is that of the substrate. The Poisson's ratio $\nu = (\nu_0 + \nu_s)/2$ where ν_0 and ν_s are the Poisson's ratio of the overlayer and substrate respectively. For all the cases that we have considered $a_x = a_y = a$; $b_x = b_y = b$ and $f_x = f_y = f$, we get from (1)

$$E_2^e = \frac{2\mu b_z \, a^2 \, f^2(1 + \nu)}{(1 - \nu)} \tag{2}$$

where b_z (= 3.004 Å) is the thickness of the overlayer. The quantity E_2^e is the homogeneous strain energy per interfacial atom.

This approach is applicable to Hg overlayers which generally grow pseudomorphically (adopting the crystallography of the substrate) or at least in a structure which is largely coincident with the substrate [2]. We shall treat, however, strain as result of compression from equilibrium (repulsive strain) separately from strain which is the result of expansion from equilibrium (attractive strain).

The measure of electronic structure we shall employ, for comparison with the adlayer, strain is the energy separation of states that normally appear to be degenerate for the free atom, as determined by photoemission. For Hg the $5d_{5/2}$ level was chosen.

Hg Overlayers

A variety of ordered Hg overlayers have been studied ([1,2] and the references therein) all with square lattices and covering a wide range of lattice constants. The very large range of lattice constants, observed with Hg overlayers, makes the Hg monolayer films rather unusual when compared with other metal overlayers [1,2]. It has been shown [3,4] that the metal-non metal phase transition in a monolayer of Hg occurs for the hexagonal lattice configuration. On all the substrates that we have considered, a monolayer of Hg has a square lattice and no metal-non metal transition is expected with variations in lattice constant [3,4].

In this study, the strain energy for a variety of Hg overlayers has been calculated. From these results, the repulsive interaction between Hg adatoms can be inspected and compared with the electronic structure of the Hg overlayer. Furthermore, since the energetics of Hg adsorption on Fe(100) [8], W(100) [9,10], Ni(100) [14], and for bulk Hg [15] have been characterized, the homogeneous strain energy can be compared with experimentally derived repulsive energies, as shown in Fig. 1.

From (2) we have calculated the homogeneous strain energy. The results are summarized in Figure 1. The values of μ_s and ν_s have been obtained from Voigt averages [16]. The results are proportional to the adatom-adatom interaction energy for Hg overlayers (Fig. 1).

This calculated strain energy E_2^e does not account for the energy difference (approximately 0.01 eV/atom) between the natural rhombohedral Hg lattice and the f.c.c. lattice adopted by the overlayer Hg thin films estimated by Worster and March [17].

By treating the homogeneous strain energy for attractive adatom-adatom interactions separately from the strain energy for repulsive adatom-adatom interactions, we can compare the strain energy with the electronic structure of the Hg overlayer thin film. For Hg adsorption on $Cu_3Au(100)$ [1], Ag(100) [5,6], W(100) [18], Ni(100) [19], Cu(100) [1] as well as for bulk Hg [20], the Hg $5d_{3/2}$ and $5d_{5/2}$ spin-orbit split doublet is observed at binding energies of approximately 9.8 and 7.8 eV below the metal Fermi energy. For ordered Hg overlayers on $Cu_3Au(100)$ [1], Ag(100) [5,6] W(100) [18], and bulk Hg [20] a new 5d-like Hg band has been observed, and discussed in detail elsewhere [1, 2, 5-7]. This new band is "split off" from the $5d_{5/2}$ level and appears at smaller binding energies. The energy separation between the new 5d-like Hg state and the $5d_{5/2}$ state is strongly related to the lattice constant [1,2] as shown in Fig. 1. It has been proposed [1,2] that as the Hg lattice parameter

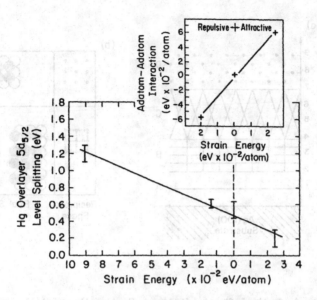

Figure. 1. The overlayer repulsive and attractive strain energies plotted against the Hg $5d_{5/2}$ level splitting. The band separations are derived from photo-emission spectra as indicated in the text. Least squares fits to the data have been indicated. The insert shows a plot of the strain energy E_2^e versus the plot of the experimentally derived adatom-adatom interactions. The regimes of the attractive and repulsive adatom interactions are indicated.

decreases, the energy separation ΔE_{5d}^B of the $5d_{5/2}$-like features increases due to the greater overlap of adjacent orbitals, thus leading to correspondingly greater hybridization. Similar observations have been made for halogen and iron overlayers as well [2,21].

Within the repulsive Hg-Hg adatom interaction regime, there appears to be a linear relationship between the observed energy separation ΔE_{5d}^B of the new Hg state and the Hg $5d_{5/2}$ state and the two dimensional homogeneous strain energy E_2^e as shown in Fig. 1. To clearly establish that ΔE_{5d}^B is proportional to E_2^e within the repulsive regime will require investigating the electronic structure of more systems. Nonetheless, we believe that this result, plotted in Fig. 1, demonstrates the correlation between the strain energy and electronic structure.

For Hg adsorption on Ag(100) at 90K, layer by layer growth occurs (Frank-van der Merwe growth) [5-7]. We can calculate for this system the critical thickness, h_c, for pseudomorphism (from the shear modulus of Ag

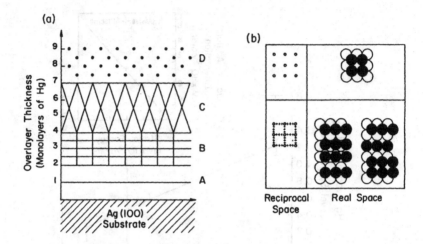

Figure 2. On the left, the diagram indicating the crystalline phase with thickness of Hg on Ag(100) is shown. For 1-2 monolayers, Hg grows pseudomorphically on Ag (A). For 2-4 monolayers thick Hg film no sharp LEED pattern could be seen (B). Beyond four monolayers, new LEED pattern is gradually established (C). For 5 to 7 monolayers, the observed diffraction pattern is a superposition of p(1x4) and p(4x1) [6] (see Fig. on right). Adsorption of Hg following 7 monolayers of Hg on Ag(100) results in polycrystalline bulk Hg (D). At the right the schematic reciprocal space nets for one and two monolayers of Hg on Ag(100) (top) as well as the reciprocal space nets for five to seven monolayers of Hg on Ag(100) (bottom) are shown. The 1x1 overlayer (top) for Hg has been drawn with the Hg atoms (black) of the outermost layer resting in the fourfold hollow of the layer below-though no evidence as to site yet exists. Two possible solutions for five to seven monolayers of Hg are shown (black) with 1x1 lattice indicated in white.

[22] and Hg [23,24]) by [25]

$$h_c = \frac{-\mu_{Ag}\left[\frac{2ab}{a+b}\right]^2 \ln\left[\frac{4\pi\mu_{Hg}\mu_{Ag} \quad (a-b)}{(\mu_{Hg}+\mu_{Ag})\left[\frac{2\mu_{Hg}\mu_{Ag}}{\mu_{Hg}+\mu_{Ag}}\right](1-\nu)(a+b)}\right]}{2\pi a\left[\frac{b-a}{a}\right](1+2\nu)(\mu_{Hg}+\mu_{Ag})}. \tag{3}$$

It has been observed (via low energy electron diffraction) that the pseudomorphic growth of Hg on Ag(100) ceases after two monolayers, with a disordered thin film structure observed with three and four monolayer thick films followed by the formation of a p(4x1) structure for 5-7 monolayer thick films [6]. In good agreement with this result, we calculate that $h_c = 9.8$ Å, suggesting this treatment of strain, based on the shear modulus, agrees with structural studies.

The p(4x1) structure can be [6] interpreted as consistent with a structure with periodic misfit dislocations as a result of missing rows, as shown in Fig. 2. As pointed out elsewhere [26] there is no reason why the free energy (as a result of the homogeneous strain) cannot be lowered by a non-uniform deformation of the lattice. Certainly the postulated solutions of the p(4x1) low energy electron diffraction pattern do relieve strain.

There has been theoretical work suggesting that as a strained overlayer film becomes thicker, an intermediate structure can be present between the coherent pseudomorphic structure occurring with the thin films (following initial growth) and polycrystalline bulk structure (the natural structure of the overlayer material) observed with very thick films [26]. The observation of the p(4x1) structure for 5,6 and 7 monolayers with indications of the bulk mercury structure observed at 9 and 10 monolayer thick films [6,7] is consistent with such a model.

Curiously, the electronic density of states is not substantially altered by this relief of the homogeneous strain energy that occurs at h_c [6]. This suggests, despite the above results correlating homogeneous strain energy (for monolayer thick films) with electronic structure, that strain alone cannot be related with electronic structure - clearly structure is an important factor in determining the density of states.

We have noted elsewhere [27-29] that the Hg $5d_{5/2}$ strain induced splitting is sensitive to long range crystallographic order and will be destroyed by a large defect density. Similar results have been observed for silicon-GaAs heterojunctions [30]. Nonetheless, the demonstrated empirical relationship between overlayer strain and electronic structure of Hg allows one to predict the Hg $5d_{5/2}$ level splitting for selected Hg overlayers for which no measurement currently exists. For example, we expect the Hg $5d_{5/2}$ level splitting for the 1 x 1 Hg overlayer on Fe(100) to be approximately 0.65 eV.

CONCLUSION

For mercury overlayers, we observe that the greater the lattice constant, the smaller the separation between states that are degenerate, by photoemission, for the free atom. By treating attractive strain separately from repulsive strain, we find that the homogeneous strain is proportional to this measure of the valence electronic structure for a range of lattice constants near the equilibrium lattice constant of the overlayer i.e. $E \propto (b-a)^2$ [2]. Changes in electronic binding energy splittings and the strain energy both arise from changes in the overlap integral of the wavefunctions on adjacent adatoms sites. The effect of crystallographic order upon electronic structure cannot, however, be considered separate from the effect of the homogeneous strain energy. Furthermore, strain relief normal to the plane is not equivalent to in-plane strain relief.

ACKNOWLEDGEMENTS

The authors are indebted to J. Weinberg, and A. Srivastava. This work was supported, in part, by the Office of Naval Research through grant # N-00014-87K-0673 and by AFOSR TASK 2305J9 monitored under RADC postdoctoral program contract number F30602-88-0027. We would like to thank the staff of

90

the Synchrotron Radiation Center in Stoughton Wisconsin. The SRC is supported by the National Science Foundation grant no. DMR-83-04368.

References

1. M. Onellion, Y. J. Kime, P. A. Dowben and N. Tache, J. Phys. C. Solid State Phys. 20, L633 (1987).
2. P. A. Dowben, Shikha Varma, Y. J. Kime, D. R. Mueller, and M. Onellion, Z. Physik, (1988).
3. H. J. F. Jansen, A. J. Freeman, M. Weinert and E. Wimmer, Phys. Rev. B28 593 (1983).
4. A. R. Miedema, J. W. F. Dorleijn, Phil. Mag. B43, 251 (1981).
5. M. Onellion, J. L. Erskine, Y. J. Kime, S. Varma and P. A. Dowben, Phys. Rev. B33, 8833 (1986).
6. P. A. Dowben, Y. J. Kime, S. Varma, M. Onellion and J. L. Erskine, Phys. Rev. B36, 2519 (1987).
7. P. A. Dowben, M. Onellion and Y. J. Kime, Scanning Microsco. 2, 177 (1988).
8. R. G. Jones and D. L. Perry, Vacuum, 31, 493 (1981).
9. R. G. Jones and D. L. Perry, Surf. Sci. 71, 59 (1979).
10. R. G. Jones, D. L. Perry, Surf. Sci. 82, 540 (1979).
11. J. H. van der Merwe, J. App. Phys. 41, 4725 (1970).
12. J. H. van der Merwe, Treastise on Material Science and Technology, vol 2, edited by H. Herman, Academic Press, New York, p. 90 (1973).
13. I. Markov and S. Stoyanov, Contemp. Phys. 28, 267 (1987).
14. R. G. Jones and A. W-L. Tong, Surf. Sci. 188, 87 (1987).
15. Handbook of Chemistry and Physics, 51st Edition, R. C. Weast, Ed.: Chemical Rubber Co. Cleveland, OH p. D-146 (1971).
16. G. Simmons and H. Wang, Single Crystal Elastic Const. and Calculated Aggregate Properties: A Handbook, (MIT Press, Cambridge, MA, 1971).
17. J. Worster and N. H. March, Solid State Commun. 2, 245 (1964).
18. W. F. Egelhoff, D. L. Perry and J. W. Linnett, Surf. Sci. 54, 670 (1976).
19. G. E. Becker and H. G. Hagstrum, J. Vac. Sci. Technol. 10, 31 (1973).
20. S. Svensson, N. Mortensson, E. Basilier, P. A. Malmquist, U. Gelius and K. Siegbahn, J. Electron Spectrosc. Relat. Phenom. 9, 51, (1976).
21. P. A. Dowben, CRC Critical Reviews in Solid State and Material Science 13, 191 (1987).
22. J. R. Neighbours and G. A. Alers, Phys. Rev. 111, 707 (1958). The value of μ_{Ag} and Poisson's Ratio of Ag are taken at 100° K.
23. R. F. S. Hearmon, Adv. in Phys. 5, 323 (1956). (μ_{Hg}, E and Poisson's ratio are at 83° K and are determined by Voigt averages.)
24. A. G. Crocker and G. A. A. M. Singleton, Phys. Stat. Sol.(a) 6, 635 (1971).
25. W. A. Jesser, D. Kuhlmann Wilsdorf, Phys. Stat. Sol. 19, 95 (1967).
26. R. Bruinsma and A. Zangwill, J. Physique 47, 2055 (1986).
27. M. Onellion, P. A. Dowben and J. L. Erskine, Phys. Lett. A. 130, 171 (1988).
28. Shikha Varma, Y. J. Kime, P. A. Dowben, M. Onellion and J. L. Erskine, J. Phys. C., Solid State, submitted.
29. P. A. Dowben, Y. J. Kime, Shikha Varma, M. Onellion and J. L. Erskine, J. Vac. Sci. Technol.,. Submitted.
30. R. S. List, J. C. Woicik, I. Lindau, W. E. Spicer, J. Vac. Sci. Technol. B5, 1279 (1987).

BOND SELECTIVE CHEMISTRY
WITH PHOTON-STIMULATED DESORPTION

J.A. YARMOFF AND S.A. JOYCE
 National Institute of Standards and Technology,* Surface Science
Division, Gaithersburg, MD 20899

*Formerly the National Bureau of Standards

ABSTRACT

Photon stimulated desorption of fluorine ions from silicon surfaces was studied via excitation of the Si 2p core level with synchrotron radiation. These results showed that the process is chemically selective in that the removal of a fluorine ion from a silicon species in a given oxidation state can be enhanced by tuning the photon energy to the excitation wavelength corresponding to a transition from the 2p core level of the bonding atom to the conduction band minimum. This process was studied as a possible means for the production of surfaces with selected compositions of species. The results of selective exposures of fluorinated surfaces to monochromatized radiation indicated that secondary desorption processes and the inherent chemistry of the surface reactions can override the effects of selective desorption. Other possibilities for selective surface reactions via core-level excitations are discussed.

INTRODUCTION

Electron and/or photon beams provide a possible means of inducing dissociation of reactant species on surfaces at low temperatures via core-level excitations.[1] Since these excitations are chemically specific, they may provide a means for selectively enhancing a particular dissociation reaction on a surface. By studying the desorption of ions from surfaces, the mechanism by which this chemical selectivity occurs can be ascertained. In a recent letter,[2] we showed that stimulated desorption of fluorine ions from silicon at the 2p core level is selective with respect to the oxidation state of the bonding silicon atom. This paper presents a study of the photon stimulated desorption (PSD) of F^+ from fluorine bound to Si (111) in the energy regime of the Si 2p core level, and comments on the usefulness of employing core-level excited stimulated desorption processes for the purposes of bond-selective surface chemistry.

EXPERIMENTAL

These experiments were performed on beamline UV-8b at the National Synchrotron Light Source at Brookhaven National Laboratory, using a 6m toroidal grating monochromator (TGM). Si (111) wafers, cut within 0.25° of the low-index plane, were cleaned via standard methods in ultra-high vacuum (UHV). Cleanliness was checked by monitoring the surface core level shifts and the valence band surface states measured with photoemission.[3]

The surfaces were exposed to fluorine by transferring them under UHV into a separate reaction chamber (base pressure 1×10^{-9} torr), and then exposing them to XeF_2. Following exposure, the samples were transferred back into the spectrometer chamber (base pressure 1×10^{-10} torr). The composition of the surfaces was measured via Si 2p core level photoemission, employing an ellipsoidal mirror analyzer[4] with a typical energy resolution of 0.1 eV. PSD was measured by reversing the polarity of the optics so as to be sensitive to ions instead of electrons, and then biasing the entrance to the multichannelplate array by -600 volts to provide additional acceleration of the ions. In this manner it was possible to collect both energy distributions of the emitted ions and the dependence of the ion yield on the incident photon energy.

RESULTS AND DISCUSSION

Core-level soft x-ray photoemission of the 2p level of silicon surfaces exposed to XeF$_2$ have provided information on the type and amount of fluoride species resident on the surface.[5] This is facilitated by the observed shift of the 2p level of approximately 1 eV per attached fluorine atom. For a Si (111)-7x7 surface, exposure to 50 L (1 L = 1x10^{-6} torr-sec) of XeF$_2$ results in the formation of mono- and trifluorides (with a small amount of difluoride). The mono- and trifluorides are present in almost equal abundance. Annealing of this surface to 300°C removes the higher fluorides, leaving a surface terminated solely by SiF species.[2] The coverage at this point is slightly less than one monolayer.

In Fig. 1, absorption and PSD spectra are shown for both the annealed (SiF only) and the unannealed (SiF and SiF$_3$) surface compositions. The upper curve shows the absorption spectrum obtained by monitoring the yield of 20 eV secondary electrons as a function of the photon energy. This electron energy was chosen so as to provide the best possible surface sensitivity, since 20 eV is roughly the energy at which the mean free path for electrons traveling through silicon is at a minimum. There is an edge seen at a photon energy of approximately 100 eV in the absorption spectrum which corresponds to transitions from the bulk Si 2p level to the conduction band minimum (CBM). This edge shows structure due to the spin-orbit splitting of the Si 2p initial state. The positions of the onsets for these transitions (determined with photoemission) have been marked by vertical lines in the figure. The length of these lines is proportional to the occupation of the spin-orbit components.

The middle curve (Fig. 1(b)) shows the PSD obtained from the annealed surface as a function of photon energy. The contributions to this spectrum from small impurities of higher fluorides have been removed by measuring the ions desorbed primarily in normal emission direction. This is possible since PSD is known to occur in a direction related to the initial bond angle,[6] and the Si-F bonds in the monofluoride species are directed normal to the surface while the bonds in the higher fluorides are not.[2,7]

The main feature in the PSD spectrum of the monofluoride species occurs at approximately 1 eV higher energy than the bulk absorption feature, and shows the same spin-orbit structure as the bulk feature. This energy correlates with the transition from the 2p level of an SiF unit to the CBM. The onsets for these transitions are also shown as vertical lines in the figure. This large feature in the PSD spectrum is thus a direct desorption process in which a photon is absorbed by the bonding Si atom, decays via an interatomic Auger mechanism, and results in the emission of an F$^+$ ion.

There is a small feature in the PSD spectrum at the energy of the bulk Si 2p absorption. This small feature represents ions which have been desorbed via an indirect process in which the secondary electrons produced when a bulk Si atom absorbs a photon act to excite a lower lying core-level (e.g. the F 2s) in the adsorbate, thereby initiating an electron stimulated desorption (ESD) process. Note also that the PSD intensity just before the edge is non-zero. In fact, the ion intensity at the peak is about twice that of the intensity just before the edge. This background is also due to secondary electrons induced ESD. The threshold for ESD for fluorine from silicon has been measured by Bozack, et al. at 28 eV, which correlates with the F 2s level.[8]

The spectrum shown in Fig. 1(c) is the PSD yield obtained from the unannealed surface (a mixture of SiF and SiF$_3$ species) collected by applying a negative bias voltage to the first grid in the analyzer in order to collect ions emitted at all angles. This spectrum shows both the indirect and direct features which were present in the monofluoride PSD spectrum, and shows additional structure at higher photon energies. There is an edge seen at the energy which is required to excite a 2p electron in an SiF$_3$ species to the CBM. The position of this onset is marked in the figure, and although it is not completely resolved from the monofluoride edge (and from whatever contribution a small amount of SiF$_2$ may have), it is clear that there is an edge

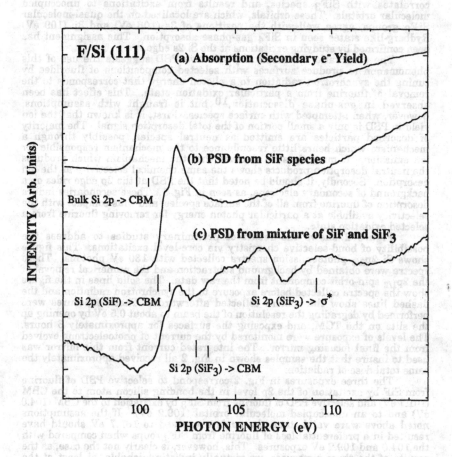

Figure 1. (a) Absorption as a function of photon energy for sample after exposure to 50 L of XeF_2 at room temperature and annealing to 300°C. The absorption was monitored via secondary electrons with a kinetic energy of 20 eV. (b) PSD yield of F^+ ions as a function of photon energy for sample prepared in same manner as (a). This spectrum represents ions observed primarily in the normal emission direction. (c) PSD yield as a function of photon energy for a surface prepared by exposure of Si (111)-7x7 to 50 L of XeF_2 without annealing. This spectrum represents ions emitted in all directions.

at this photon energy. The additional structure seen above the edge also correlates with SiF$_3$ species, and results from excitations to unoccupied molecular orbitals. These orbitals, which are localized on the quasi-molecular SiF$_3$ groups, agree well with the positions of 3s (106 eV) and 3p (109 eV) Rydberg-like states seen in SiF$_4$ gas-phase absorption. This assignment has been confirmed by studying excitations at the Si 2s edge.[9]

The observation of chemically selective PSD suggests the use of this phenomenon to produce surfaces with selected compositions of fluorides by tuning the synchrotron radiation to a photon energy that corresponds to the removal of fluorine from a particular oxidation state. This effect has been observed in gas-phase dissociation,[10] but is fraught with assumptions, however, when attempted with surface species. First, it is known that the ion yield in PSD is only a small portion of the total desorption signal. The majority of desorbed particles are emitted as neutral species, possibly through a mechanism which bears little resemblance to the mechanism responsible for ion emission. It must thus be assumed that the mechanism which produces the neutral desorption products shows the same chemical selectivity as the ion desorption. Secondly, it should be noted that the PSD at the 2p edge rides on a background of secondary electrons, as seen in Fig. 1. These secondaries induce desorption of fluorine from all of the surface species, and thus compete with the selectivity available at a particular photon energy for removing fluorine from a selected oxidation state.

Figure 2 shows the result of preliminary studies to address the possibility of bond-selective chemistry via core-level excitations. This figure shows Si 2p$_{3/2}$ photoemission spectra collected with 130 eV photons. These spectra were obtained by background subtraction and the numerical removal of the 2p$_{1/2}$ spin-orbit component from the raw data. The solid lines in the figure show the spectra collected before exposure to the synchrotron radiation and the dashed lines show the spectra collected afterwards. These exposures were performed by degrading the resolution of the beam to about 0.5 eV by opening up the slits on the TGM, and exposing the surfaces for approximately 6 hours. The levels of exposure were monitored by the current of photoelectrons evolved from the final focusing mirror. The integrated current from this mirror was used to insure that the samples shown in Fig. 2 all received approximately the same total dose of radiation.

The three exposures in Fig. 2 correspond to selective PSD of fluorine from SiF by excitation of the 2p level in the bonding silicon atom to the CBM (101.7 eV) and selective PSD of fluorine from SiF$_3$ by excitation to the CBM (104.0 eV) and to an unoccupied molecular orbital (109.2 eV). If the assumptions noted above were valid, then the surface exposed to 101.7 eV should have resulted in a preferential loss of fluorine from SiF groups when compared with the 104.0 and 109.2 eV exposures. This, however, is clearly not the case, as the results of the three exposures are virtually indistinguishable, at least at the levels of exposure employed here.

These results imply that in the stimulated desorption of neutral species at the silicon 2p core level, the major process contributing to the loss of fluorine does not involve a direct excitation of the bonding Si atom. Thus, secondary induced ESD processes leading to excitation of the F 2s or of valence levels may be more important than the direct excitations seen in ion desorption. A second possible interpretation of the results is that the composition of a surface containing a given number of chemisorbed fluorine atoms is dependent on the chemistry of the Si-F system, and not on the history of how the fluorine atoms arrived at the surface. This does not eliminate the possibility for bond-selective chemistry, but suggests performing experiments in which the surfaces are held at cryogenic temperatures during the radiation exposures and subsequent photoemission measurements in order to inhibit the restructuring of the fluoride species.

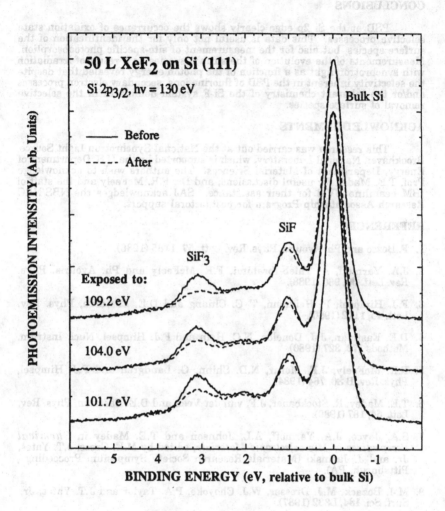

Figure 2. Si 2p$_{3/2}$ photoemission for Si (111)-7x7 surfaces exposed to 50 L of XeF$_2$ followed by irradiation with 101.7, 104.0 and 109.2 eV synchrotron light. The solid lines show the surfaces prior to irradiation and the dashed lines show the surfaces after irradiation. These spectra are shown after background subtraction and numerical removal of the 2p$_{1/2}$ spin-orbit component.

CONCLUSIONS

PSD at the Si 2p edge clearly shows the occurrence of oxidation state selective processes. This data is useful not only for the identification of the surface species, but also for the measurement of site-specific photoabsorption. Measurements of the evolution of the surface species as a result of irradiation with synchrotron light as a function of the photon energy revealed that despite the selectivity inherent in the PSD of fluorine ions, secondary electron processes and/or the inherent chemistry of the Si-F system dominates in the selective removal of surface species.

ACKNOWLEDGEMENTS

This research was carried out at the National Synchrotron Light Source, Brookhaven National Laboratory, which is supported by the U.S. Department of Energy, Department of Material Sciences. The authors wish to acknowledge Prof. T.E. Madey for useful discussions, and Dr. F.R. McFeely and the staff of IBM beamline UV-8 for their assistance. SAJ acknowledges the NBS/NRC Research Associateship Program for postdoctoral support.

REFERENCES

1. F. Bozso and Ph. Avouris, Phys. Rev. Lett. **57**, 1185 (1986).

2. J.A. Yarmoff, A. Taleb-Ibrahimi, F.R. McFeely and Ph. Avouris, Phys. Rev. Lett. **60**, 960 (1988).

3. F.J. Himpsel, P. Heimann, T.-C. Chiang and D.E. Eastman, Phys. Rev. Lett. **45**, 1112 (1980).

4. D.E. Eastman, J.J. Donelon, N.C. Hein and F.J. Himpsel, Nucl. Instrum. Methods **172**, 327 (1980).

5. F.R. McFeely, J.F. Morar, N.D. Shinn, G. Landgren and F.J. Himpsel, Phys. Rev. B **30**, 764 (1984).

6. T.E. Madey, R. Stockbauer, J.F. van der Veen and D.E. Eastman, Phys. Rev. Lett. **54**, 187 (1980).

7. S.A. Joyce, J.A. Yarmoff, A.L. Johnson and T.E. Madey in *Chemical Perspectives of Microelectronic Materials*, edited by M.E. Gross, J.T. Yates, Jr. and J. Jasinski (Materials Research Society Symposium Proceedings, Pittsburgh, PA).

8. M.J. Bozack, M.J. Dresser, W.J. Choyoke, P.A. Taylor and J.T. Yates, Jr., Surf. Sci. **184**, L332 (1987).

9. J.A. Yarmoff and S.A. Joyce, unpublished.

10. W. Eberhardt, T.K. Sham, R. Carr, S. Krummacher, M. Strongin, S.L. Weng and D. Wesner, Phys. Rev. Lett. **50**, 1038 (1983).

POLARIZED X-RAY ABSORPTION STUDIES OF OXIDE SUPERCONDUCTORS

E. Ercan Alp, S.M. Mini, M. Ramanathan, B.W. Veal, L. Soderholm, G.L. Goodman, B. Dabrowski, G. K. Shenoy, *Argonne National Laboratory*, Argonne, Illinois 60439, USA, J. Guo, D.E. Ellis, *Northwestern University, Evanston, Illinois 60601*, A. Bommanavar, *Brooklyn College of CUNY, Brooklyn, New York, 11210*, O.B. Hyun, *Iowa State University, Ames, Iowa 50011-3020*

ABSTRACT

Polarized X-ray absorption studies have been carried out at the Cu K-edge to study the effect of Sr doping in La_2CuO_4 and oxygen doping in $YBa_2Cu_3O_{6+x}$. These measurements help to elucidate the transitions giving rise to the absorption edges. We offer an explanation of the polarization shifts of features in terms of the results of our embedded cluster calculations of electronic structure.

INTRODUCTION

Polarized X-ray absorption spectroscopy has been successfully used to describe the nature of electronic transitions in square planer copper complexes (1), as well as in magnetically oriented powders of the oxide superconductors (2). In this paper, the results of polarized XANES spectroscopy measurements on magnetically oriented powders of $La_{2-x}Sr_xCuO_4$ and $YBa_2Cu_3O_{6+x}$ are described, and the full angular dependence of absorption cross-sections are given. The results indicate the strong anisotropic nature of the electronic structure in both classes of compounds. The polarized nature of synchrotron radiation, can thus be advantageously used to gain further insight into the local charge distribution around a particular kind of atom in a solid. These results give an additional dimension to XANES spectroscopy - one of the few techniques that nondestructively measure charge on an atom in a solid .

EXPERIMENTAL

The x-ray absorption spectra have been collected at room temperature, in the transmission mode at beamlines X-18B, and X-11A of NSLS at Brookhaven National Laboratory, and also at CHESS facility of Cornell University. Si (220) crystals are used to monochromatize the synchtrotron radiation. Cu-metal spectra have been recorded simultaneously to provide an energy calibration between different beamlines, and between separate runs. The inflection point in the first resolved peak of copper metal has been chosen to be the zero of the energy scale. This is relatively easy to locate and reproduce, since the derivative of the spectrum has a sharp peak at that point. The energy calibration is estimated to be better than 0.25 eV.

Mat. Res. Soc. Symp. Proc. Vol. 143. ©1989 Materials Research Society

La$_{2-x}$Sr$_x$CuO$_{4-y}$ samples are difficult to orient when both x=0 and y=0. A sample in which x=0, and y=0.04 yielded the best orientation. Finely ground samples were cast into disks by mixing with fast curing epoxy, and were held under 8.2 T magnetic field. Both samples orient such that the crystallographic c-axis is along the magnetic field direction.

RESULTS AND DISCUSSION

La$_{2-x}$Sr$_x$CuO$_4$ is orthorombic when x=0.0, and it is tetragonal and superconducting for x=0.15 (3). The copper atoms in both structures occupy a distorted octahedral site, surrounded by 4 planar oxygen atoms at a distance of 1.89 Å, and two axial oxygens at 2.43 Å. The XANES spectrum of unoriented powder sample of La$_2$CuO$_4$ is shown in Fig. 1, along with that of CuO, and KCuO$_2$ for comparison. The characteristic energy of absorption, determined based on a recently developed method (4) yields 3.58 for Cu$_2$O, 6.49 for CuO, 7.54 for La$_2$CuO$_4$, 7.94 for La$_{1.85}$Sr$_{0.15}$CuO$_4$, 6.46 for YBa$_2$Cu$_3$O$_{6.5}$, 7.66 for YBa$_2$Cu$_3$O$_{6.9}$, and 8.25 for KCuO$_2$. (All energies are in electron volts). This "characteristic energy" is defined in terms of the energy-weigthed area of an absorption cross section divided by its ordinary area and gives a quantitative location for the steeply rising portion of the absorption curve. Any shift here in characteristic energies as a function of Sr-doping is of the same order as our experimental uncertainity, about 0.5 eV. This relative insensitivity of Cu K-edge features to the Sr-doping has led investigators to believe that the primary effect of Sr-doping is to create electronic holes in the oxygen 2p states, rather than in the Cu 3d-states (5,6).

In Fig. 2, we compare the Cu K-edge XANES spectra for unoriented powder samples of Cu$_2$O and YBa$_2$Cu$_3$O$_{6+x}$ for x=0.5, and 0.9. The structure of YBa$_2$Cu$_3$O$_{6+x}$ has been shown to depend on the value of x. For 0.0 < x < 0.5 the compound is tetragonal, while for higher oxygen concentrations, 0.5 < x < 1.0 the material is orthorombic (10). In both structures, there are two crystallographically inequivalent

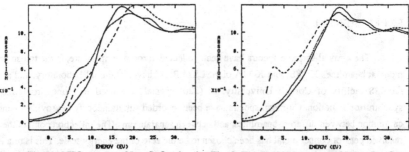

Fig. 1. The XANES spectra of La$_2$CuO$_4$ (——), CuO (·····), and KCuO$_2$ (---) at the Cu K-edge.

Fig. 2. The XANES spectra of YBa$_2$Cu$_3$O$_{6.9}$ (——), YBa$_2$Cu$_3$O$_{6.5}$ (·····) , and Cu$_2$O (---)

copper sites. Cu 2 (in the nomenclature of Ref. 10) is coordinated to 5 oxygen atoms four of which form a slightly distorted plane perpendicular to the crystal's c-axis. The fifth capping oxygen, O4, is located along the c-direction and is also coordinated to Cu 1. The major changes in the structure as a function x occur in

the Cu 1 environment. At x=0, Cu 1 has only two near neighbor, O 4 atoms along the c-direction. This linear arrangement is similar to that of Cu in the monovalent standard Cu_2O. As x increases, the Cu 2 environment remains essentially unchanged, but the Cu 1 site changes significantly. For x=1, Cu 1 has four near neighbor oxygen atoms, arranged in plane containing c-axis. Since there are two different Cu sites, one of which changes significantly with x, it is difficult to understand the composite features in the unoriented powder spectra of $YBa_2Cu_3O_{6+x}$ depicted in Fig. 2.

Using the technique of magnetically oriented powders cast into disks (2), we have obtained polarized XANES spectra of $La_{2-x}Sr_xCuO_4$ and $YBa_2Cu_3O_{6+x}$ shown in Figs. 3 and 4. The spectra obtained with the electric vector parallel to the crystallographic c axis we term as "along the c direction", and spectra with the electric vector perpendicular to the c axis we term " in the ab plane." The samples are not ordered with respect to the a or b axis.

The noticable shift in energy of the steeply rising portion of the cross section curve as a function of the polarization direction in Fig. 3 makes it practical to identify features and to assign them to final state characteristics for the single Cu site environment in the $La_{2-x}Sr_xCuO_4$ lattice. The shifts in energy as the polarization changes are clearly much greater for this 2-1-4 system than any changes that are associated with the Sr-doping. These observations can be made quantitative in terms

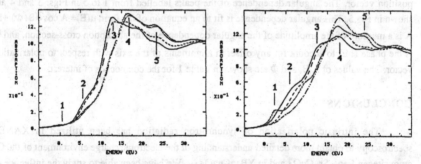

Fig. 3. XANES spectra of $La_{2-x}Sr_xCuO_4$ for x=0.0 along the c-direction (——) and along the ab-planes (····) and for x=0.15 along the c direction(----), and along ab-planes (– ·— ··) at the Cu K-edge.

Fig. 4. XANES spectra of $YBa_2Cu_3O_{6+x}$. For x=0.1 along the c-direction (——) and along the ab-planes (····) and for x=0.91 along the c direction (----), and along ab-planes (– ·— ··) at the Cu K-edge.

of the characteristic energies for the four absorption edges shown in Fig. 3. These energies for x=0 are 6.07 along the c direction and 8.05 in the ab plane; and the corresponding values for x=0.15 are 6.60 and 8.42, all in eV's. Since the uncertainty for these characteristic energies is about 0.5 eV, we are confident only of the approximately 2 eV shift in the edge position as a function of polarization. In contrast, the situation depicted in Fig. 4 shows energy shifts of comparable magnitude for both changes in polarization and changes in oxygen content.

Multiple scattering (MS) calculations of the near edge spectra have previously been reported for clusters of representative symmetry (5,6,7). The interpretation presented here is based on a new approach in which the MS potential has been derived from ground and transition state calculations by the self-consistent discrete variational X-α method with the cluster embedded in an extended crystal lattice (8). The details of the simulation and the analysis of the spectra are presented elsewhere (9). The orbital energies calculated for an 11 atom cluster in the Cu-1s transition states include three excited, virtual orbitals derived predominantly from Cu-4p basis functions on the transition state atom. We believe that the energy separation and orientations of these virtual orbitals provide a good understanding of the observed shift of the Cu K-edge in the 2-1-4 compound as a function of polarization. We find an orbital with 80% Cu $4p_z$ character directed along the c axis lies some 3.4 eV below two orbitals with 91% Cu $4p_{x,y}$ character in the a,b plane. The 1s ---> $4p_z$ transition is excited by light polarized along the c axis, and the 1s ---> $4p_{x,y}$ by light polarized in the a,b plane. The difference in orbital energies is comparable to the observed energy shift. The corresponding analysis for the 1-2-3 compound is more difficult because of the two crystallographically distinct Cu atoms within perpendicularly oriented local coordinate planes. Work is in progress on identifying the final state features for XANES of this 1-2-3 system as well.

The electric dipole nature of the x-ray absorption process dictates that the transition intensities should have a $\cos^2\theta$ dependency on the angle θ between the polarization and the position vector. The angular dependence of the peaks labelled from 1 to 5 in Figs. 3 and 4 are shown in Fig.5. This angular dependence is fit to an equation of the form $\mu(E)= A \cos^2(\theta+\Phi) + B$. A is a measure of the amplitude of the angular dependence of the absorption cross-section, and Φ is the phase shift to account for any offset in alignments of the axis with respect to polarization vector. The values of A,B, and Φ are shown in Table 1 for the compounds of interest.

CONCLUSIONS

The intrinsic polarization of synchrotron radiation has been utilized in XANES spectroscopy to obtain more detailed understanding of the electronic charge environment of the Cu atom sites in $La_{2-x}Sr_xCuO_4$ and in $YBa_2Cu_3O_{6+x}$. We have been able to study the influence of geometric characteristic of the final state orbitals on XANES. We find that for the 2-1-4 system any changes in charge environment of the Cu due to Sr-doping are within our experimental uncertainty, and are dwarfed by the polarization shifts in the K edge energy. For the 1-2-3 system we find the changes in edge shape and position due to O atom doping and those due to polarization are of comparable magnitude. Because of the two crystallographically distinct types of Cu atoms present in the 1-2-3 system, the pattern of edge shifts for this case is more complicated than that for the 2-1-4 system.

Table 1. Results of angular dependence of normalized x-ray absorption coefficient at the Cu-K edge of La_2CuO_4 and $YBa_2Cu_3O_x$ based superconductors. The angular dependence is fit to an equation of the form $\mu(E) = A \cos^2(\theta+\phi) + B$, where θ is the angle between polarization direction of the synchrotron radiation and c-axis of the oriented samples, ϕ is the phase shift. A is a measure of the amplitude of the angular dependence of the absorption cross-section in, and B is the absolute value of the absorption cross-section at $\theta+\phi=0.0$, in normalized units.

| Sample | Coefficient | PEAK NUMBER | | | | |
		1	2	3	4	5
La_2CuO_4						
	E (eV)	1.78	6.28	11.817	16.82	22.82
	A	0.003	-0.12	-0.48	0.16	0.20
	B	0.047	0.346	1.227	1.210	1.035
	φ (deg)	73	100	100	92	92
$La_{1.85}Sr_{0.15}CuO_4$						
	E (eV)	1.78	6.28	11.78	17.79	23.29
	A	0.003	-0.10	-0.35	0.10	0.13
	B	0.047	0.305	1.107	1.212	1.072
	φ (deg)	108	101	100	102	102
$YBa_2Cu_3O_{6.9}$						
	E (eV)	-0.86	7.63	12.13	17.13	23.63
	A	0.01	-0.21	-0.22	0.25	0.18
	B	0.037	0.525	0.961	1.145	1.039
	φ (deg)	105	98	99	99	99
$YBa_2Cu_3O_{6.1}$						
	E (eV)	1.78	6.28	11.79	17.28	21.786
	A	0.06	-0.18	-0.27	0.13	0.18
	B	0.150	0.515	0.985	1.188	1.018
	φ (deg)	88	91	92	85	88

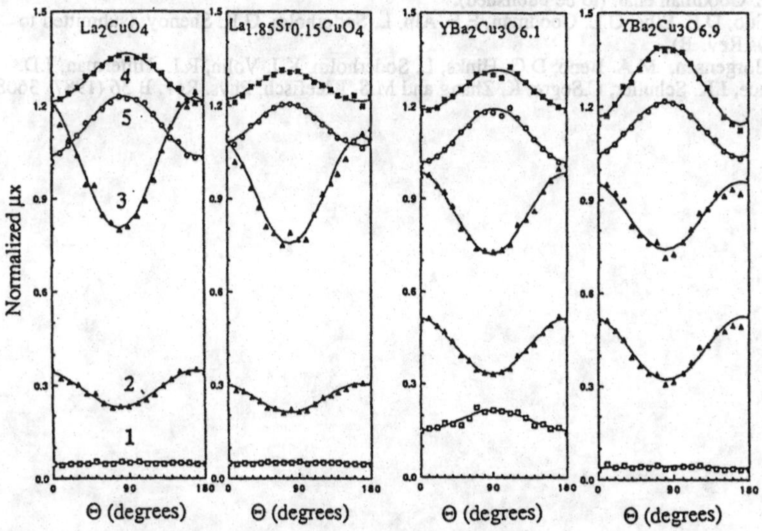

Fig. 5. Angular dependence of $\mu(E)$ for peaks numbered 1-5 in Figs. 3 and 4. The quantitative details are given in Table 1. The solid curves are least square fitted curves to the equation $\mu(E) = A \cos^2(\theta+\Phi) + B$.

ACKNOWLEDGMENTS

One of us (EEA) acknowledge help of D. Finnemore for making the right connection for orienting our samples. We also acknowledge the support of US DOE, BES Materials and Chemical Sciences under contract # W-31-109-ENG-38, and its role in the development and operation of X-11 beamline at the National Synchrotron Light Source under contract # DE-AS05-ER107-42. The NSLS is supported by US DOE, BES Materials and Chemical Sciences under contract # DE-AC02-76CH00016.

REFERENCES :

1. T.A. Smith, J.E. Penner-Hahn, M.A. Berding, S. Doniach, K.O. Hodgson, J.Am.Chem.Soc.107 (1985) 5945.
2. S. Heald, J. Tranquada, Phys.Rev.B 38 (1988) 761.
3. J.D. Jorgensen, H.-B. Scüttler, D.G. Hinks, D.W. Capone, K. Zhang, M.B. Brodsky, and D.J. Scalapino, Phy.Rev.Lett. 58 (1987) 1024.
4. E. E. Alp, G.L. Goodman, L. Soderholm, S.M. Mini, M. Ramanathan, G.K. Shenoy, A.S. Bommanavar, (submitted to Phys. Rev. Lett.)
5. V. J. Emery, Phys. Rev. B 38 (1988) 4547.
6. J.C. Fuggle, P.J.W. Weijs, R. Schoorl, G.A. Sawatzky, J. Fink, N. Nücker, P.J. Durham, W.M. Temmerman, Phys. Rev. B 37 (1988) 123.
7. E.E. Alp, G.K. Shenoy, D.G. Hinks, D.W. Capone, L. Soderholm, H. -B. Schüttler, J. Guo, D.E. Ellis, P.A. Montano, M. Ramanathan, Phys. Rev. B 35 (1987) 7199.
8. G.L. Goodman et.al. (to be published).
9. J. Guo, D.E. Ellis, G.L. Goodman, E.E. Alp, L. Soderholm, G.K. Shenoy, (submitted to Phys.Rev. B)
10. J. Jorgensen, M.A. Beno, D.G. Hinks, L. Soderholm, K.J. Volin, R.L. Hitterman, J.D. Grace, I.K. Schuller, C.Segre, K. Zhang and M.S. Kleefisch, Phys. Rev. B 36 (1987) 3608.

INTERACTION OF THIN Ga OVERLAYER WITH InP(110):
A PHOTOEMISSION STUDY

R. CAO, K. MIYANO, T. KENDELEWICZ, I. LINDAU, and W. E. SPICER
Stanford Electronics Laboratories, Stanford University, Stanford, CA 94305

ABSTRACT

Photoemission study of the Ga/InP(110) interface, in particular at the In 4d cooper minimum (CM) reveals that the growth of the deposited Ga on InP(110) at room temperature (RT) has two modes: chemisorption at low coverage and metallic island formation at high coverage, whereas the Ga overlayer is much more uniform at 80K low temperature (LT). A replacement reaction between Ga and InP is found to take place only underneath the Ga islands. Metal screening from the Ga islands is suggested to weaken the substrate bonds and enhance the replacement reaction. Distinct behavior of Fermi level pinning has been observed at different temperatures. This is correlated with the temperature dependence of the overlayer morphology as well as the interfacial reaction.

INTRODUCTION

Study of column III metals/III-V semiconductor interfaces is of particular interest in understanding morphological, chemical, electronic, and electrical properties of metal/III-V interfaces. Ga/InP is an important member of this family. Despite previous studies on this interface prepared at RT, two important questions have still remained unanswered [1,2]. The first is the growth of Ga on InP(110). Although metallic islands at the high coverage have been reported the initial stage growth is not clear. For instance, is there an intermediate stage of small cluster formation in this case as that at the interfaces of Al/GaAs(110) and Al/InP(110)? [1-3] This is crucial in understanding other interfacial phenomena since small cluster formation was suggested to have a large influence on the chemical and electrical properties at the interfaces [4]. The second question is about the reaction at this interface. A replacement reaction between Ga and InP has been suggested by all the previous studies on the basis of observed elemental In component using photoelectron spectroscopy (PES) [1,2]. However, the mechanism which initiates this reaction is still unknown. Similar heats of formation of InP (-21.2 KCal/mol) and GaP (-21 KCal/mol) [5] urge one to find out a mechanism of how the activation barrier is overcome. We noticed that all the previous PES studies of this interface were performed with a photon energy around 80 eV (synchrotron) or 40.8 eV (He II) for Ga 3d and In 4d. Overlap of Ga 3d and In 4d at those photon energies made it difficult to obtain the detailed information at the interface, in particular at the early stage. In order to overcome this problem we tuned the photon energy to the In 4d CM (150 eV) [6] to enhance Ga 3d feature.

Recent studies of the Fermi level movement at the metal/III-V interfaces at reduced temperatures have provided valuable information on the mechanisms of Schottky barrier formation [7-10]. Particularly, it has been found that for metal/InP interfaces, the Fermi level pinning at the interface is closely correlated with the interfacial reaction and/or InP surface disruption [10]. Interfacial reaction and InP surface disruption were detected for most RT prepared and some LT prepared metal/InP interfaces, and the Fermi levels were found to be pinned by defects [11]. In contrast, for some metals (e.g. Ag), LT lead to nearly ideal interfaces. As a result, the metal induced gap states (MIGS) [12] appear to be dominant in the Fermi level pinning. In this work we performed a study of the Ga/InP interfaces at LT, which provides another example to strongly support our conclusion obtained from previous noble metal/InP studies [10].

EXPERIMENT

The experiments were performed in a standard UHV chamber (base pressure in 10^{-11} torr range) at the Beam Line I-1 at the Stanford Synchrotron Radiation Laboratory and at the Stanford Electronics Laboratories using a monochromatic He discharge lamp. The Ga/InP interfaces were prepared by cleaving single n-type (Sn doped, n= 5 x 10^{17} atoms/cm³) and p-type (Zn doped, p = 5 x 10^{17} atoms/cm³) InP crystal in (110) orientation followed by stepwise Ga deposition. The samples were held at a constant temperature (RT or 80K LT) during the interface preparation and PES measurements. Ga was evaporated from a heated tungsten basket. The source was well outgassed prior to evaporation and the pressure during the Ga deposition was in low 10^{-10} torr range. A low coverage evaporator was used to perform low Ga coverage deposition with better accuracy of coverage and less thermal disturbance to the sample surfaces. The coverage was monitored by an *in situ* crystal thickness monitor and the unity sticking coefficient was assumed. The synchrotron radiation was tuned to 150 eV to perform a study around the In 4d CM. The spectra of In 4d & Ga 3d at 80 eV and P 2p at 170 eV were also taken where best surface sensitivity is obtained. In the case of the He discharge lamp, In 4d & Ga 3d core level spectra were taken at He II ($h\nu = 40.8$eV) and the valence band spectra were taken at He I ($h\nu = 21.2$eV).

Ga/n-InP(110)   RT

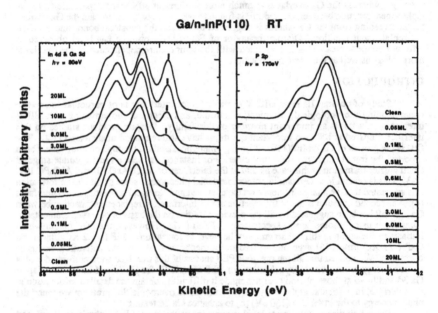

Figure 1.   Photoemission spectra of P 2p and In 4d & Ga 3d core levels of Ga/n-InP(110) as a function of Ga coverages in ML. The small bars in the left panel indicate the segregated In components.

RESULTS

Figure 1 shows spectra of P 2p (hv = 170 eV) and In 4d & Ga 3d (hv = 80 eV) of the RT Ga/n-InP(110) interfaces as a function of Ga coverage in monolayers (ML: 1ML = 8.4 x 10^{14}atoms/cm^2). P 2p (left panel) exhibits shift and attenuation without appreciable change of lineshape as Ga coverage increases. On the other hand, many more features appear in the spectra of In 4d & Ga 3d (right panel). In the low coverage regime (<0.3ML), no Ga signal is detected, and In 4d undergoes a shift without lineshape change. The same amount of shift is also observed from P 2p and the valence band spectra in this region. This is interpreted as the substrate InP band bending due to Ga deposition. Starting from 0.3ML, an additional peak, marked by small bar, appears on the high kinetic energy side of the spectra. This has been identified as segregated elemental In and can be used as a fingerprint that a replacement reaction is taking place at the interface. This component grows with increasing Ga coverages, implying that more reaction occurs. At high coverage, another doublet feature emerges on the low kinetic energy side. This is Ga 3d from the deposited Ga and same as that of metallic Ga.

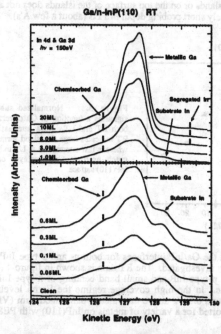

Figure 2. In 4d & Ga 3d core levels taken at the In 4d Cooper minimum (hv = 150 eV) as a function of Ga coverages. Different features are marked. The spectra taken with low Ga coverages are magnified and shown in the bottom panel, and those with high Ga coverages are shown in the top panel.

Overlap of Ga 3d and In $4d_{7/2}$ makes the data analysis very difficult, in particular the behavior of Ga at the early stage. In order to get more information discrimination against In 4d is crucial. Theoretically and experimentally it has been established that the photoionization cross section of the 4d-shell changes with incident photon energy and passes through a minimum value (so-called Cooper minimum) [6]. For In 4d, the CM appears around photon energy 150 eV. The cross section at the CM is about 100 times lower than that at 80 eV. On the other hand, the cross section of the 3d-shell does not have this property. In the same photon energy range there is only slight change in the Ga 3d cross section. Therefore, taking the In 4d & Ga 3d spectra at the In 4d CM will enhance the Ga 3d signal relative to that of In 4d. The selected spectra taken at 150 eV are presented in figure 2. In the low coverage regime (<0.3ML) the dominant Ga 3d feature is the peak around 126.3 eV kinetic energy. This comes from the chemisorbed Ga on InP. With increasing Ga coverage a peak around 127.4 eV kinetic energy emerges, which is the same as that from metallic Ga and overlapping with In $4d_{7/2}$. The important finding is that there is no intermediate structure between the chemisorbed Ga and metallic Ga, whereas such a feature is clearly seen in the case of Al growth on GaAs and InP [1-3]. This feature indicates a step transition from isolated Al atoms to small cluster formation. Missing of the intermediate structure between the chemisorbed Ga and metallic Ga can be interpreted as gradual formation of clusters, in contrast to step transition in the case of Al. Gradual cluster formation was also reported at the RT In/GaAs interface [13]. As will be discussed later, it is the difference in the overlayer growth which leads to different mechanisms of the interface reaction.

More information on the overlayer growth on the surface can be obtained by following the substrate signal intensity attenuation as a function of the coverage of the deposited material [14]. In figure 3 the normalized substrate signal intensity against that of the clean surface is plotted. At RT (open data points) slow attenuation of the substrate signal indicates formation of sparse and large Ga islands. This is consistent with the appearance of the metallic Ga feature in figure 2 even at low Ga coverages. Here we measure the areas underneath P 2p peak and the substrate In 4d (not including reacted component). They fit with each other quite well. It is not surprising because for this system large islands are formed so that the normalized intensity essentially represents portion of the uncovered areas of the InP surfaces. The reaction which leaves products either underneath the islands or on the top surface of the islands does not affect the intensity analysis due to the extremely short probing depth of PES (about a few Å's).

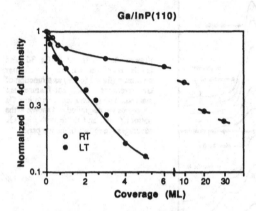

Figure 3. Normalized substrate signal as a function of Ga coverage at both RT and LT. Slow attenuation at RT indicates large Ga island formation; whereas fast attenuation at LT indicates much uniform Ga overlayer on the InP(110) surface.

The Fermi level movement at the Ga/InP interfaces for both n- and p-type InP as a function of Ga coverages were closely investigated. The RT data is shown in figure 4 (open data points). Large band bending of p-type InP and small band bending of n-type InP are observed in the low coverage regime. In the high coverage regime the Fermi levels are approaching to the same position around 0.9 eV above the valence band maximum (VBM). This value agrees well with those reported for a variety of metals on InP(110) with PES and electrical measurements [15,16].

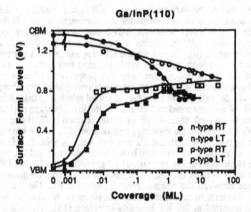

Figure 4. Evolution of the surface Fermi levels at the RT and LT Ga/InP(110) interfaces as a function of Ga coverages. Temperature dependence on the Fermi level pinning behavior is observed. At RT the Fermi levels are pinned around 0.9 eV above the VBM, which is due to the defects. At LT, however, the Fermi levels are pinned around 0.75 eV above the VBM due to the MIGS.

The same interfaces prepared at 80K LT were studied using ultra violet PES. The spectra will not be shown here for sake of brevity. We did not observe the segregated In component at the LT interfaces, indicating that chemical reaction is largely suppressed at these interfaces. The normalized substrate signal intensity is also plotted in figure 3 as a function of Ga coverage (closed data points). Much faster attenuation at LT with respect to that at the RT interfaces suggests that the Ga overlayer is much more uniform at LT. The Fermi level movement at these interfaces is shown in figure 4 (closed data points). Despite the similarity in the Fermi level movement in the low coverages regime, the Fermi level pinning pattern is quite different at LT, compared with that RT. A lower pinning position (around 0.75 eV above the VBM) is observed, which is about 0.15 eV lower than that at the RT interface.

DISCUSSION

By studying the Ca/InP interfaces at the In 4d CM, we have established that at RT the growth of Ga on the InP(110) surfaces has two modes, that is, the predominant chemisorption at low coverages and metallic island formation at high coverages. There is no step transition from the isolated atoms to small clusters. At reduced temperature a much more uniform overlayer is formed. Change of growth mode upon the temperature is essential due to the change of the surface mobility of the metal atoms. The same phenomenon has been observed at many other metal/III-V interfaces[7,10].

The replacement reaction indicated by the appearance of segregated In is observed only at the RT interfaces starting from 0.3ML, where Ga islands are dominant. Full metallicity has been established in these islands justified by the peak position of Ga 3d and the density of states around the Fermi cutoff, where a well defined lineshape is seen with the Fermi level bisecting the cutoff. However, P 2p does not show any change of lineshape. These findings suggest that the replacement reaction takes place only underneath the Ga islands with the reacted phosphorus remaining where it was and In segregating to the surface of the Ga islands.

The product of the replacement reaction between Ga and InP is most likely GaP plus free In. We notice that the difference between the heat of formation of InP and that of GaP is small so that extra energy is needed to initiate the reaction between Ga and InP although such a reaction may be thermodynamically favorable (notice that the heats of formation quoted in this paper may not be accurate, and they are also based on bulk thermodynamics estimation). For some column III metals on III-V semiconductor surfaces, specifically Al/GaAs and Al/InP interfaces, previous studies revealed that the onset of the interface reaction coincide with small Al cluster formation on the the semiconductor surfaces. It was suggested by Zunger that the reaction is initiated by the energy released from exothermic small Al cluster formation [4]. Can one use the same mechanism to explain the reaction at the Ga/InP interfaces? There are two distinct differences between the Ga/InP(110) and Al/GaAs and Al/InP interfaces. First, in contrast to the step transition from isolated atoms to small cluster at the Al/III-V interface, Ga on InP only shows gradual cluster formation., One does not expect large amount of energy released in this case. Secondly, the reaction becomes stronger with increasing Ga coverages; whereas the reaction at the Al/III-V interfaces seems to takes place only at the relative low coverage, where small clusters are formed. Therefore using the cluster model to explain the reaction in our system seems questionable. In fact, lack of small cluster formation of In on GaAs is suggested to be the key point of why In does not react with GaAs [13]. On the other hand, we notice that the reaction is taking place when metallic Ga islands become dominant. Besides, it is clear, spectroscopically, that the reaction occurs under the metallic Ga islands. It seems reasonable to link the overlayer metallicity with the interfacial reaction. It is interesting to recall some of the early studies of metal/Si interfaces. It was found that metal-Si reaction did not start until a critical coverage was reached, and the overlayer was metallic then. Hiraki proposed that metal screening from the overlayer weakened the substrate Si-Si bond and initiated the reaction [17]. Our results seem to agree with this model. With increasing Ga coverage, more surface areas are covered by the metallic islands and conceivably more reaction product (segregated In) is seen. At LT the reaction is largely inhibited, in particular in the low coverage regime. This is ascribed to higher activation barrier and maybe more importantly the non-metallic characteristics of the uniform Ga overlayer in this coverage regime.

As far as the Fermi level stabilization at the Ga/InP(110) interfaces two different pinning position are observed (0.9 and 0.75 eV above the VBM at RT and LT, respectively). They appear to be associated with the interfacial reaction. Reaction and InP surface disruption are detected even in the submonolayer regime at the RT interface; whereas they are largely inhibited at LT. We argue that the pinning behavior at this interface can be well explained in terms of the defect and MIGS mechanisms [11,12]. Large amount of defects created due to reaction and InP surface disruption at RT pin the surface Fermi level around 0.95 eV, which is associated with defect level [9,10]. In contrast, much less defects are expected at the nearly ideal interface formed at LT. Moreover, the uniform metallic overlayer assures that the MIGS will play an important role in Fermi level pinning. In fact, the pinning position, which is far away from the defect level but close to calculated charge neutrality level due to MIGS (0.76 eV above the VBM [18]), reinforces that the MIGS but not defects are responsible for the Fermi level pinning at the LT interface. The similar behavior of the Fermi level movement was observed for noble metals on InP(110) by Cao et al and the correlation has been discussed in detail [10]. Here the Ga/InP interface provides another example showing that both defects and MIGS can pin the Fermi level. This is different from the previous point of view that a single mechanism explains everything. Moreover, this shows that the pinning mechanism can be controlled by varying the temperature at which the interface is formed.

CONCLUSION

We have shown the growth of the deposited Ga on InP(110), the interfacial reaction and the Fermi level pinning at both RT and LT prepared Ga/InP(110) interfaces. At RT Ga atoms are predominantly chemisorbed on the InP(110) surfaces at ultra low coverages and form metallic islands at high coverages; whereas they tend to lay down uniformly at reduced temperature. Instead of small cluster formation, metal screening is suggested to enhance the replacement reaction occurring underneath the metallic Ga islands. The Fermi level pinning is found to correlate with interface perfection and can be well explained in terms defects (disrupted interface) and the MIGS (nearly perfect interface).

ACKNOWLEDGEMENT

This work is supported by DARPA and ONR under contract No. N00014-83-K-0073 and ONR under contract No. N00014-86-K-0376. Part of the experiments was performed at the Stanford Synchrotron Radiation Laboratory which is supported by the Department of Energy under contract No. DE-AC03-82ER-13000, Office of Basic Energy Sciences, Division of Chemical/Material Sciences.

REFERENCES

1. T. Kendelewicz, M. D. Williams, W. G. Petro, I. Lindau, and W. E. Spicer, Phys. Rev. B 31, 6503 (1985).
2. R. H. Williams, A. McKinley, G. J. Hughes, and T. P. Humphreys, J. Vac. Sci. Technol. B 2, 56 (1984).
3. R. R. Daniels, A. D. Katnani, T. -X. Zhao, G. Margaritondo, and A. Zunger, Phys. Rev. Lett. 49, 895 (1982).
4. A. Zunger, Phys. Rev. B 24, 4372 (1981).
5. A. Kahn, C. R. Bonapace, C. B. Duke, and A. Paton, J. Vac. Sci. Technol. B 1, 613 (1983).
6. J. J. Yeh and I. Lindau, Atomic Data and Nuclear Data Table 32, 1 (1985) and references therein.
7. R. Cao, K. Miyano, T. Kendelewicz, K. K. Chin, I. Lindau, and W. E. Spicer, J. Vac. Sci. Technol. B 5, 998 (1987).
8. K. Stiles, A. Kahn, D. Kilday, and G. Margaritondo, J. Vac. Sci. Technol. B 5, 987 (1987).
9. R. Cao, K. Miyano, T. Kendelewicz, I. Lindau, and W. E. Spicer, Appl. Phys. Lett. 53, 210 (1988).
10. R. Cao, K. Miyano, I. Lindau, and W. E. Spicer, J. Vac. Sci. Technol. A (to be published).
11. W. E. Spicer, I. Lindau, P. Skeath, C. Y. Su, and P. Chye, Phys. Rev. Lett. 44, 420 (1980).
12. J. Tersoff, Phys. Rev. Lett. 52, 465 (1984).
13. R. R. Daniels, T. -X. Zhao, and G. Margaritondo, J. Vac. Sci. Technol. A 2, 831 (1984).
14. R. Cao, K. Miyano, K. K. Chin, I. Lindau, and W. E. Spicer, SPIE proceeding 946, 219 (1988).
15. T. Kendelewicz, N. Newman, R. S. List, I. Lindau, and W. E. Spicer, J. Vac. Sci. Technol. B 3, 1206 (1985).
16. N. Newman, W. E. Spicer, T. Kendelewicz, and I. Lindau, J. Vac. Sci. Technol. B 4, 931 (1986).
17. A. Hiraki, Surf. Sci. Rep. 3, 357 (1984) and references therein.
18. J. Tersoff, Phys. Rev. B 30, 4874 (1984).

Absorption Spectroscopy:
Structural Measurements

X-RAY ABSORPTION SPECTROSCOPIC STUDIES OF CATALYTIC MATERIALS

G. H. VIA*, J. H. SINFELT*, G. MEITZNER* AND F. W. LYTLE**
*Exxon Research and Engineering Company, Rt. 22 E, Annandale, NJ 08801
**The Boeing Company, Seattle, WA 98124

ABSTRACT

X-ray absorption spectra (XAS) contain information in the L_{III} near-edge region on filling of the absorber d-band, and in the extended fine-structure region on the physical environment of the absorber. We report here an evaluation of the effect on platinum L_{III} edges of preparation in clusters with a high fraction of Pt atoms at the surface. We also report the effects on platinum and rhenium L_{III} edges from addition of copper. These effects are surprisingly small.

We have also re-evaluated extended x-ray absorption fine-structure spectra (EXAFS) of platinum and rhenium in alumina-supported platinum-rhenium bimetallic catalysts. A novel feature of this new analysis was the requirement that interatomic distances, coordination numbers, and Debye-Waller type factors maintain certain physically necessary relationships among themselves. This procedure decreased the number of free variables and increased the amount of information returned by the analysis.

INTRODUCTION

Heterogeneous catalysts consisting of metal clusters dispersed on a refractory oxide support are employed in the manufacture of many chemicals and fuels. The metallic part typically comprises less than one weight percent of the catalyst and may include more than one metal. The support is a refractory oxide or mixed-oxide. A variety of non-metallic modifiers are also added to the catalyst to enhance or to suppress catalysis of particular reactions. The determination of the physical and electronic structures of sites associated with catalytic activity is a necessary step in the continuing refinement of supported metal catalysts.

X-ray absorption spectroscopy is a tool of tremendous power and versatility for the study of complex materials, especially those with extremely small domains or possessing order on a scale of only a few Angstroms. In heterogeneous catalysts, for example, the oxide support is typically amorphous and the metal particles may be smaller on average than 10 Angstroms in diameter. The information in a spectrum separates into the near-edge extended x-ray absorption fine-structure (NEXAFS) region and the extended x-ray absorption fine-structure (EXAFS) region as shown in figure 1.

In the case of the L_{III} absorption edges of transition metals, a 2p electron excited by a photon of energy close to its binding energy can become trapped by a vacant bound state of d-type symmetry. The L_{III} edges thus exhibit a resonance (a "white line") that decreases in intensity as the valence d-band fills [1,2,3]. This is shown in figure 2a for a series of reduced metals, and in figure 2b for metals in various oxidation states. In this report we present results that indicate that d-band vacancies are only slightly affected by formation of the reduced metal into supported clusters, or by addition of copper.

The EXAFS region of an x-ray absorption spectrum usually exhibits an oscillation in the value of the absorbance, shown in figure 1. This oscillation may be regarded as a consequence of scattering of the ejected photoelectron by atoms in the vicinity of the absorber atom. Analysis of EXAFS can yield numbers and types of neighboring atoms to a radius of about 5 Angstroms, and distances and mean-square relative displacements to these atoms.

Typical XAFS Spectrum (Rhodium Metal)

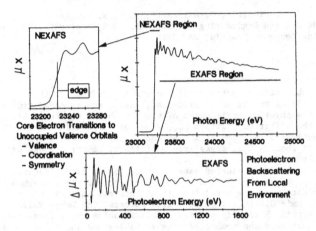

FIG. 1. A rhodium metal x-ray absorption spectrum is shown separated into its near edge x-ray absorption fine-structure (NEXAFS) region, which is regarded as a source of electronic structural information, and its extended x-ray absorption fine-structure (EXAFS) region, which yields physical structural information on the environment of the absorber.

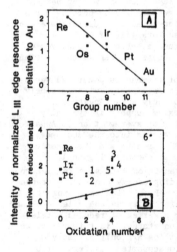

FIG. 2. The upper field (A) shows a plot of the intensity, i.e. the height, of the normalized L_{III} white lines of a series of metals against group number. The value of the intensity at the corresponding position on the gold L_{III} edge, which has no white line, has been subtracted from each. The three points for osmium and the two points for iridium illustrate typical uncertainty in the determination of line intensities due to thickness effects and contamination.

The lower field (B) shows a plot of white line intensity versus oxidation state for a series of compounds. The value of white line intensity for the corresponding reduced metal has been subtracted from each. The plot shows that removal of one electron by oxidation results in about the same increment of increase in edge intensity regardless of the metal.

Key:
1. $Pt(acac)$
2. $PtCl_2$
3. H_2PtCl_6
4. IrO_2
5. H_2PtCl_6
6. $NaReO_4$

One strength of XAS is its broad applicability. Synchrotron radiation sources now provide an intense continuous spectrum from the vacuum ultraviolet (VUV) through very high-energy x-rays. X-ray absorption spectra can in principle be measured from most of the elements. Interferences and optical problems prevent these measurements on elements lighter than carbon. Spectra from several elements near the middle of the periodic table are hard to use because the K absorption edges occur at unreachable energies above 3C KeV, and the accessible L edges are very close together and interfere with each other. In addition to high information content and broad applicability, the x-rays used for measurements of most edges are penetrating so that powdered or liquid samples, and controlled atmospheres and pressures, are acceptable.

During the last several years we have used XAS to determine the structures of supported platinum-copper, rhenium-copper [4] and platinum-rhenium [5] bimetallic clusters. Platinum and rhenium were two in a series of metals we combined with copper in supported-metal catalysts to investigate structural and catalytic trends [4]. The platinum-rhenium sample was similar to commercial-type reforming catalysts used in the manufacture of aromatic hydrocarbons from C_6 through C_9 aliphatic hydrocarbons [6]. It differed from the commercial-type reforming catalysts in that the metals content was higher.

EXPERIMENTAL

The dispersed platinum samples consisted of 1 wt.% platinum supported on SiO_2 (Cabot Cab-o-sil HS-5) or on η-Al_2O_3. They were prepared by a simple impregnation procedure, wherein a quantity of the oxide was wetted with an aqueous solution of the appropriate concentration of hexachloroplatinic acid. Volumes of 2.2 ml or 0.65 ml of solution were used per gram of silica or alumina, respectively. The samples were dried at 380K, then reduced in flowing hydrogen at 775K for the silica-supported catalyst and 750K for the alumina-supported catalyst. Dispersions, defined as the ratio of surface platinum atoms to total platinum atoms in the samples, were determined from hydrogen chemisorptions. These were 66% for the silica-supported catalyst and 88% for the alumina-supported catalyst. The chemisorption procedure has been described elsewhere [7].

The silica-supported platinum-copper and rhenium-copper catalysts, and the alumina-supported platinum-rhenium catalyst were prepared by the impregnation procedure described above. Two metals (as hexachlorplatinic acid, perrhenic acid and copper nitrate) were in one solution to give 1 wt.% of platinum and 0.3 wt.% copper, or 2 wt.% of rhenium and 0.6 wt.% of copper on the dried catalysts. The platinum-rhenium catalyst contained 1 wt.% each of platinum and rhenium. These metals concentrations were chosen to give 1:1 atomic ratios of metals in the catalysts.

The bimetallic catalyst precursors were dried at 380K, then reduced in a stream of hydrogen at 775K for 3 h. The samples were passivated by slow exposure to air. Before collection of spectra the samples were re-reduced at about 700K, and cooled in hydrogen to liquid nitrogen temperature.

Spectra were collected at SSRL during 1985 to 1987.

RESULTS

Absorption edges of platinum and rhenium

The first part of this report is concerned with the effect on L_{III} absorption edges of high metal dispersion and of the presence of copper.

These effects were assessed by fitting the absorption edge spectrum from a well characterized bulk or dispersed sample to the absorption edge from the same metal in the dispersed state or in bimetallic clusters with copper. The differences between edges were surprisingly small, and a systematic approach to the fitting correspondingly important. Figure 3a shows how the edge spectrum from a standard was fit to the edge spectrum from a sample.

Three variables were adjusted in sequence. An offset, or additive constant, was used to shift the reference spectrum vertically with respect to the sample spectrum. This fit was performed through the range from 50 eV to 20 eV in front of the edge. (In this work the edge position, $\underline{i.e.}$ the zero, was defined as the inflection point in the rising portion of the spectrum.) Next a scalar multiplier called the amplitude was used to adjust the normalization of the reference spectrum to match the sample spectrum. This fit was performed on the range from 20 eV to 100 eV beyond the absorption edge. Finally an energy (horizontal) shift was applied to the reference spectrum. This fit was performed on the range from 20 eV before to 20 eV beyond the absorption edge. Figure 3b shows the difference spectrum. The location and magnitude of the difference between two spectra was extremely sensitive to the method of the fit. This procedure minimized the difference.

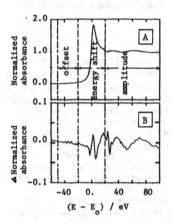

FIG. 3. The upper field (A) illustrates the parameters that were used to perform the edge fits for this work. The offset was a vertical shift of the reference spectrum relative to the sample spectrum. Its value was determined by a fit in the pre-edge region. The amplitude was a multiplier used to renormalize the reference spectrum to match the normalization of the sample spectrum. Its value was determined in the relatively flat region beyond the edge. The energy shift was a horizontal shift of the reference spectrum. Its value was determined in the region from 20 eV before to 20 eV beyond the edge. These three parameters were fit sequentially.

The lower field (B) shows the value of (sample spectrum - reference spectrum). Note the change of scales from the upper field to the lower field.

The sample consisted of 1 wt.%Ir - 0.3 wt.% Cu / SiO_2. The reference spectrum was from a sample of 1 wt.% Ir/SiO_2 with dispersion of 100%.

The fits of an edge spectrum from bulk platinum to spectra from a Pt/SiO_2 catalyst with dispersion of 66% and a Pt/Al_2O_3 catalyst with dispersion of 88% are shown in the left fields of figure 4. The right fields of figure 4 plot the differences between the sample and reference spectra. Figure 5 shows fits of platinum and rhenium reference spectra to platinum or rhenium spectra from the PtCu and ReCu samples. Difference spectra are also shown on the right of figure 5. The platinum reference spectrum for figure 5 is from the dispersed Pt/SiO_2 sample that was fit in figure 4. The dispersed reference was used to isolate the effect of copper on the platinum absorption edge. Note the change in scales between the fields on the left showing fits, and the difference spectra on the right in figures 4 and 5.

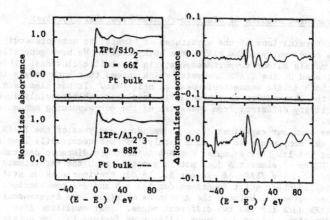

FIG. 4. Fits to L_{III} absorption edges of platinum in 1% Pt/SiO$_2$, with platinum dispersion equal to 66%, and of 1% Pt/Al$_2$O$_3$, with platinum dispersion of 88%. Dispersion is defined as the ratio of surface platinum atoms to total platinum atoms in the sample. The fits were to determine the effect of high dispersion on filling of the d-band of reduced platinum. The fields on the right show the value of the difference (sample - reference). The reference in both cases was 10% Pt/SiO$_2$ prepared to give large platinum crystallites that approximate bulk platinum.

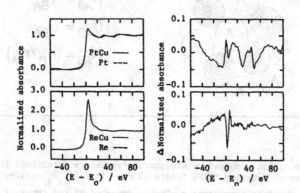

FIG. 5. Fits to L_{III} absorption edges of platinum (upper fields) and of rhenium (lower fields) in 1% Pt - 0.3% Cu/SiO$_2$ and 2% Re - 0.6% Cu/SiO$_2$ catalysts. The platinum standard for the upper fit was 1% Pt/SiO$_2$ with dispersion of 66%. The rhenium standard for the lower fit was 1% Re/Al$_2$O$_3$, dispersion undetermined. The fields at right show the differences from point-by-point subtractions of the reference spectra from the sample spectra. Note the change of scales from the fields on the left to the fields on the right.

EXAFS Structural studies of platinum-rhenium bimetallic catalyst

Our determinations of the structures of bimetallic heterogeneous catalysts have employed the EXAFS from both metals. We have generally found that two metals on one support associate in bimetallic clusters. This has been indicated by the EXAFS analysis, which showed that both metals included atoms of both metals among their nearest neighbors. It has also been indicated by the observation that bimetallic catalysts exhibit catalytic activity and selectivity different from a mixture of the corresponding monometallic catalysts.

It has been our experience that independent analyses of the EXAFS spectra from the two components of the same catalyst do not necessarily lead to a consistent result. For example, the A-B interatomic distance determined from the EXAFS of element A does not in general match the interatomic distance determined from the EXAFS of element B. In our previous work we systematically adjusted phase shift functions for the A-B and B-A combinations until we obtained agreement between the distances obtained from independent fits of the EXAFS data for the two different edges. That condition also affected the coordination numbers and Debye-Waller type factors that we determined to describe the material. In other work a constraint was placed on the Debye-Waller type factor determined from either of the different atoms in the pair [8]. The A-B and B-A coordination numbers should also be equal if the atomic ratio of the components A and B is 1:1 [9]. Correlations between structural parameters for fitting two spectra simultaneously are summarized in figure 6. The analysis of multiple spectra with more than one set of correlated variables has been an intractable problem.

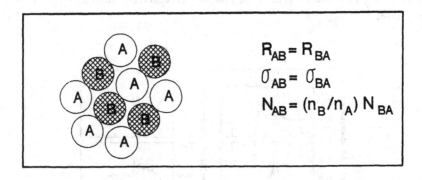

$$R_{AB} = R_{BA}$$
$$\sigma_{AB} = \sigma_{BA}$$
$$N_{AB} = (n_B/n_A) \, N_{BA}$$

FIG. 6. Conditions imposed on structural parameters determined from analysis of platinum and rhenium EXAFS from a 1% Pt - 1% Re/Al$_2$O$_3$ catalyst. The figure on the left represents a cluster composed of two types of atoms. R, σ, and N are interatomic distances, rms relative displacements, and coordination numbers, respectively. The first letter in the subscript refers to the absorber element, and the second letter refers to the backscattering element. n_A and n_B are atomic fractions of elements A and B, respectively, in the catalyst. Note that the conditions shown on the right are satisfied by any real system, and are not arbitrary boundaries.

We report here the results of an EXAFS study of the structure formed by platinum and rhenium together on alumina. EXAFS is an appropriate tool since the metal clusters were too small to produce a useful diffraction pattern, and even electron microscopy was of marginal value. Also, interpretation of chemisorption results is unclear due to the different abilities of platinum and rhenium to adsorb gases. Results were reported previously [5] of an analysis of these same spectra. In that analysis the EXAFS data for the two edges were fitted independently and the only constraint on results was that the platinum-rhenium interatomic distance equal the rhenium-platinum interatomic distance. The interatomic distances and values of Debye-Waller type factor $\Delta\sigma^2$ are shown in table 1. Coordination numbers are shown in table 2.

TABLE I. Platinum-rhenium interatomic distances and Debye-Waller factors $\Delta\sigma^2$ for 1% Pt - 1% Re/Al$_2$O$_3$ catalyst. The only condition on the results from the previous analysis [5] was the requirement that the Pt-Re interatomic distance equal the Re-Pt interatomic distance. The present analysis retains that condition, and requires as well that the value of σ^2 (mean-square relative displacement), and platinum-rhenium coordination number, have the same values whether determined from platinum or from rhenium.

Atom Pair	Previous Analysis		Present Analysis	
	R / Å	$\Delta\sigma^2/Å^2$(a)	R / Å	$\Delta\sigma^2/Å^2$(a)
Pt-Re Catalyst				
Pt-Pt	2.75	(b)	2.75	0.003
Pt-Re = Re-Pt	2.64	(b)	2.64	0.005
Re-Re	2.73	(b)	2.73	0.003

(a) Values are expressed relative to the value for the rhenium reference material.
(b) Not previously determined.

TABLE II. Coordination numbers describing platinum-rhenium bimetallic clusters in 1% Pt - 1% Re/Al$_2$O$_3$ catalyst. A strong correlation between values of $\Delta\sigma^2$ and coordination numbers prevented their independent determination in the previous analysis [5]. The coordination of Pt by Re and of Re by Pt were constrained to be equal for the present analysis.

Atom Pair	Previous Analysis		Present Analysis
Pt-Re Catalyst	N		N
Pt-Pt	(a)		4.8
Pt-Re	(a)		2.9
Re-Pt	(a)		2.9
Re-Re	(a)		4.2
% Pt neighbors of:	Pt	65	62
	Re	44	41

(a) Compositions were determined, but not coordination numbers.

The present analysis used an iterative non-linear regression procedure like before, but the EXAFS spectra for the two elements were fitted simultaneously rather than independently, as a result of several advances we have made recently in the computer programs used in the analysis. Moreover, the relationships shown in figure 6 between coordination numbers, interatomic distances, and Debye-Waller type factors were always maintained. This work also differs from the previous analysis in that we have substituted the back-scattering phase functions calculated by McCale, et al. [10] for those calculated by Teo and Lee [11]. The results of the re-analysis are shown in the right-hand columns of tables 1 and 2. The fits we obtained to platinum and rhenium Fourier-filtered EXAFS are shown in figure 7. The calculation of back-transforms shown in figure 7 have been described previously [5]. The fits were as good as we obtained previously, although we were attempting to determine values for 9 adjustable parameters instead of 12 as in our previous analysis. We gratefully acknowledge the contribution of Dr. Kenneth F. Drake, currently at Los Alamos National Laboratory, who wrote the computer programs that made this work possible.

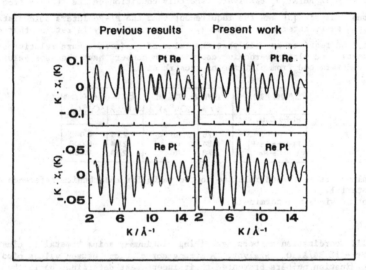

FIG. 7. Fits to Fourier-filtered EXAFS of platinum (upper fields) and of rhenium (lower fields) in 1% Pt - 1% Re/Al$_2$O$_3$ catalyst. The calculation of the back-transforms has been described elsewhere [5]. The fits achieved in the previous work [5] and in the present work were of similar quality although the present work employed only 9, instead of 12, independent variables.

CONCLUSIONS

Edges

The edge analyses reported here were part of a systematic evaluation of absorption edges. A surprising conclusion concerned the low magnitude of the changes in L$_{III}$ edge structures that resulted from high dispersion or from addition of copper. Such changes as occurred were neither large nor discrete.

Most of the features in the difference spectra in figures 3, 4, and 5 are EXAFS oscillations. These appear in the difference spectra because EXAFS is sensitive to physical structure, and the physical structures were different in the dispersed or copper-containing samples and in the reference materials. Most of the features in the difference spectra at the position of the white-lines may also have been due to differences between physical structures. The small changes we found in intensities of white-lines suggest that for these systems the change in valence d-band occupancy due to high dispersion or to contact with copper was a small fraction of one electron. It is another possibility that the work only shows that this type of spectroscopy is insensitive to electronic rearrangements of the sort induced by these configurations of metals. We caution that accurate observation of these effects requires careful attention to quality of standard materials, and a consistent approach to the analysis of the edge spectra.

EXAFS

The benefits from increasing the information to solve an analytical problem are often difficult to realize in practice. We have shown that more than one EXAFS spectrum can be analyized simultaneously, resulting in a smaller number of adjustable parameters. This was accomplished by introducing new information that is not model dependent. In the case of the platinum and rhenium EXAFS analysis the result did not change from the previous one, except that we were able to determine individual coordination numbers and Debye-Waller factors.

The picture that emerges of the platinum-rhenium/alumina catalyst includes bimetallic clusters that contain similar numbers each of platinum and rhenium atoms. Rhenium may have a slight tendency to occupy low-coordination sites, based on its slightly lower total coordination number. Our recent analysis agrees with the previous one, in that both find a short platinum-rhenium interatomic distance. This may indicate a strong bond between atoms of the two metals.

Some of this work was done at SSRL which is supported by the NSF through the Division of Materials Research and the NIH through the Biotechnology Resources Program in the Division of Research Resources. (in cooperation with the Department of Energy).

[1] F. W. Lytle, P. S. P. Wei, R. B. Greegor, G. H. Via, and J. H. Sinfelt, J. Chem. Phys., 70, 7849(1979).

[2] J. A. Horsley, J. Chem. Phys., 76, 1451(1982).

[3] F. W. Lytle, R. B. Greegor, V. A. Biebesheimer, G. Meitzner, J. A. Horsley, G. H. Via, and J. H. Sinfelt, Murphree Award Symposium, 191st ACS National Meeting, New York, NY, 1986.

[4] G. Meitzner, G. H. Via, F. W. Lytle, and J. H. Sinfelt, J. Chem. Phys., 83, 353(1985).

[5] G. Meitzner, G. H. Via, F. W. Lytle, and J. H. Sinfelt, J. Chem. Phys., 87, 6354(1987).

[6] J. L. Carter, G. B. McVicker, W. Weissman, W. S. Kmak, and J. H. Sinfelt, Appl. Catal., 3, 327(1982).

[7] D. J. C. Yates and J. H. Sinfelt, J. Catal., 8, 348(1967).

[8] S. Sakellson, M. McMillan, and G. L. Haller, Paper 70g, American Institute of Chemical Engineers, 1984 Annual Meeting, San Francisco(1984).

[9] E. Zschech, W. Blau, K. Kleinstuck, H. Hermann, N. Mattern and M. A. Kozlov, J. Non-Cryst. Sol., $\underline{86}$, 336(1986).

[10] A. G. McKale, B. W. Veal, A. P. Paulikas, S. -K. Chan and G. S. Knapp, J. Am. Chem. Soc., $\underline{110}$, 3763(1988).

[11] B. -K. Teo and P. A. Lee, J. Am. Chem. Soc., $\underline{101}$, 2815(1988).

TIME—RESOLVED X-RAY ABSORPTION SPECTROSCOPY: STRENGTHS AND LIMITATIONS

A. FONTAINE, E. DARTYGE, J.P. ITIE[+], A. JUCHA, A. POLIAN[+], H. TOLENTINO,
G. TOURILLON
LURE (Lab. CNRS, CEA, MEN) Bât. 209D, 91405 ORSAY Cedex, FRANCE
[+]Physique des Milieux Condensés (C.N.R.S. - U.A. 782),
Université P&M Curie, T13 E4, 4 place Jussieu, F-75252 Paris Cedex 05

X-ray Absorption Spectroscopy has proved to be a powerful tool to elucidate a huge number of questions in materials science. Great interest exists in time-resolved experiments achieved with extreme energy resolution and energy scale stability to take a full benefit of the strong correlation between the stereochemical environment of the absorbing atom and the exact shape and position of the absorption edge.

Fast energy dispersive X-ray spectroscopy allows in-situ observations with data collected in a short time. Nowadays the main limitation concerns very low-concentration samples since it is no longer possible to use the dispersive geometry because detection of the signal via the decay channels is no more possible.

I - INTRODUCTION

Using the combination of X-ray energy dispersive optics and a position-sensitive detector able to work under high flux conditions we are able to proceed with both _in-situ_ and _time-dependent_ investigations[1]. Several papers have evidenced already what are the major benefits one can expect for chemistry [2], physics[3], material science[4], and biophyics[5,6,7].

This new tool is herein illustrated by three experiments. The first one deals with an in-situ time-resolved electrochemical reaction. The second illustration draws a great benefit of the energy scale stability and the extreme sensitivity of the detection scheme for the in-situ investigation of oxygen uptake and removal of oxygen in high Tc superconductor. The last example points out the definitive advantage of the smallness of the beam size at the sample position for high pressure experiments which is a large piece of new science of our spectroscopy.

II—ENERGY DISPERSIVE SCHEME

4 ms only are needed to collect a full EXAFS spectrum spread over 500 eV. Each of the 1024 sensing elements of the photodiode array transforms an average of 10^5 8 keV-Xray photons into 8.8×10^7 electron-hole pairs. In total, each frame is made of 10^8 photons which can be repeated at 550 Hz or 220 Hz according to the used analog-digital converter (10 or 12 bits). The cooled photodiode array is the key tool which associates a good spatial resolution and a large dynamics. Advantages and limitations of this scheme are:

a) The copper XANES spectra (fig. 1) have been recorded using a Si 311 crystal to give an account of the **energy resolution which is as good** as it is with a step by step scan using a double detuned crystal optics.

b) Because of the lack of mechanical movements, once the optics is tuned for a given absorption edge, **the energy scale is very stable.** This enables an extreme sensitivity meaning that very minute energy shift of the absorption threshold induced by chemical change can be detected accurately.

c) The linearity and the dynamics of the photodiode array give very good signal/noise ratio. The signal is proved to be essentially statistically limited.

d) Combined to time-resolved capability the in-situ observation permits a systematic investigation of the time-dependent system and an a-posteriori decision about stages of interest.

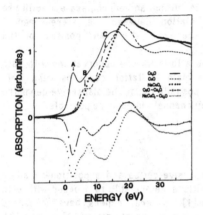

<u>Figure 1</u> : Copper K-edge XAS spectra of Cu_2O, CuO and $NaCuO_2$ whose formal oxidation states are Cu(I), Cu(II) and Cu(III). The differences between the monovalent spectrum and the others are plotted, showing that the **A** peak ($1s^2$ $3d^{10}$ $4p^0$ → $1s^1$ $3d^{10}$ $4p^1$) is very well separated from the other features in the spectra.

Energy resolution

The energy resolution δE of an X-ray reflector comes from the derivative of the Bragg's law, $\delta E = E \cot g(\theta) \delta\theta$ where E is the nominal energy and $\delta\theta$ is the overall angular resolution. This includes:

i) the spatial resolution of the position sensitive detector ρ , which was measured to be almost two pixels, i.e., $\rho \simeq 50\mu m$.

ii) the intrinsec Darwin width of the rocking curve of the perfect crystal;

iii) the penetration depth of the incident X-ray into the crystal;

iv) the size of the X-ray source whose contribution can be negligible under the conditions discussed below.

Besides the polychromatic imaging at the focus point, which is the optimized position for the sample, there exists another type of focalization, called herein monochromatic focalization, which strongly controls the achievable energy resolution. The extended source provides, for each energy, a span of incident rays passing through the same point at the local Rowland circle. Therefore, if the detector is put at this monochromatic focus point, all the photons with this energy fall into a single pixel of the detector and the energy resolution is no longer limited by the extended source.

Owing to the good spatial resolution of our position sensitive detector, the dominant term is usually the Darwin width [8]. Nevertheless, the Darwin width can be reduced

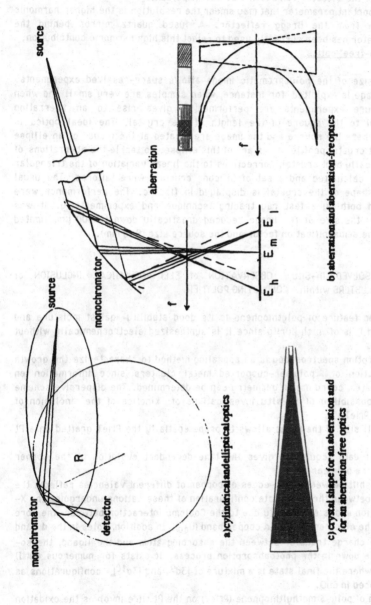

Fig. 2 Main features of cylindrical and elliptical optics. The elliptical optics is the ideal one although the cylindrical gives rise to the U-like aberration.

a) cylindrical and elliptical optics

b) aberration and aberration-free optics

c) crystal shape for an aberration and for an aberration-free optics by going to higher order and/or asymmetric reflections and thence the spatial resolution of the detector can become as well important.

Another important parameter that may smear the resolution is the higher harmonic contribution from the Bragg reflector. A fused quartz mirror behind the monochromator has been currently used to reject this high harmonic contribution.

"Aberration-free" optics

The small size of the polychromatic image allows space-resolved experiments. This advantage is exploited, for instance, when samples are very small and when high pressure experiments are performed[9], gives rise to an aberration proportional to the square of the length of the crystal. The ideal optics is elliptical, where the source and the image are located at the focus of an ellipse and the bent crystal profile fits an arc of this ellipse. To tackle the aberrations of the cylindrically bent crystal, corrections to the linear variation of the triangular shape were calculated and a set of Silicon crystals were tailored. The usual corrected shape of the crystal is displayed in fig.2-c. The performance were investigated both by a fast ray tracing technique and experimentally. It was verified that the size of the image reduced drastically down to 400 μm, limited mostly by the demagnification factor and the source size (S~6 mm).

III-TIME-RESOLVED In-Situ OBSERVATION of ELECTROCHEMICAL INCLUSION of METALLIC CLUSTERS within a CONDUCTING POLYMER.

An appealing feature of polythiophene is its good stability against moisture and oxygen. and it is of high purity since it is synthesized electrochemically without any calayst.

X-ray absorption spectroscopy is an appealing method to characterize the growth and interaction of polymer-supported metal clusters since information on oxidation states, coordination geometry can be determined. The dispersive scheme offers the possibility of **in-situ** investigations of kinetics of the inclusion of copper into PMeT .

-i) The small size of the focus allows to probe spatially the PMeT grafted on a Pt wire.

-ii) The fast data acquisition gives the time-dependent evolution of the copper clusters in the polymer.

The energy shift between the K-edges of copper of different valencies reflects the difference between the final state configuration of these cations undergoing the X-ray absorption and is essentially due to the Coulomb interaction between the core hole c and the electrons in the d copper band (U_{cd}). In addition, holes in the d band enables the charge transfer between the absorbing atom and the ligand, the so-called shake down in the photoabsorption process. It exists for numerous Cu(II) compounds where the final state is a mixture of $|3d^9>$ and $|3d^{10}L>$ configurations as well-evidenced in CuO.

The grafting of poly-3 methylthiophene (PMeT) on the Pt wire involves the oxidation of the monomer, 3-methylthiophene, 0.5M in CH_3CN + 0.5M $N(C_4H_9)_4SO_3CF_3$ at 1.35 V/SCE (satured calomel electrode). The polymer is formed directly in its doped conducting state. This modified electrode is put in a 3 mm-thick electrochemical cell composed of a Teflon ring covered by two Kapton windows. An aqueous copper

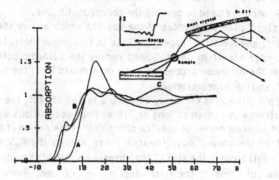

FIG. 3 Cu K-edge absorption spectra of three copper species obtained with a dispersive EXAFS using a Si(311) crystal: curve A, 50-mM aqueous CuCl₂ solution; curve B, Cu¹⁺-bipyridine complex in acetonitrile; curve C, copper foil (photon energy 8978 eV is taken to be the reference energy 0). Data has been collected in 9600 ms for curve A, 300 ms for curve B, and 460 ms for curve C. Inset: Scheme of the dispersive EXAFS experiment.

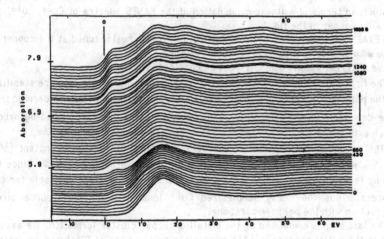

FIG. 4 *In situ* measurements of the evolution of the Cu K edge when PMeT is cathodically polarized in an H₂O-CuCl₂ 50-mM electrolytic medium.

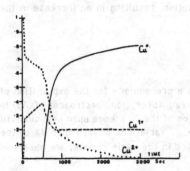

FIG. 5 Variations of the concentration of Cu²⁺, Cu¹⁺, and Cu⁰ inside the polymer vs the polarization time.

solution (H_2O + 50 mM $CuCl_2$, pH = 6) is then added to the cell. The grafted electrode and another Pt wire are used to develop the copper inclusions.

Fig.3 shows clearly the differences of the spectra which allows the estimation of the Cu^{2+}, Cu^{1+}, and metallic copper contents in this experiment. The evolution of the copper K-edge spectrum was followed versus the cathodic polarization time. The kinetics of the inclusion processes were determined from the time dependence of Cu^{2+}, Cu^{1+}, and Cu^0 concentrations.

Since the sample attenuates the X-ray beam by a factor of 200, the collection time for each spectrum was chosen as long as 3.6 s. Data acquisitions were spaced 7.2 s apart. Fig.4 shows a series of spectra of the near-edge region of the Cu K- edge. A rapid shift of the copper edge towards lower energy (by 8.5 eV) appears, in addition to a small bump in the rise of the absorption just at the edge: the Cu^{2+} ions transform into Cu^{1+} ions at the first step of the reaction. Since Cu^{1+} ions are unstable in aqueous solution, they must be stabilized by the polymer.

If the cathodic polarization is continuously applied, the bump in the rise of the absorption is shifted from a value consistent with Cu^{1+} to the Cu^0 value (fig.4).

The time-dependent concentration of species (fig.5) can be determined by a simple analysis in terms of a linear combination of the XANES spectra of Cu^{2+}, Cu^{1+}, and metallic copper of the fig.3.

The EXAFS data have been collected using a Si(111) crystal tuned at the copper K-edge with a wide energy band pass[10] (~500 eV.).

Three different kinetic domains must be considered :

1) The first one is a fast $Cu^{2+} \rightarrow Cu^{1+}$ transformation where Cu^{1+} ions are stabilized by the polymer backbone and form complexes with the $SO_3CF_3^-$ ions as derived from time-dependent EXAFS data, which show that Cu^{1+} ions are surrounded by oxygen which exists into $SO_3CF_3^-$. The time constant of this rapid fixation is 27 s.

2) The second kinetic domain is characterized by a longer time constant [600s (fig.5)], where the Cu^{1+} ion concentration increases form 25 % to 40 %. Since the doping level of PMeT is equal to 25 %, no more $SO_3CF_3^-$ ions are available for this process. Thus the newly synthesized Cu^{1+} ions should come from a direct interaction with the polymer backbone.

3) The last step is dominated by the metallic-copper cluster formation. We assume that in the initial process, all the accessible sulfur sites of PMeT are saturated. Then, in the absence of stabilizing agent in aqueous solution, the monovalent copper ions undergo disproportionation to produce Cu^{2+} ions and metallic copper. Additional Cu^{2+} is then drained from the solution, resulting in an increase in the absolute copper content.

IV –NEW HIGH TC SUPERCONDUCTORS

The knowledge of the electronic structure is a prerequisite for the explanation of the mechanism of superconductivity. Using X-ray Absorption Spectroscopy (XAS) in dispersive mode, we have studied the evolution of the Cu K edge upon changing **in situ** the oxygen stoichiometry of $Y_1Ba_2Cu_3O_{7-\delta}$. Variations of less than 1% of the absorption spectrum could be reliably detected in the same sample. We observed directly the transformation of trivalent ($|3d^9\underline{L}\rangle$ configuration) into divalent copper ($|3d^9\rangle$ configuration) upon increasing δ within the high oxygenated superconducting phase. When the compound becomes semiconducting a dramatic amount of monovalent copper, i.e. a $|3d^{10}\rangle$ configuration, is produced. Even for the transformation in the range of δ from 0.2 to 0.3 the ocurrence of a small fraction of this $|3d^{10}\rangle$ configuration is unambiguously observed, meaning that the hole concentration in that range is greater than that given by the chemical formula.

High energy spectroscopies, including XPS (mainly from Cu 2p and O 1s levels) [11-14], XAS at Cu K and L_3 edges [15,18], resonant photoemission [19] and EELS [20,21] at oxygen K-edge, have lead to the following clear-cut conclusions:

i) the electronic ground state is highly correlated,

ii) the trivalent copper exists mainly as a $|3d^9\underline{L}\rangle$ configuration in the ground state ($|3d^9\underline{L}\rangle$ meaning the occurence of a hole in the copper **d** band and a hole in the oxygen **p** band), while no evidence for $|3d^8\rangle$ configuration (two holes in the Cu **d** band) is found,

iii) empty states are present in the oxygen **p** band in the metallic phase.

Fig.1 shows the copper K edge spectra of **Cu$_2$O, CuO** and **NaCuO$_2$** whose formal oxidation states are Cu(I), Cu(II) and Cu(III), respectively. Differences between them (**CuO–Cu$_2$O** and **NaCuO$_2$–Cu$_2$O**) are also displayed. The structure A at 2.5 eV present in the Cu(I) spectrum reflects the dipolar allowed transition from the **1s** state to the **p$_x$** and **p$_y$** degenerated empty states. According to a well-documented and systematic survey of XAS in many copper compounds, this A structure is always observed in linearly coordinated copper systems and its position comes from the $|3d^{10}\rangle$ configuration in the ground state. As can be seen, this structure is well separated from the others appearing in the Cu(II) and Cu(III) spectra.

The stoichiometry δ has been changed by controlling both the oxygen partial pressure and the temperature in a furnace able to keep a temperature stability of $\delta T \simeq \pm 1°C$ from room temperature up to 1000°C. The values of δ have been determined from the thermogravimetry measurements of Kishio et al[22)].

Here the emphasis is put on the difference of XAS spectra between two states corresponding to different values of δ. The stability of the energy scale which is better than 50 meV in this range of energy, yields a great sensitivity and reliability in the difference signals.

In the first step (path **a** in fig.6), a well loaded sample ($\delta \simeq 0.05$) with a narrow superconducting transition at 92 K was progressively heated from room temperature up to 600 °C under 1 atmosphere of O$_2$ (i.e. up to $\delta \simeq 0.20$). Notice that in this range, the sample remains in the 90 K phase. The difference of XAS spectra (fig.7-a) displays two maxima at about 4.5 and 10.2 eV, which increase progressively with the temperature. This difference compares very well with the derivative of the XAS spectrum from a $Y_1Ba_2Cu_3O_{6.95}$ sample (fig.8). This similarity can be easily explained if we assume that the absorption spectrum undergoes a slight shift δE towards the lower energy side. The most straightforward explanation is that, when δ increases, a small fraction of the ligand holes disappears, corresponding to a partial evolution from the $|3d^9\underline{L}\rangle$ configuration into the $|3d^9\rangle$ one in the ground state. A rough simulation of this

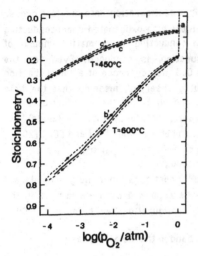

Figure 6: Nonstoichiometry δ of $YBa_2Cu_3O_{7-\delta}$ sample as a function of the oxygen partial pressure for several temperatures. The paths (dashed lines) represent the conditions the sample was submitted during the in-situ investigation. The values of δ are taken from the reference [22].

Figure 7: The difference between copper K-edge XAS spectra of $YBa_2Cu_3O_{7-\delta}$ annealed under the following conditions

    a) sample annealed from room temp. to 600°C under 1 atm of O_2;

    b) sample annealed with oxygen partial pressure varying from 1 atm to 10^{-4} Torr for sample at 600°C

    c) id at 450°C.

The first step (**a**) yields essentially the transformation $|3d^9\underline{L}\rangle \rightarrow |3d^9\rangle$ and the second one (**b**) essentially $|3d^9\rangle \rightarrow |3d^{10}\rangle$. The intermediate step (**c**) yields a mixture of both transformation. The A peak transition at 1.7 eV is clearly seen in the last two steps. An additional fact is that no variation in the quadrupolar pre-edge transition Q is observed.

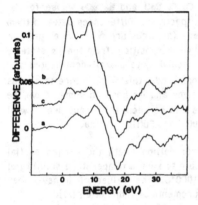

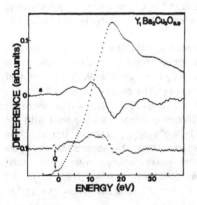

Figure 8: Copper K-edge XAS spectrum of $YBa_2Cu_3O_{6.95}$ and its derivative (dots). The difference signal obtained by annealing under oxygen atmosphere is a derivative-like one and is due to a shift in the spectrum to the lower energy side.

difference spectra reveals a shift $\delta E \simeq 1.2$ eV, consistent with the difference of binding energy between these two configurations ($\delta E \simeq 2$ eV) evaluated by Cu 2p XPS. By changing the pressure from 1 to 10^{-4} atmospheres of oxygen at 600 °C. Thus, the stoichiometry δ was changed from $\delta \simeq 0.2$ to $\delta \simeq 0.8$ (paths **b** and **b'** in fig. 6). Now, the difference between XAS spectra evidences unambiguously the appearence of a strong additional feature at 1.7 eV (fig. 7b), shifted by about -0.8 eV compared to the **A** peak of Cu_2O, which is interpreted as a signal coming from a $|3d^{10}\rangle$ configuration. This result shows clearly up that the transformation of Cu(I) species becomes predominant in that range of δ. The removal of oxygen consists essentially in the transformation of the $|3d^9\rangle$, into the $|3d^{10}\rangle$ configuration. Nevertheless, it appears that partial transformation of the $|3d^9\underline{L}\rangle$ into the $|3d^9\rangle$ configuration is still present, as indicated by the remaining features at about 4.5 and 10.2 eV.

After the 600°C cycling the sample was cooled under oxygen atmosphere down to 450° and the same cycling of oxygen atmosphere was accomplished. Here, the oxygen deficiency varies from $\delta \simeq 0.1$ to $\delta \simeq 0.3$ (path **c** and **c'** in fig.6). The difference between the XAS spectra (fig.7c) shows the additional feature at about 1.7 eV, even if a great contribution is due to the derivative-like signal of the first step. In the considered range of δ, the main contribution comes from the transformation of the $|3d^9\underline{L}\rangle$ into the $|3d^9\rangle$ configuration. However, there exists some contribution coming from monovalent copper, i.e. $|3d^{10}\rangle$ configuration. This evidence of monovalent species is in apparent contradiction with the neutrality rule that predicts the appearence of Cu(I) only when δ is larger than 0.5. It should be noted that the oxygen uptake and removal is perfectly reversible for both temperatures.

The presence of the $|3d^{10}\rangle$ configuration, which is most likely localized on Cu(I) sites, in the $Y_1Ba_2Cu_3O_{7-\delta}$ sample well beyond the tetragonal→orthorhombic transition, implies that i) the hole concentration in the oxygen **p** band is higher than that given by the chemical formula and ii) that the tetra→ortho transition should take place for δ larger than 0.5. Indeed, precise correlated magnetic and cristallographic measurements [23] have shown that this phase transition occurs at $\delta \simeq 0.6$. Moreover, the initial oxygen uptake ($\delta > 0.75$), which drains electrons from the Cu(1) reservoir, does not change the Neel temperature. On the contrary, when the hole creation starts ($\delta < 0.75$) the antiferromagnetic order is strongly affected and the Neel temperature decreases, vanishing at $\delta \simeq 0.4$.

V – HIGH PRESSURE ON SOLID KRYPTON

Because of the spherical electronic shells of their constituent atoms, rare gas solids (R.G.S.) are the best candidates to compare experimental results with theoretical predictions using various interaction potentials. XAS gives informations about the interatomic distances and their mean square deviation which can be compared to data obtained by calculation or other experimental techniques. This section presents results on solid krypton under high pressure up to 20 GPa.

The high pressure cell was a classical Block–Piermarini diamond anvil cell, cryogenically loaded with Kr. The pressure was measured using the power 5 law ruby scale. Because of the small size of the sample in the cell, 300 μm in diameter, only a part of the beam is usable.

In order to determine the nearest neighbour interatomic distance and the Debye-Waller factor σ, the classical EXAFS fitting procedure was used. Since at ambiant pressure and room temperature krypton is a gas, the backscattering amplitude and the phase shift have been obtained from the data at 15.7 GPa where the lattice parameter was known from X ray diffraction [24]. Only the variation of the Debye-Waller factor can be measured.

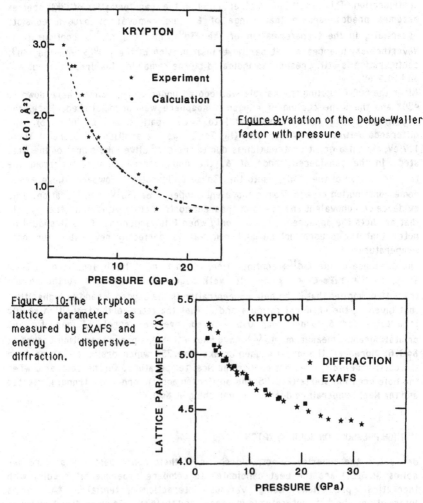

Figure 9:Vaiation of the Debye–Waller factor with pressure

Figure 10:The krypton lattice parameter as measured by EXAFS and energy dispersive-diffraction.

The variation with pressure of the first neighbour interatomic distance obtained from the present EXAFS measurement agrees with x ray diffraction results (fig.10). The thermal contribution to the Debye–Waller factor is determined by the mean square displacement of the central absorbing atom relative to its neighbours. Using self consistent harmonic theory plus a cubic anharmonic term[25], Loubeyre calculated the mean displacement amplitude of the phonons in krypton as a function of the pressure, using Aziz's HFD-B pair potential. Results of these calculations are shown in fig. 9 (points). Since EXAFS data determine only changes in σ^2, a constant value of 0.0069 A^2 was added to experimental results (stars in fig.10). The comparison shows the remarkable agreement between both sets of data. The very fast variation of σ^2 with pressure at low pressure relates to the approach of melting and to the weak Van der Waals interatomic potentials.Our set of data can provide useful informations to deduce the density of the krypton bubbles formed after implantation in metals.

V-CONCLUSION

For concentrated systems, in-situ time-dependent experiments have been achieved. The mechanisms and the kinetics of the electrochemical inclusion of copper particles inside an organic conducting polymer yield a complete illustration of the typical experiments one can consider.The time scale achievable of the order of the tenth or the second restricts this time-resolved spectroscopy to materials science where mass transportation is involved. ESRF can open the millisecond time scale giving access to more dilute samples and perhaps visualisation of changes of conformation of large molecules as found in biophysical dedicated cases.An additionnal advantage of the dispersive scheme comes from the focus of the beam which allows the use ofvery small apertures to carry out high pressure experiments.

REFERENCES

1) E. DARTYGE, C. DEPAUTEX, J.M. DUBUISSON, A. FONTAINE, A. JUCHA, P.LEBOUCHER G.TOURILLON,NIM A246 (1986) p. 452
2) G. TOURILLON, E. DARTYGE, A. FONTAINE, A. JUCHA, PRL (1986) 57, 5, 506
3) E. DARTYGE, A. FONTAINE, G. TOURILLON, R. CORTES, A. JUCHA, Phys.LettersA, 113A,7,p.384
4) G. MAIRE, F. GARIN, P. BERNHARDT, P. GIRARD, J.L. SCHMITT, E. DARTYGE, H. DEXPERT, A. FONTAINE, A. JUCHA, P. LAGARDE Applied Catalysis 1986,26,305
5) T. MATSUSHITA, H. OYANAGI, S. SAIGO, H. KIHARA, U. KAMINAGA, EXAFS and Near Edge Structure III, ed.K.O. HODGSON, B. HEDMAN, J.E. PENNER-HAHN.
6) I. ASCONE, A.BIANCONI, E.DARTYGE, S.DELLA LONGA, A.FONTAINE, M.MOMENTEAU.Biochimica & Biophisica Acta915 (1987),168. 7) J.S.ROHRER, M-S.JOO, E.DARTYGE, D.E.SAYERS, A.FONTAINE, E.C.THEIL. J. of Bio. Chem. 262,13385,(1987).
8) H.TOLENTINO, E. DARTYGE, A. FONTAINE, G. TOURILLON J.App.Crys. 21,15-21(1988);
9) A.POLIAN, J.P. ITIE, E.DARTYGE, A.FONTAINE, G.TOURILLON Acc.Phys. RevB..
10) E.DARTYGE, A.FONTAINE,G. TOURILLON, A.JUCHA, J. de Phys. C8sup 12,42, 1986

11) T.GOURIEUX, G.KRILL, M.MAURER, M-F.RAVET, A.MENNY, H.TOLENTINO,
A.FONTAINE; Phys.Rev.B37,7516(1988)

12) J.C.FUGGLE, P.J.W.WEIJS, R.SCHOORL, G.A.SAWATZKY, J.FINK,   N.NUCKER,
P.J.DURHAM, W.M.TEMMERMAN; Phys.Rev.B 37, 123 (1988)

13) A.FUJIMORI, E.TAKAYAMA-MUROMACHI,Y.USCHIDA, B.GKAI Phys.Rev.B, 35, 6,
8814 (1987)

14) D.VAN DER MAREL, J.VAN ELP, G.A.SAWATZKY, D.HEITMANN; Phys.Rev. B 37,
5136 (1988)

15) F.BAUDELET, G. COLLIN, E.DARTYGE, A.FONTAINE, J.P.KAPPLER, G.KRILL,
J.P.ITIE, J.JEGOUDEZ, M.MAURER, Ph.MONOD, A.REVCOLEVSCHI,   H.TOLENTINO,
G.TOURILLON, M.VERDAGUER Z. für Physik B, 69,141 (1987)

16) A.BIANCONI,A.CONGIUCASTELLANO,M.DESANTIS,P.RUDOLF,P.LAGARDE,
A.M.FLANK, A.MARCELLI,Sol.St.Comm.63,11,1009 (1987)

17) A. BIANCONI J. BUDNICK, A. M. FLANK, A. FONTAINE, P. LAGARDE, A.MARCELLI,
H.TOLENTINO, B.CHAMBERLAND, G.DEMAZEAU, C.MICHEL, B.RAVEAU Phys.
Let.A,127, 5,285(1988)

18) A. BIANCONI, M.DE SANTIS, A.DI CICCO, A.M.FLANK, A.FONTAINE, P.LAGARDE,
H.KATAYAMA-YOSHIDA, A.KOTANI, A.MARCELLI. Phys.Rev.B Rapid Com.(1988)

19) P.THIRY, G.ROSSI, Y.PETROFF, A.REVCOLEVSKCHI, J.JEGOUDEZ; Europhys.Lett. 5
55(1988)

20) P.E.BATSON,M.F.CHISHOLM;Phys.Rev.B37,635(1988)

21) N.NUCKER, H.ROMBERG, X.X.XI, J.FINK, B.GEGENHEIMER,Z.X.ZHAO; Submitted
1988 to Phys.Rev.B

22) K.KISHIO, J.SHIMOYAMA, T.HASEGAWA, K.KITAZAWA, K.FUEKI;
Jap.J.Appl.Phys. 26L1228 (1987)

23) J.ROSSAT-MIGNOD,J.Y.HENRY; RCP Caen Mai 1988 and Private Com.

24) A. POLIAN, J.M. BESSON, M. GRIMSDITCH and W.A. GROSSHANS unpublished

25) P. LOUBEYRE, D. LEVESQUEand J.J. WEISS Phys. Rev. B33, 318 (1986)

EXAFS Studies of Multilayer Interfaces

S. M. Heald and G. M. Lamble

Brookhaven Laboratory, Upton, NY 11973, USA

ABSTRACT

Important for the understanding of multilayer materials is a determination of their interface structure. The extended x-ray absorption fine structure (EXAFS) technique can be useful, particularly for interfaces with a high degree of structural disorder. This paper reviews the application of EXAFS to multilayers, and describes the standing wave enhancement of the EXAFS from multilayer interfaces. Examples are given for W-C and Ni-Ti multilayers.

INTRODUCTION

Multilayer materials are increasingly finding technological uses, particularly in optical and electronic applications. The key in preparing improved multilayers is an understanding of their interface structure. For many applications EXAFS is well suited for probing interface structures in multilayers. This paper reviews some past work in this area, and introduces a new method for obtaining EXAFS data from multilayers: standing wave assisted EXAFS measurements.

The EXAFS technique looks at the local environment of the absorbing atoms, and results from interference between the outgoing photoelectron caused by an x-ray absorption event, and backscattering by the surrounding atoms. A useful parameterization is the single scattering expression[1],

$$\chi(k) = \sum_j \frac{A_j(k)}{R_j^2} e^{-2k^2\sigma_j^2} \sin(2kR_j + \Phi(k)). \tag{1}$$

$\chi(k)$ is the EXAFS interference function after the smooth absorption background is removed, and the energy has been converted to the photoelectron wave vector k. The EXAFS contains information about the interatomic distances R_j for the first few atomic shells about the absorbing atom. Each shell give rise to a characteristic frequency, and can often be isolated by Fourier transform techniques. The amplitude functions $A(k)$ and σ_j give information about the types of atoms in a shell, and the disorder in the bond lengths. The advantage of EXAFS is that no long range order is required, a situation which often in interfaces.

EXAFS APPLIED TO MULTILAYERS

The simplest method for measuring the EXAFS from a multilayer is the transmission technique. This has been applied to systems such as Cu-Hf[2] and Nb-Zr[3]. However, a transmission measurement requires a thick sample ($\geq 1\mu m$) on a x-ray transparent substrate. More generally applicable are fluorescence and electron yield detection methods. These can be applied to thin samples(only a few multilayer periods) prepared on thick substrates. For fluorescence measurements the sensitivity can be further enhanced by employing small incidence angles. Results obtained on the W-C system[4] are a good example. In this case only 10 bilayers were needed to obtain high quality data for W layer thicknesses of $\simeq 30\text{Å}$.

The fluorescence technique has high sensitivity, but still has problems for single crystal substrates. If the multilayers are thin such that the X rays penetrate to the substrate, then Bragg reflections can enter the fluorescence detector and distort the data. For the above mentioned W-C results this problem was avoided by using amorphous substrates. Bragg peaks can also minimized by using glancing angles to reduce the penetration of the X rays to the substrate. However, since the components in multilayers are generally present in concentrated amounts, self absorption will then give significant reductions of the EXAFS amplitudes as will be discussed below.

A better technique is to detect the electron yield from the sample. Since the electrons can be collected from an $\simeq 1000\text{Å}$ surface layer, the technique is well matched to typical multilayer samples. In addition, self absorption corrections are eliminated, and Bragg peak interference is greatly reduced. Figure 1 shows examples of data collected using electron detection for a W-C multilayer consisting of 10 bilayers with a W thickness of 43Å and a C thickness of 40Å. The

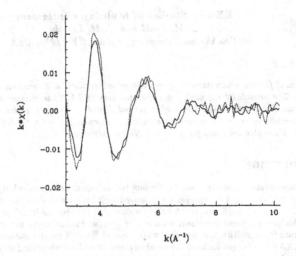

Figure 1: Comparison of EXAFS $\chi(k)$ for a W(43Å)-C(40Å) multilayer before(solid line) and after(dashed line) exposure to NSLS bending magnet radiation for 12 days.

two spectra compare the W L_3 edge EXAFS before and after exposure to the NSLS bending magnet radiation for approximately 12 days. In this sample the EXAFS signal is very small resulting somewhat noisy data, but it can be seen that there is very little change in the sample with exposure. No indication of the formation of W_2C was observed. In contrast, annealing at $350°$ C has previously been shown to convert an identical sample almost completely to W_2C.

An enhancement of the fluorescence technique can be made by taking advantage of the standing wave fields set up in a multilayer structure when the multilayer Bragg reflections are excited[5]. The basic principle is shown in Fig. 2. This shows a calculation of the electric field intensity in a Ni-Ti multilayer for incident angles near the first order Bragg peak. Strong modulations of the electric field are produced, which means that the fluorescence signal is no longer being excited uniformly throughout the sample. Thus, the EXAFS measurement can be enhanced from selected regions of the sample by an appropriate choice of incidence angles. As the data show this is possible even if there is substantial interfacial roughness. Fig. 2(a) shows the reflectivity from an actual Ni-Ti multilayer prepared by sputtering[6]. The reflectivity is substantially reduced by interfacial roughness. This reduction can be modeled by $\simeq17$Å of roughness. The standing wave modulation for this amount of roughness, however, is only slightly affected as the calculation in fig. 2(b) shows.

Figure 3 shows the results of a calculation of the contribution to the EXAFS signal from interfacial regions under standing wave conditions. For the case shown the total interface contribution can be enhanced by almost a factor of two for the appropriate choice of incidence angle. Of equal importance is the fact that the signal from interface type (1) can be almost completely eliminated. This makes it possible to look for differences in the two types of interfaces present in multilayer samples.

The major disadvantage of the standing wave technique is additional experimental complexity. In an EXAFS measurement the energy must be scanned which causes a corresponding shift in the Bragg peak angle. This means that the incidence angle must be varied as the scan proceeds in order to maintain constant standing wave conditions. Another problem is the self absorption distortion mentioned above. The measured fluorescence signal from a concentrated sample is related to the

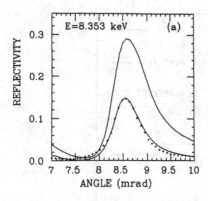
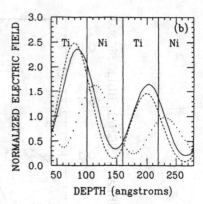

Figure 2: a) Reflectivity near the first order Bragg peak for the Ni-Ti multilayer described in the text (crosses). The solid lines are calculated reflectivities assuming no roughness (large amplitude) and 17Å rms roughness (smaller amplitude). b) Calculated electric field intensity for the near surface region in a Ni-Ti multilayer with d=120Å at 8.353 keV. The interface roughness is 17Å rms, and the curves are calculated for 8.23 mrad (solid line), and 9.0 mrad (points). The dashed line is the result for 8.23 mrad assuming no roughness.

absorption coefficient by the relation[1]

$$\mu' = \frac{I_f}{I_0} = \frac{\mu sin(\theta)}{\mu_t/sin(\theta) + \mu_f/sin(\phi)} \qquad (2)$$

where $\mu_T = \mu + \mu_B$ is the total absorption of the sample including the other components, μ_B, and μ_f is the absorption for the fluorescence photons. The incidence angle is denoted by θ, and the fluorescence photon exit angle by ϕ. For glancing angles $\theta \ll \phi$ and the μ_f term can be ignored. Thus, eq. 2 can be simplified to $\mu' = \mu/(\mu + \mu_B)$ where it is easily seen that if $\mu \simeq \mu_B$, modulations in the measured μ' will be strongly reduced in amplitude. EXAFS spectra, however, are normalized relative to the step in the absorption edge which is also reduced. Taking this into account the reduction in the normalized EXAFS signal can be calculated for a two component system as

$$\frac{\chi'}{\chi} = \frac{\mu_{1L} + f\mu_2}{\mu_{1K} + f\mu_2}, \quad f = \frac{1-x}{x} \qquad (3)$$

where μ_{1L} and μ_{1K} are the absorption of the element being measured below and above its absorption edge, μ_2 is the absorption of the other component, and x is the concentration of element 1.

When standing waves are excited in a multilayer structure, the concentration must be adjusted to account for the fact that the components no longer contribute to the absorption in direct relation to their average concentration. This can be done by integrating the electric field intensity over the individual layers and summing to determine the contribution of each element. The variation of the EXAFS amplitude with the standing wave pattern can also be used to verify that standing wave fields are being excited, as will be shown in the next section.

RESULTS FOR NI-TI

Figure 4 shows results obtained for the Ni K-edge in the same sample for which the reflectivity data is shown in Fig. 2 (60Å Ti and 60Å Ni layers). In this case the comparisons are most easily made using the Fourier transforms of the EXAFS. A strong amplitude variation is seen indicating that standing wave affects are being observed, but the overall shape of the spectra are very similar to data obtained on a bulk Ni sample. This immediately indicates that there is not substantial

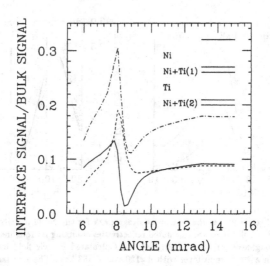

Figure 3: The ratio of the interface signal to the bulk Ni signal for the multilayer shown in the inset. Compared are the signals for interface (1) (solid line), interface (2) (dashed line), and the total interface signal (dot-dashed line)

intermixing between the Ni and Ti. Further analysis indicates that Ni-Ni bond length in the multilayer is the same as for the bulk metal to 0.01Å, and the EXAFS Debye-Waller factor is increased over the metal by $\simeq 0.0015$ Å^2.

The situation for the Ti K-edge is quite different as shown in Fig. 5. Again amplitude variations indicative of standing wave affects are present, although the increased absorption near the Ti edge reduces their amplitude. The most striking feature is the presence of a low-R peak not seen in the pure standard. Detailed analysis indicates that this peak is due to low-Z neighbors such as C or O located at distance of 2.15±0.05Å. Subsequent Auger analysis identified these to be due to C and O in about a 2:1 ratio. The reactivity of Ti is well known, and the contaminates were undoubtedly picked during the deposition process. The observed distance is in good agreement with the first neighbor distance in TiO at 2.09Å and TiC at 2.16Å. TiO$_2$ with a Ti-O distance of 1.92Å seems to be ruled out. In addition by carrying out calculations similar to those shown in Fig. 3, it was determined that the low-Z impurity is distributed relatively uniformly throughout the Ti layers.

To further investigate the presence of the standing wave fields, the amplitude of the Ni EXAFS was determined at a variety of angles. The results are shown in Fig. 6, and compared to two calculations using equation 3. The solid line assumes 17Å rms roughness and ideal Ni and Ti layers. The dashed line includes the fact the density of the Ti layer is reduced by the presence of low-Z impurities. The results clearly verify the presence of standing wave fields in about the expected amplitude. The only discrepancy is small and systematic reduction in the overall amplitude. This could be caused by a small amount of intermixing (1-2 monolayers) at the interfaces. Model calculations indicate that since the Ni and Ti backscattering functions are similar, such a small amount of mixing would not result in an observable Ni-Ti bond, but would manifest itself mainly in a reduction of Ni-Ni signal. Note, however, that mixing on the 17Å level as indicated by the interface roughness would be clearly observable, and would cause serious disagreement between the Ni-Ni EXAFS amplitude and the calculations. Thus, we conclude that the reflectivity reduction is due to a rough interface with only small amount of intermixing.

CONCLUSIONS

This paper has reviewed the application of EXAFS techniques to multilayer systems. EXAFS

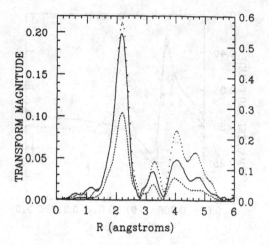

Figure 4: Fourier transform of the Ni EXAFS for a Ni-Ti multilayer at 8.02 mrad (solid line) and 9.41 mrad (dashed line) compared to data obtained on a pure Ni foil (points). The scale for the multilayer is on the left, and for the foil on the right.

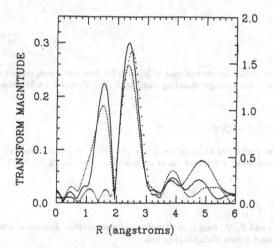

Figure 5: Fourier transform of the Ti EXAFS for a Ni-Ti multilayer at 14.6 mrad (solid line) and 15.0 mrad (dashed line) compared to dat obtained on a pure Ti foil (points). The scale for the multilayer is on the left, and for the foil on the right.

is seen to be valuable complement to other techniques, and is particularly suited for studying interfacial mixing or reactions between layers. It can also determine the role of impurities in modifying the layer structure as the results on the Ti edge demonstrated. The paper also introduced the standing wave enhancement of EXAFS measurements on multilayers, and has shown that they can be applied to multilayers with substantial interface roughness. This technique can substantially enhance the sensitivity to interface regions, and offers the possibility of probing structure within

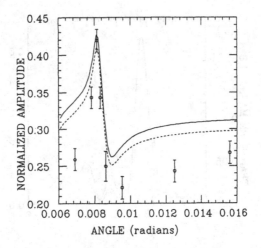

Figure 6: Reduction of the EXAFS amplitude for the Ni first shell in the multilayer as compared to Ni foil. The points ar the data and the curves are calculated reduction factors for a 120Å period multilayer with equal thickness Ni and Ti layers, and 17Å rms roughness. The solid line assumes that the Ni and Ti densities are the normal metal values, while the dashed curve is for a Ti density of 90% of the metal.

the multilayer unit cell. The present case of Ni-Ti is far from optimum, since x-ray optical systems such as W-C have much stronger standing wave fields, and W and C backscattering are much easier to distinguish.

ACKNOWLEDGEMENTS

This work was performed at beamline X-ll at the NSLS, which is supported in part by the U.S. DOE office of Basic Energy Sciences under contract Nos. DE-AC02-76CH00016 and DE-AS05-80-ER10742.

References

[1] E. A. Stern and S. M. Heald, in *Handbook of Synchrotron Radiation*, edited by E. E. Koch (North-Holland,Amsterdam,1983) p. 955.

[2] S. M. Heald, J. M. Tranquada, B. M. Clemens, and J. P. Stec, J. de Physique 47,C8-1061 (1986).

[3] T. Claeson, J. B. Boyce, W. P. Lowe, and T. H. Geballe, Phys. Rev. **B29**, 4969 (1984).

[4] G. M. Lamble, S. M. Heald in *Multilayers, Synthesis, Properties, and Non-electronic Applications*, edited by T. W. Barbee, F. Spaepen, and L. Greer (Mat. Res. Soc. Proc. **103**, Pittsburgh, PA 1988) pp. 101-07.

[5] S. M. Heald and J. M. Tranquada, J. Appl. Phys., in press.

[6] A. M. Saxena and C. F. Majkrzak, Proc. AIP Conf., **89**, 193 (1982).

FLUORESCENCE YIELD NEAR EDGE SPECTROSCOPY (FYNES) FOR ULTRA LOW Z MATERIALS:
AN IN-SITU PROBE OF REACTION RATES AND LOCAL STRUCTURE

D. A. FISCHER,* J. L. GLAND,** AND G. MEITZNER†
*Exxon PRT, Bldg. 510E, Brookhaven National Laboratory, Upton, NY 11973
**Chemistry Dept., University of Michigan, Ann Arbor, MI 48109
†Exxon Research and Engineering Co., Annandale, NJ 08801

ABSTRACT

Fluorescence Yield Near Edge Spectroscopy (FYNES) of ultra low Z materials represents a synchrotron radiation milestone for in-situ determination of local structure for a range of materials problems from monolayers to bulk samples even in the presence of a reactive atmosphere. Two examples will be presented highlighting the broad range of materials problems addressed by the FYNES technique. First a study of the kinetics of CO displacement on Ni(100) by hydrogen at pressures up to .1 torr. These kinetic results highlight the unique capabilities of FYNES to directly characterize surface reaction rates in the presence of reactive gases. Second, a pioneering fluorescence EXAFS study characterizing low concentration fluorine materials will be discussed. Finally new opportunities in FYNES presented by increased photon flux from insertion-device-based sources will be explored.

INTRODUCTION

X-ray absorption spectroscopy has proven to be a very useful tool in determining local structure in a range of materials problems. Although historically a hard x-ray technique, recently Near Edge X-Ray Absorption Fine Structure (NEXAFS) (in the soft x-ray region using electron yield) has been developed as a method for obtaining local structure of adsorbed species.[1] We discuss here a new detection method for NEXAFS extending traditional fluorescence yield (FY) methods to the soft x-ray region. In fact, we have demonstrated that such a FY technique is feasible, and we have reported results on studies for thiophene adsorbed on Ni(100) [sulfur K edge] [2] and ehtylene on Cu(100) [carbon K edge].[3] We have also performed prototype synchrotron experiments coupling a soft x-ray proportional counter,[4] reaction cell, and suitable soft x-ray entrance window to form a system allowing the experimenter to perform sample characterization experiments under real reaction conditions encountered in catalysis.[5] The initial series of experiments described briefly in this paper were focussed on coadsorption of CO and hydrogen on the Ni(100) surface. We found that chemisorbed CO can be rapidly displaced from the Ni(100) surface by hydrogen pressures above 10^{-4} torr in the 290 to 330 K temperature range. This unexpected displacement occurs despite the fact that CO is adsorbed with a heat of 30 kcal/mole,[6,7] while hydrogen is adsorbed with a heat of 23 kcal/mole.[8]

We describe a second series of experiments demonstrating the feasibility of performing low Z EXAFS for characterizing low concentration (<5%) fluorine containing materials. This study yielded EXAFS of CaF and fluoroapetite as a preliminary result in characterizing fluorine in teeth.

APPARATUS

The experimental apparatus consists of a multiple level vacuum chamber with a small high pressure reaction chamber on top where the FYNES experiments are performed. The primary vacuum chamber was equipped with standard surface science instrumentation including facilities for Auger electron spectroscopy, thermal desorption, low energy electron diffraction, and sputtering. A long travel manipulator was used to transfer the sample to the reaction chamber at the upper level. The reaction chamber could be isolated from the main chamber by a gate valve and pumped independently using a turbomolecular pump as indicated in Figure 1. The reaction chamber could also be isolated from the synchrotron ring by a combination of two thin windows and a ballast region. The windows were 1,000 Å boron and tin films, supported on a high transmission nickel grid mounted on the center of two 2-3/4" gate valves. Thus each window could be inserted or removed

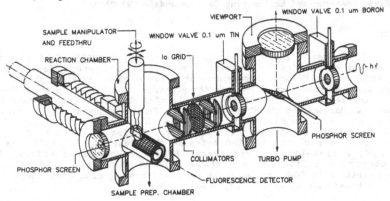

Fig. 1. Apparatus for soft x-ray absorption measurements under a reactive atmosphere.

independently. The windows transmit over 50% of the radiation in the 300 eV energy range. The tin window has an absorption edge around 490 eV which reduces second order radiation coming from the monochromator for primary energies above 245 eV in the carbon edge region. Each window could withstand a hydrogen pressure differential over 10 torr without any significant leakage to the vacuum side.

Fluorescence radiation from the sample was collected by a differentially pumped ultra-high vacuum compatible proportional counter mounted below the sample orthogonal to the plane of incidence (to minimize scattered light see Ref. 3) as shown in the figure. The detector assembly is mounted via a welded bellows and for these experiments was positioned ~1 cm from the sample, providing a solid angle of collection of 10% of $4\pi Sr$. Radiation from the sample passes into the detector through two 1 μm thick polypropylene windows (each 83% transmitting for C-K_α radiation) [9] with a differentially pumped region in between. The polypropylene windows are mounted on a 90% transmitting stainless steel grid and are capable of withstanding atmospheric pressure.

The energy discrimination characteristics of the proportional counter detector have proven to be crucial for our application, since excitation of the Ni-L_α edge by third order light occurs at about the same monochromator position needed for C-K_α electron excitation. In order to perform carbon-edge FYNES experiments we had to reject the signal coming from the nickel sample substrate by using a pulse height discriminator and setting a window around the adsorbate carbon peak of interest.

Energy dispersive fluorescence yield detection for our second series of experiments above the fluorine K edge was particularly useful in increasing the signal to background or edge jump. This was accomplished by setting a window on the high energy tail of the fluorine K_α peak sharply reducing the contribution of oxygen background fluorescence from the fluoroapetite $(Ca_{10}(PO_4)_6F_2)$ sample. As shown in Fig. 2, we observe a fluorine edge jump increase of four for this sample as compared to total FY (oxygen K_α and fluorine K_α).

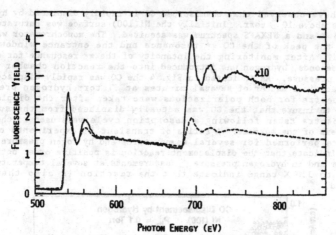

Fig. 2. Fluorescence yield from a pressed powder wafer of fluoroapetite, dashed curve is total fluorescence yield and solid curve uses energy discrimination to emphasize the fluorine K_α.

RESULTS AND DISCUSSION

Spectra at glancing (30°) and normal incident photon geometries for a CO saturated Ni(100) crystal in vacuum at 100K are shown in Fig. 3. These spectra were taken with about 300 mA in the ring (100 points, 4 secs/pt). They have been normalized by spectra from the clean surface, but a count rate of about 4000 c/s was observed at the π resonance peak for normal incidence ($h\nu$ - 287.6 eV) in the original data. These spectra illustrate the high sensitivity of our FY detection techniques. The spectra are qualitatively identical to those obtained using electron detection techniques.[10] However, the signal-to-background (STB), which is the ultimate parameter in determining the sensitivity of the technique, is close to two for the edge jump of a CO saturated nickel surface, and goes as high as 10 for the π resonance peak at normal incidence. These STB are better than those obtained by any electron yield technique by more than an order of magnitude.[3] If we establish a criteria where structure determinations are possible for STB larger than 10%, thise means that our fluorescence detection scheme should be able to detect coverages as low as 1% of a chemisorbed CO monolayer.

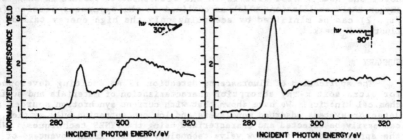

Fig. 3. The normalized fluorescence yield NEXAFS from a 1/2 monolayer CO on Ni(100) are shown for glancing (left) and normal (right) incidence.

We have observed that chemisorbed CO is rapidly displaced by hydrogen pressures above 10^{-4} torr. Initially the Ni(100) surface was saturated with CO at 85 K and a NEXAFS spectrum was acquired. The monochromator was then tuned to the peak of the CO π^* resonance and the entrance windows were inserted. After monitoring the intensity of the π^* resonance for several hundred seconds, hydrogen was introduced into the reaction chamber to the desired pressure. As indicated in Fig. 4 the CO was rapidly displaced from the surface in a matter of several minutes at .1 torr hydrogen pressure. FYNES spectra for both polarizations were taken after the displacement reaction to insure that the CO was actually displaced from the surface. Auger spectra taken following a desorption cycle were used to check the cleanliness of the surface. A series of transient FY experiments of this type were performed for several temperatures and hydrogen pressures. The results indicate that the displacement reaction is positive order both in CO coverage and in hydrogen pressure. Measurements at several temperatures in the 290 to 330 K range indicate that the reaction is also thermally activated.

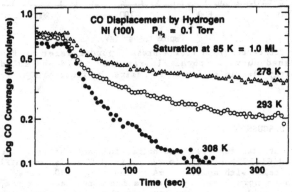

Fig. 4. The log of CO coverage taken from transient FYNES experiments, as a function of displacement time for several temperatures at .1 torr H_2 pressure.

Our second series of FY experiments was dedicated to demonstrating the feasibility of performing low Z EXAFS at low concentration (<5%). To that end we proposed to study a biological problem, that of characterizing the local structure of fluorine in teeth (treated topically or ingested). For purposes of the demonstration we obtained pressed powder wafers of CaF_2 and Fluoroapetite ($Ca_{10}(PO_4)_6F_6$) cycling them in our reaction/sample chamber (sample change over and pump down to high vacuum in 15 min.). Fig. 5 shows the high quality EXAFS data obtainable for these two samples in run times of about one hour. Also visible in the Fluoroapetite spectra is background oxygen EXAFS oscillations which as outlined in the apparatus section (see Fig. 2) can be minimized by accepting only the high energy tail of the fluorine K_α peak.

SUMMARY

The application of fluorescence detection is an exciting development for ultra soft x-ray absorption characterization of materials and surface chemical kinetics. We have shown that with current synchrotron sources and detector systems adsorbed monolayers and low concentration bulk samples, FY can easily be detected and characterized using the FYNES techniques. With the application of a window valve technology we have taken advantage of the photon-in/photon-out principle to bridge the gap between UHV surface science and the real reaction conditions of catalysis.

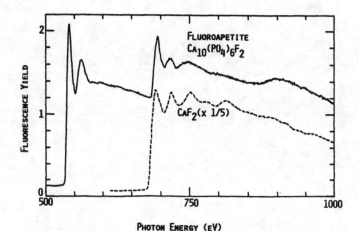

Fig. 5. Fluorescence yield from pressed powder wafers of fluoroapetite (solid) and CaF_2 (dashed).

We have shown the need for energy dispersion detection in FYNES in order to discriminate fluorescence of interest from background. To that end the focusing multilayer mirror collector holds much promise for the future of FYNES.[4] In fact with the advent of very high brightness insertion device beamlines at the Advanced Light Source, multilayer mirror collection will be the method of choice for the future.

REFERENCES

1. J. Stohr, in "Principles, Techniques, and Applications of EXAFS, SEXAFS, and XANES," R. Prins and D. C. Konigsberger, eds., John Wiley, New York, 1985.

2. J. Stohr, E. B. Kollin, D. A. Fischer, J. B. Hastings, F. Zaera, and F. Sette, Phys. Rev. Lett. 55. 1468 (1985).

3. D. A. Fischer, U. Dobler, D. Arvanitis, L. Wenzel, K. Baberschke and J. Stohr, Surf. Sci. 177. 144 (1986).

4. D. A. Fischer, J. Colbert, and J. L. Gland, Rev. Sci. Instr., March/April (1989).

5. F. Zaera, D. A. Fischer, S. Shen, J. L. Gland, Surf. Sci. 194, 205 (1988).

6. J. C. Tracy, J. Chem. Phys. 56, 2736 (1971).

7. J. C. Tracy and J. M. Burkstrand, CRS Crit, Rev. in Solid St. Sci. 4, (1974), Issue 3, 381.

8. K. Christmann, O. Schober, G. Ertle, M. Neumann, J. Chem. Phys. 60, 4528 (1974).

9. B. Henke and M. Tester in Advances in X-ray Analysis, Vol. 18, Plenum Press (1975).

10. J. Stohr and R. Jaeger, Phys. Rev. B 26, 4111 (1982).

EXAFS STUDIES OF COBALT SILICIDE FORMATION PRODUCED BY HIGH DOSE ION IMPLANTATION

Z. TAN, J. I. BUDNICK, F. SANCHEZ+, G. TOURILLON*, F. NAMAVAR, H. HAYDEN, AND A. F. FASIHUDDIN
University of Connecticut, Dept. of Physics and Institute of Materials Science, Storrs, CT 06268.
+ Permanent address: Departamento de Fisica, Universidad Nacional de La Plata, C. C. 67, 1900 La Plata, Argentina.
* LURE, Bat. 209D, 91405 Orsay Cedex, France.

ABSTRACT

The early stages of cobalt silicide formation in high dose (1.0 to $8.0 \times 10^{17} Co/cm^2$) cobalt implanted Si(100) are studied by extended X-ray absorption fine structure (EXAFS), X-ray diffraction (XRD) and Rutherford backscattering spectroscopy (RBS). Locally ordered silicide that is not detectable in XRD has been observed with EXAFS in the as-implanted samples. Long-range ordered phases are observed in the $3 \times 10^{17} Co/cm^2$ samples. After thermal annealing at 700-750°C, single phase $CoSi_2$ with (400) orientation is formed in all implants.

INTRODUCTION

In the past few years we have carried out systematic studies of silicide formation by high dose ion implantation of transition metals into silicon [1,2]. Other workers have also produced silicides by ion implantation technique (for example Ref. 3 and Ref. 4). We report here a study of silicide formation by high dose cobalt implantation into Si(100) at elevated temperatures (100-400°C). The Short-range order and long-range order of the silicide structure are investigated by extended X-ray absorption fine structure (EXAFS), X-ray diffraction and Rutherford backscattering spectroscopy (RBS). Short-range ordered phases are observed in the as-implanted samples with EXAFS for the first time.

EXPERIMENTAL

Polished single crystal silicon with (100) orientation was uniformly implanted (scanning beam) at substrate temperatures of 100 to 400 °C with 150 keV to 165 keV Co^+ ions at a current density of 10-20 $\mu A/cm^2$. The implantation was carried out in a vacuum of 10^{-6} torr. The sample holder was surrounded by a liquid nitrogen trap and warmed by a heater with temperature monitored and stablized during the implantation. Further description of the implantation equipment set up has appeared elsewhere [5]. The EXAFS measurements were carried out at beam line X-11 at the National Synchrotron Light Source (NSLS) and at L. U. R. E. in France. The NSLS data on the implanted samples were measured by fluorescence method with a 45° incident-45° exit set up. Data on $CoSi_2$ and cobalt foil were measured in transmission. Electron yield detection of EXAFS [6] was used in measurements carried out at L. U. R. E..

RESULTS

1. RBS measurements

Mat. Res. Soc. Symp. Proc. Vol. 143. ©1989 Materials Research Society

Fig.1 presents the 1.9 MeV ^{4}He RBS yield of samples implanted with a total dose of 1.0, 3.0, and 8.0× 10^{17}Co/cm^2 respectively and Fig. 2 shows the corresponding Co concentration profile. At a dose of 1.0×10^{17}Co/cm^2 the Co atoms are buried with a peak concentration of near 12% at about 1000 Å (dots in Fig. 2). As the dose is increased to 3× 10^{17}Co/cm^2, the Co atoms begin to appear at the surface, mainly because of erosion effects induced by sputtering. The Co atom concentration increases steadily to near 40% in the nearly flat top 1000 Å layer. When the dose is increased further to 8× 10^{17}Co/cm^2, the Co concentration only increases slightly to 44%. In the meantime, the Co retention decreases to 42% from the 90% observed in the samples implanted with lower doses. Thus at a dose of about 3× 10^{17}Co/cm^2, the Co atom concentration starts to saturate under the present condition of implantation. The RBS yield and concentration profile of the 1×10^{17}Co/cm^2 sample before and after annealing are compared in Fig. 1 and Fig. 2. Upon annealing at 700 °C, the Co atoms diffuse towards the Co concentration peak region and sharpen the peak. This suggests a further separation of the buried silicide and the surface silicon layer.   On the contrary, the Co atoms in the annealed 8 × 10^{17}Co/cm^2 sample (Fig. 3) diffuse towards the Si substrate giving rise to a square-like distribution with a Co atomic concentration of about 33%.

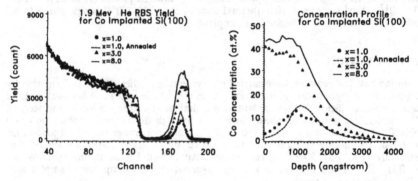

Fig. 1 (left)   RBS spectra of Co implanted Si(100). Doses (X) are indicated in units of 1×10^{17}Co/cm^2.   The implantation was performed at temperatures between 350°C and 400°C. The sample implanted with a dose of 1×10^{17}Co/cm^2 has been annealed at 700°C for 2 hours.

Fig. 2 (right) Co atom concentration profile obtained by assuming the density as a linear combination of the bulk atomic densities of Si and Co. The concentration profiles were not corrected for the effects of the energy straggling and finite detector resolution.

Fig. 3 RBS spectra of the as-implanted and annealed 8×10^{17}Co/cm^2 sample.

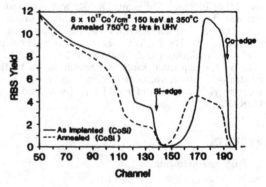

2. X-ray diffraction and EXAFS measurements

All silicides in the implanted samples observed by X-ray diffraction are highly textured. We first discuss the as implanted samples. Several weak but definite lines that can be indexed with the orthorhombic Co_2Si phase were observed in the X-ray diffraction of the $1 \times 10^{17}Co/cm^2$ sample. No diffraction lines can be associated with the CoSi phase. The EXAFS determines that the dominant silicide phase in this sample is $CoSi_2$ (details will follow shortly). CoSi and $CoSi_2$ with (400) orientation were found by XRD in the $3 \times 10^{17}Co/cm^2$ sample. There are additional diffraction lines from this sample that can be indexed with Co_2Si phase. For the $8 \times 10^{17}Co/cm^2$ sample, only one weak and broad line was observed in the diffraction pattern. This line cannot be associated with the Co_2Si or $CoSi_2$ but only possibly with CoSi of a lattice parameter expanded by 0.10 ± 0.04Å. This is to be compared with the EXAFS results below. After thermal annealing at typically 700°C, all implanted samples form single phase and almost completely oriented $CoSi_2$ (400). There are results [7] that prove that the $CoSi_2$ phase in some of our annealed samples is single-crystal epitaxially grown with the silicon substrate.

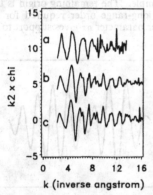 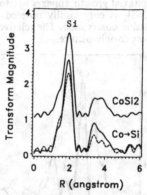

Fig. 4 (left)   Co K-edge k^2 weighed EXAFS function for the as-implanted samples. (a) 1 $\times 10^{17}Co/cm^2$, implanted at $T_i = 400$°C, data taken at room temperature; (b) $7.5 \times 10^{17}Co/cm^2$, $T_i = 100$°C , 77 K data; (c) $8.0 \times 10^{17}Co/cm^2$, $T_i = 350$°C , 77 K data.
Fig. 5 (right) Fourier transform magnitude of $k^2 \cdot \chi$ for the $1 \times 10^{17}Co/cm^2$ sample. Dashed line: as-implanted; Solid line: annealed at 700°C for 2 hours. The data for $CoSi_2$ is also included.

EXAFS with its element selectivity and short-range nature is an ideal local probe for the implanted samples. Tuning the X-ray energy to the Co K-absorption-edge allows the study of the Co local environment no matter if the silicides are at the surface, buried or imbedded in the Si matrix. The EXAFS data were analyzed using the standard method [8]. The EXAFS functions obtained after background subtraction were normalized to the stepsize at the edge energy. Fig. 4 presents k^2 weighed EXAFS data for the as-implanted samples. The Fourier transform magnitudes are shown in Fig. 5 for the as-implanted and annealed $1 \times 10^{17}Co/cm^2$ sample together with that of $CoSi_2$ powder. From the Fourier transform magnitudes, it is immediately clear that $CoSi_2$ is formed in both the as-implanted and annealed samples. The first coordination shell about the Co atom has 8 Si atoms at $R = 2.32$ Å (the peak at 2 Å in Fig.5).

The first shell data is further analyzed by methods of the log-ratio and non-linear least square fitting based on the single electron scattering description of EXAFS [9]. $CoSi_2$ was used as a reference compound for backscattering amplitude and phase. The $CoSi_2$ in the annealed sample is well ordered and the Co-Si distance R, coordination number N and the mean squared relative displacement σ^2 are, to within the experimental uncertainties, the same as those of the bulk $CoSi_2$. This result is in good agreement with the diffraction measurement which revealed a single phase $CoSi_2$. The as-implanted sample shows some disorder (a larger σ^2) as compared to the annealed sample. This disorder is also visible from Fig. 5 where the Fourier transform magnitude of the as-implanted sample is apparently smaller than the annealed one. Co-Co pairs in the 2.53 – 2.67Å range, expected in Co_2Si, were not found in the EXAFS data. Therefore, if as suggested by the diffraction experiments on the as-implanted sample, some Co_2Si phase formed during the implantation its relative amount should be very small.

The dominant silicide phase $CoSi_2$ in the as-implanted sample is not detected by the X-ray diffraction. There are two possible origins for this. One is that the $CoSi_2$ grows in the Si(100) wafer epitaxially with a perfect match between the lattice parameters, therefore the only diffraction peak (400) from the $CoSi_2$ overlaps with the (400) peak from the Si(100) substrate. This epitaxial growth would lead to a Co-Si distance of 2.35 Å, but the Co-Si distance determined in the EXAFS data is R = 2.32± 0.01 Å. Thus the epitaxial growth appears not to exist in the sample. The remaining origin is that the $CoSi_2$ is only locally ordered and lacks the long-range order required for the diffraction observation. The relatively large disorder parameter σ^2 gives support to the local-order-only picture.

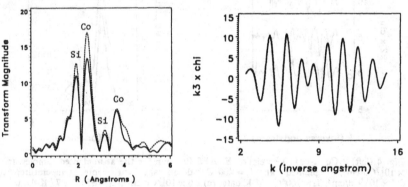

Fig. 6 (left) Fourier transform magnitude of $k^3 \cdot \chi$ for the as-implanted samples. The data were taken at 77 K by a fluorescence method. Solid line: sample A ($7.5 \times 10^{17} Co/cm^2$, $T_i = 100°C$); Dashed line: sample B ($8.0 \times 10^{17} Co/cm^2$, $T_i = 350°C$).
Fig. 7 (right) Filtered (R window: 1.0-3.0 Å) $k^3 \cdot \chi$ data and fit (dashed line) for sample A. The fit is in the 2.8-12.0 Å^{-1} range.

We now turn to the local structure of the samples implanted with high nominal doses; sample A with a dose $7.5 \times 10^{17} Co/cm^2$ implanted at 100 °C and sample B with a dose $8.0 \times 10^{17} Co/cm^2$ implanted at 350 °C. Fig. 4 and Fig. 6 respectively shows their EXAFS function and Fourier transform. The near-neighbor shells are analyzed up to 4.0 Å from the Co absorber using the $CoSi_2$ and Co foil data as references for Co-Si and Co-Co pairs respectively. The 3rd and 4th shell are analyzed after subtracting out the 2nd Co-Co shell. The peaks in between 1.0-3.0 Å fit excellently to a Si shell plus a Co shell. Extreme care was taken to avoid a false fit. The data and the best fit for sample A are shown in Fig. 7.

The structure of sample A and sample B are similar, so the following results apply to both of the two. All observed near neighbor distances from the Co atom in the implanted samples correspond to those of CoSi with a small lattice expansion.

However the 3 Si atoms at 2.47 Å expected in the CoSi structure are totally missing in the implanted samples (attempts to fit the data including 3 Si atoms at 2.47 Å were not successful). There are 7 to 8 Si atoms observed at 2.34 − 2.36 Å in stead of 4 atoms expected in a CoSi structure. To check possible polarization effects in EXAFS, we have calculated the EXAFS contribution from atoms in each coordination shell. The result indicates that there is no polarization effect for the CoSi structure although it is a somewhat peculiar (all atoms are at the interior of the unit cell) cubic structure. The observed number of Co atoms at 2.74 Å is about 3 which is significantly less than the 6 expected in a CoSi structure. Note that the effect of the inelastic loss is small for this second shell. If the silicide is uniform in the implanted region, the missing Co atoms may indicate a Si concentration richer than the stoichiometry of CoSi, which is consistent with the concentration profile by RBS (Fig. 2). Sample A and sample B resemble but differ in detail from the CoSi structure. The nearest Si neighbor configuration is more like the 8 Si at 2.32 Å in the $CoSi_2$ structure than that in CoSi, a point to be emphasized further. Other ordered silicide phases such as Co_2Si and $CoSi_2$ are excluded. The amount of isolated Co atoms in the unreacted Si matrix is also small. The local order around the Co atom is short-range in nature since no diffraction peak (except a very weak one) associated with Co silicides was observed.

DISCUSSION AND SUMMARY

We discuss the silicide formation in the implanted samples in terms of their crystal structures. It is interesting to observe that a short-range ordered $CoSi_2$ phase is readily formed in the as implanted $1 \times 10^{17} Co/cm^2$ sample. The peak Co concentration (12%) at this dose is much less than the stoichiometry of $CoSi_2$, and the observed $CoSi_2$ phase does not appear to be Co atom poor. It is very likely that the $CoSi_2$ clusters are isolated in the mostly unreacted Si grains because the lack of enough Co atoms limits the $CoSi_2$ cluster growth. During annealing, these isolated $CoSi_2$ clusters act as nucleation seeds for further $CoSi_2$ growth and Co atoms diffuse towards the more seeded region, i.e. the region with higher Co concentration. This diffusion behavior is probably responsible for the RBS yield observed on the annealed sample (Fig. 1). The (400) orientation of the $CoSi_2$ should be due to the Si(100) substrate and the close match between the lattice parameters of Si and $CoSi_2$ (1.2% mismatch). As the implant dose increases, the $CoSi_2$ clusters grow because of more Co atom supply. In the meantime, silicide of Co concentration higher than $CoSi_2$ starts to form in regions with Co atoms in excess of that for $CoSi_2$. CoSi is indeed detected in addition to the $CoSi_2$ phase in the as-implanted $3 \times 10^{17} Co/cm^2$ sample. When the dose is further increased, the total Co atoms retained only increase slowly. The implantation probably creates a larger amount of damage to the CoSi and $CoSi_2$ structures in the slightly Co richer material. Even if the ordered $CoSi_2$ is destroyed, the $(CoSi_8)$ core seems to be unaffected presumably because there are fewer 2.72 Å Co-Co pairs (only 3 was observed in EXAFS of sample A and sample B) than are needed to bring the Si near-neighbors into the configuration of CoSi phase. As the Co concentration approaches the stoichiometry of CoSi, the Co and Si atoms have a tendency to locally order as CoSi structure, resulting in the observed more distant Co-Si and Co-Co distances. The under stoichiomtry and limited diffusion during the implantation may be among the reasons for the lack of long-range order in the very high dose implants. Evidences of large disorder and defects in similar implants were also observed [3] from resistivity measurements.

In summary, we have observed short-range ordered and long-range ordered silicides in the high dose cobalt implanted Si(100). We have shown the usefulness of the EXAFS technique in determining the locally ordered structure in such an implanted system. The silicide formation and their dependence on implantation doses and the evolution upon thermal annealing has been discussed based on the crystal structures.

We note finally that the silicide formation process is affected by many factors whose role requires various future investigations to understand.

One of us (ZT) would like to thank Dr. D. Pease for helpful discussions. We acknowledge the support of the U. S. Department of Energy, Division of Materials Sciences, under contract No. DE-AS05-80-ER10742 for it's role in development and operation of Beam Line X-11 at the National Synchrotron Light Source. The NSLS is supported by the Department of Energy, Division of Materials Sciences and Division of Chemical Sciences under contract No. DE-AC02-76CH00016.

REFERENCES

1. F. H. Sanchez, F. Namavar, J. I. Budnick, A. H. Fasihuddin, and H. C. Hayden, Mater. Res. Soc. Symp. Proc. 27, 341 (1986).

2. F. Namavar, F. H. Sanchez, J. I. Budnick, A. H. Fasihuddin, and H. C. Hayden, Mater. Res. Soc. Symp. Proc. 74, 487 (1987).

3. A. E. White, K. T. Short, R. C. Dynes, J. P. Garno, and J. M. Gibson, Appl. Phys. Lett. 50(2), 95 (1987).

4. J. K. N. Lindner and E. H. te Kaat, preprint, 1988.

5. F. Namavar, J. I. Budnick, F. H. Sanchez, and F. Otter, Nucl. Instrum. Meth. B 7/8, 357 (1985).

6. G. Tourillon, E. Dartyge, A. Fontaine, M. Lemonnier, and F. Bartol, Phys. Lett. A 121, 251 (1987).

7. F. Namavar, unpublished, (SPIRE Company, 1988).

8. E. A. Stern and S. M. Heald in *Handbook of Synchrotron Radiation,* editor E. E. Koch, Vol. 1, P.955, North-Holland, Amsterdam, 1983.

9. E. A. Stern, D. E. Sayers and F. W. Lytle, Phys. Rev. B 11, 4836 (1975).

STRUCTURAL CHANGES IN Ag,Fe,Mn,AND Ge MICROCLUSTERS

J. Zhao[1], M. Ramanathan[1,2], P.A. Montano[1,3], G.K. Shenoy[2] and W. Schulze[4]

1-Department of Physics,W.Virginia University, Morgantown,WV 26506
2-Argonne National Laboratory, Argonne, ILL 60439
3-Department of Physics,Brooklyn College of CUNY,Brooklyn,NY 11212
4-Fritz Haber Institute der Max-Planck-Gesellschaft,D-1000 Berlin

ABSTRACT

The structures of Ag, Fe, Mn, and Ge microclusters were determined using EXAFS.The measurements were performed over a wide range of clusters sizes.The clusters were prepared using the gas aggregation technique and isolated in solid argon at 4.2 K. A strong contraction of the interatomic distances was observed for Ag dimers and multimers. Silver clusters larger than 17 A mean diameter show a small contraction of the nn distance and a structure consistent with an fcc lattice . By contrast smaller clusters show the presence of a small expansion and a strong reduction or absence of nnn in the EXAFS signal. This points towards a different crystallographic structure for Ag microclusters . In iron clusters we observed a gradual reduction of the nn distance as the cluster size decreases. The interatomic distance for iron dimers was determined to be 1.94 A,in good agreement with earlier measurements. The iron microclusters show a bcc structure down to a mean diameter of 9 A .Iron clusters with 9 A mean diameter show a structure consistent with an fcc or hcp lattice.Mn microclusters show a simpler crystallographic structure than α-Mn.The measurements on Ge clusters show the presence of only nearest neighbors. There was clear evidence of temporal annealing as determined by variations in the near edge structure of the K-absorption edge.Absorption edge measurements were also performed on free Ge clusters traveling perpendicular to the direction of the synchrotron radiation beam.

Introduction

The properties of small clusters or small particles lie between the well established domains of atomic and molecular physics and that of condensed matter. From a fundamental point of view ,it is crucial to understand how the electronic and structural properties change as atoms come together to form dimers , trimers and so on until finally a large unit develops that has the full characteristics of the bulk. Of particular importance is the understanding of the physical and chemical properties of microclusters. We define microclusters as small aggregates made up of about 10 to 100 atoms ,smaller units are better described as molecules , such a classification is strongly element dependent, and the borderline between the two regions is not well defined. In microclusters the surface atoms represent a large fraction of the material , and they are expected to play a fundamental role in determining the crystallographic and electronic properties of the clusters. From a technological point of view ,it is important to understand the properties of clusters and their relation to heterogeneous catalysis,nucleation,aerosol physics and chemistry ,

photographic and electronic imaging materials , and in thin films
used in information storage [1]. For the above reasons , there has
been increasing interest in metal and semiconductor clusters
during the last decade [2] .

Our interest in recent years is in the area of structural
characterization of small clusters. We have used Extended X-ray
Absorption Fine Structure (EXAFS) [3] and synchrotron radiation to
determine the structure of microclusters. EXAFS provides
information on the local structure, even when the cluster density
in the beam is well below that needed for conventional X-ray
diffraction. In addition , unlike diffraction techniques , the
structural information can be gathered from EXAFS when the
clusters have only two atoms [4,5] . By tuning the X-ray energy to
the absorption edge of different elements in the cluster , the
local structure or the partial radial distribution around each
atom can be obtained.In the present work we report an EXAFS study
of the structural size dependence of silver ,iron ,manganese ,and
germanium microclusters. We observed that the structural evolution
of the microclusters does not follow the same crystallographic
trends as in the solid. We were able to observe the X-ray
absorption signal for free Ge microclusters . In the following
paragraphs we describe the experimental procedures employed in our
measurements, and summarize the results obtained for Ag , Fe ,Mn ,
and Ge microclusters.

Experimental

The microclusters were prepared using the gas - aggregation
technique .A narrow size distribution of clusters can be obtained
using this method [6] .In this technique , the pure elements are
evaporated from a Knudsen cell in an argon atmosphere [5,7] . The
clusters are then transported by the gas stream through an
aperture into a liquid-helium cryopump,where the gas is condensed.
A collimated molecular beam passes through a second aperture on
the axis of the cryopump .The beam can be used for gas phase
studies , or deposited on a substrate.We isolate the
microclusters were in argon at 4.2 K. The metal concentration was
kept around 0.1 atomic % to avoid agglomeration in the solid
matrix. The structure and size distribution of such clusters have
been the subject of careful electron microscopy,EXAFS, and mass
spectrometric studies. The free cluster measurements were
performed using a cross beam method with the cluster beam
perpendicular to the photon beam . The fluorescence data were
collected using a proportional counter at 90 degrees with respect
to the photon and clusters beams. The X-ray absorption and
fluorescence spectra were recorded at beam line X-18 B at NSLS .
An intrinsic Ge solid state detector was used for the Ag,Mn, and
Fe fluorescence measurements . Energy calibration was performed
using ultrahigh purity samples of Ag, Fe,Mn and Ge. The standards
were cooled to liquid nitrogen temperature and were analyzed to
obtain the EXAFS phase shifts and amplitudes. For each sample
several measurements were performed until a good signal-to-noise
ratio was obtained from the added data. Details of the method of
analysis can be found in reference [7]. For the silver samples
measurements were also performed at the Stanford Synchrotron
Radiation Laboratory , this was done to test the reproducibilty of
the results, in particular below 17 A mean diameter.

Experimental Results

Silver microclusters: silver metal has an fcc lattice with a nearest neighbor (nn) distance of 2.87 A at 78 K.The EXAFS spectrum of silver at 78 K allows measurements up to the 5-th shell (6.44 A). For silver particles smaller than 14 A (mean diameter) we detect only the presence of nn and no higher shell are observable by EXAFS ,the magnitude of the Fourier transforms of $k^3\chi$ are shown in Fig.1 . Two factors can contribute to the absence of nnn peaks, one is the presence of a disordered structure, and second a strong reduction in the coordination number due to the presence of a large number of surface atoms. We detected the presence of Ag dimers and multimers (trimers and quadrumers) for samples containing the smallest particles sizes. The interatomic distance for the dimer was found to be 2.47 +/- 0.02 A .Multimers of Ag also show a significant contraction of the nn distance . For particles larger than 25 A a measurable contraction of the nn distance is observed [7] , but as the particle size decreases , a small expansion is observed instead of further contraction, see Fig.2. No contraction of the interatomic distances is observed for particles with a mean diameter less than 17 A. This might be an indication of a structural change in silver microclusters. An expansion of the lattice parameter in 15 Å Pd clusters was reported by Heinemann and Poppa [8]. They attributed this expansion to the formation of icosahedral particles . It is possible that the small expansion observed for the silver microclusters is associated with a crystallographic phase transition from an fcc to an icosahedral lattice .The mean diameter of the microclusters was determined from the average coordination number obtained from EXAFS and independently corroborated by mass spectroscopic methods [9].

Iron microclusters: Samples were prepared with mean diameter ranging from 9 A to 15 A ,the uncertainty in the diameter determination was between 1 to 2 A. α-Fe has a bcc lattice ,we were able to observe up to five shells at 78 K . We used a thin iron metal foil as our standard. The EXAFS measurements of the microclusters show the presence of 3 to 4 shells . We have also measured the dimer interatomic distance and obtained a value of 1.94 +/- 0.02 Å . In figure 3, we have plotted the nn distance vs particle size .A small contraction of about 1 % in reference to the bulk is observed for the larger particles. A larger contraction is observed for the smallest particles , about 3 %. This sample also shows the presence of an iron distance at 3.52 A such a distance does not exist in bcc iron ,see Fig.4 (Fourier transform of $k^2\chi$). The ratio of the first and second shell distances of the 9 A microclusters is 0.682 and compares well with the ratio for fcc or hcp lattices ,0.707. The structural change in the 9 A is also observable in the near edge structure.We consider that the EXAFS results are clear evidence of a crystallographic transformation in iron microclusters with a diameter around 9 A.

Manganese microclusters:α-Mn was taken as the standard for our measurements of Mn microclusters.α-Mn has a very complex structure with 58 atoms in a unit cube of edge equal to 8.894 Å.It has four different site sets , each atom has between 12 and 16 neighbors at distances from 2.24 to 3.00 Å . There are two dominant coordination shells at 2.50 and 2.65 Å respectively.The

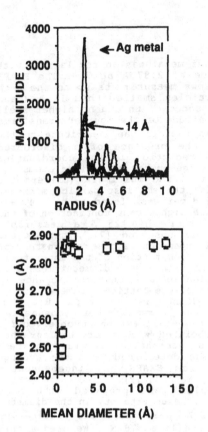

Fig. 1. Fourier Transforms of Ag bulk and 14 Å clusters.

Fig.2 NN distance vs cluster diameter

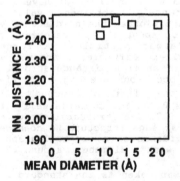

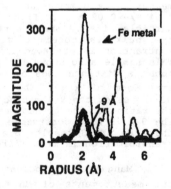

Fig.3. NN distance vs cluster size.

Fig.4. Fourier transforms of Fe metal and 9 Å clusters

Fourier transforms of the EXAFS signal ,k²χ, are shown in Fig. 5.The full analysis of the two prominent EXAFS peaks using the phase shifts of α-Fe , gives values of 2.51 and 2.66 Å .This values are very close to the position of the two prominent distances in α-Mn.So we feel confident of using the parameterized amplitude and phase shifts of α-Fe to analyze the Mn microclusters. The microcluster'mean diameter varied between 8 to 18 Å. The structure of Mn particles with less than hundred atoms is unlikely to have the same complex structure as bulk Mn.The positions of the first three shells(the first and second shells are not resolved) for clusters with mean diameters of 14 ,16 ,and 18 Å are very close to that of α-Mn. However , for smaller particles the third shell position is about 0.09 to 0.13 Å less than in the larger clusters (the EXAFS error is about 0.03 Å), and the first and second shells seems to coalesce into a single shell.The ratio of the two shells distances for 8 Å clusters is 2.65/3.64=0.723 . This ratio is close to the value of 0.707 for fcc or hcp lattices.Such results are suggestive of a simpler lattice for the Mn microclusters.The possibility of an icosahedral structure can not be totally excluded from our analysis.More measurements on smaller and well characterized clusters are neccessary to elucidate this point.

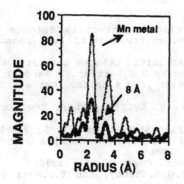

Fig.5. Fourier transforms of Mn bulk and 8 Å clusters

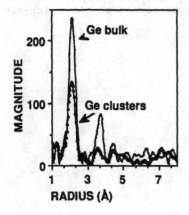

Fig. 6. Fourier transforms of bulk Ge and 14Å clusters.

Germanium microclusters:semiconductor clusters form tightly bound systems with properties that reflect the nature of their directional covalent bonds ,in great contrast to the metals. We observed a great difference in the absorption edge structure for Ge clusters deposited very fast on the matrix (10 A/sec). There wa a clear time evolution of the near edge structure,with a significant increase in the number of empty p- states. Samples prepared by slow deposition(0.5 to 1 A/sec) do not show any time dependence of the near edge structure. The EXAFS analysis of the clusters (mean sizes 7-8 A and 10-15 A) does not show the presence of nnn ; only nn are observed ,see Fig.6 .The nn distances measured were 2.42 A (diameter 7-8 A) and 2.45 A (diameters 10-15 A) ; these values are essentially equal to bulk Ge (2.44 A), within the experimental error of EXAFS.We were also able to measure the near edge structure for free Ge microclusters traveling perpendicular to the photon beam . Figure 4 shows such an spectrum. The results for the free particles were significantly different from the matrix isolated samples . However ,they show the great possibilites this technique offers in the characterization of clusters in the gas phase.

The authors aknowledge the support of the U.S. DOE and the DFG.

References

1. "Contribution of Cluster Physics to Materials Science and Technology",Eds. J.Dvenas and P.M.Rabette,Nijhoff Publ.,Boston, 1986.
2. R. E. Smalley,in Comparison of Ab Initio Quantum Chemistry with Experiment,State of the Arts,edited by R.J.Bartlett (Reidell ,NY 1985).R. L.Whetten,D.M.Cox ,D.J.Trevor ,and A.Kaldor ,Surface Sci. **156**, 8,(1985).
3. D. E. Sayers , E. A. Stern , F. W. Lyttle , Phys. Rev. Lett. **27**, 1204 , (1971).
4. P. A. Montano and G.K.Shenoy ,Solid State Comm. **35**, 53 (1980).
5. P. A. Montano ,G. K. Shenoy , E. E. Alp , W. Schulze , and J. Urban , Phys. Rev. Lett. **56**, 2076 (1986).
6.H.Abe ,W.Schulze,and B.Tesche ,Chem. Phys. 47, 95 (1980)
7.P.A.Montano ,W.Schulze,B.Tesche,G.K.Shenoy, and T.I.Morrison, Phys. Rev. **B 30** , 672 (1984).
8.K.Heinemann and H.Poppa, Surf. Sci. **156**,265 (1985).
9.Mass spectrometric measurements performed at the Fritz-Haber Institute

REFLECTION-EXTENDED-X-RAY-ABSORPTION-FINE-STRUCTURE SPECTROSCOPY AT THE CARBON K-EDGE

G. G. Long*, D. R. Black*, and D. K. Tanaka**

*National Institute of Standards and Technology
 Gaithersburg, MD 20899

**The Johns Hopkins University
 Baltimore, MD 21218

ABSTRACT

The carbon K-edge Reflection-Extended X-Ray Absorption Fine Structure (refl-EXAFS) spectra from graphite, diamond and glassy carbon have been investigated. There is good phase shift transferability between the two well-known bonding types in diamond and graphite, provided that appropriate inner potential corrections to the K-edge energy E_o are made. The model spectra from diamond and graphite were used to investigate the nature of glassy carbon. It was found that, for the particular form of glassy carbon used in this study, the bonding more closely resembled sp^3 than sp^2. This result is preliminary pending our evaluation of the influence of surface oxygen.

INTRODUCTION

There has been much interest[1] in the structural and physical properties of glassy carbon because of its unusual combination of electronic and physical characteristics. Despite numerous investigations of the structure of this material, the major questions concerning whether glassy carbon has a random three-dimensional network structure similar to that of other glasses, or whether it has a predominantly two-dimensional graphite character has not been definitively answered. In addition, several studies[2,3] have indicated that the structure of glassy carbon depends strongly on the details of the method of preparation.

In this work, we present a surface-extended x-ray absorption fine structure (surface-EXAFS) study of the structure of glassy carbon. Measurements were made by means of the reflection-EXAFS technique in which the fine structure near the carbon K-edge was obtained *via* reflectivity experiments at grazing incidence below the angle for total external reflection. This method is particularly useful in the soft x-ray region of the spectrum where samples cannot easily be made sufficiently thin for absorption measurements in a transmission geometry. Further, this technique offers the advantages of high signal intensity combined with surface sensitivity.

For the range of grazing incidence angles below the critical angle of total external reflection, the fine structure observed in reflectance spectra and the fine structure above the absorption edges of materials have been demonstrated[4] to be closely related through the optical constants of the material.

REFL-EXAFS THEORY

The reflectivity of x-rays at grazing incidence to a polished surface can be described using the Fresnel equations, which are derived from a consideration of plane waves incident on a planar boundary. In this formalism, the reflectivity is given in terms of three parameters: the reflection angle, and the real and imaginary parts of the index of refraction.

The complex dielectric permeability is given by $K = 1 - 2\delta - 2\beta i$. The dielectric permeability determines the phase velocity in the medium, $v = c/\sqrt{K}$, and the complex index of refraction is $n = \sqrt{K}$. If $\delta \ll 1$ and $\beta \ll 1$, then to first order the index of refraction is $n = 1 - \delta - \beta i$. In this approximation, $1 - \delta$ is related to the real part of the phase velocity of the wave in the medium and β is related to the absorption coefficient μ, from which the fine structure is measured. For soft x-rays, δ and β are of the order of 10^{-2} - 10^{-3}, and second order corrections to n must be included[5].

The reflectivity can be written in the small angle approximation

$$R = \frac{|E|^2}{|E_0|^2} = \frac{(\phi - a)^2 + b^2}{(\phi + a)^2 + b^2} \tag{1}$$

where ϕ is the reflection angle,

$$a^2 = \tfrac{1}{2}[\sqrt{(\phi^2 - 2\delta)^2 + 4\beta^2} + (\phi^2 - 2\delta)],$$

and

$$b^2 = \tfrac{1}{2}[\sqrt{(\phi^2 - 2\delta)^2 + 4\beta^2} - (\phi^2 - 2\delta)].$$

The values of δ and β, without including solid-state effects, can be calculated for carbon using equations that are valid close to the K-edge[6]. In addition, both δ and β exhibit fine structure related to the EXAFS from condensed matter.

Under "total" reflection conditions, there are no waves propagating into the medium. Instead, there are waves of varying amplitude (evanescent waves) present which decrease with increasing depth. These waves enter the medium and then emerge again from the same surface. The (1/e) penetration depth can be calculated from the imaginary part of the reflectivity using the relation: $d = \lambda/4\pi b$, where λ is the wavelength of the radiation. For angles smaller than 3.0 degrees, the penetration depth in carbon of x-rays of energy up to 700 eV is less than 6 nm.

EXPERIMENT

1. Samples

The glassy carbon sample was prepared using furfuryl alcohol, polyethylene glycol, Triton X-100, and paratoluene sulfonic acid. The sample was first slowly warmed in air and then gradually heat-treated in an argon environment up to 700°C. Since the final firing temperature of glassy carbon is between 700° and 3000°C, our sample is representative of material with the least heat treatment.

The standard samples used in this study were highly oriented pyrolytic graphite (HOPG) and natural diamond (type II, a). The surface of the HOPG sample was polished using tape. The diamond surface was prepared by gem polishing techniques. The glassy carbon sample was polished by means of metallographic techniques to a 0.05 μm MgO compound. Any remaining debris from the surface preparation was removed by means of solvents.

2. Refl-EXAFS Measurements

The surface reflectance measurements were performed on the U-15 (SUNY/NSLS) beamline at the National Synchrotron Light Source at Brookhaven National Laboratory, Upton, NY. This beamline is equipped with a toroidal grating monochromator (TGM) with a 600 line/mm gold-coated grating. The TGM has a constant wavelength resolution which depends on the size of the aperture in front of the grating and on the parameters of the stored electron beam in the ring. There is a non-negligible amount of harmonic contamination for photon energies up to 700 eV. The beamline has a 150 nm aluminum contamination barrier which separates the sample chamber from the storage ring vacuum. The barrier contributes a flux jump of a few percent at the oxygen K-edge due to a thin oxide layer.

The refl-EXAFS experiment requires a measurement of the intensity, I_0, of the incident soft x-rays, which is accomplished by having the incident beam pass through an 85% transparent phosphor-coated grid. The phosphor converts x-ray light into visible light, which is measured by means of a photomultiplier tube. The remainder of the beam impinges at grazing incidence on the sample, and the intensity of the reflected beam is measured by means of a phosphor screen (coated with the same phosphor) and a second photomultiplier tube. A small amount of the unreflected beam passes above the sample and permits a direct measurement of the reflection angle. During the refl-EXAFS scans, a beamstop prevents this unreflected beam from reaching the detector.

3. Refl-EXAFS Analysis

The first step in the data reduction is the removal of instrumental factors from the raw data over the range from 250 to 800 eV. A measure of the instrument function is provided by blank scans, which are obtained without the sample present. This permits a quantitative measure of carbon contamination in the optics, of the oxygen in the contamination barrier, and of the relative behavior of the phosphors and the detectors. Next, corrections to the measured intensities must be made because of the presence of second and third order harmonics. The relative second and third order content in the 250 to 700 eV part of the incident spectrum was provided[7] using data obtained by means of a variable absorber technique. Corrections to the reflected intensity are more complex because part of the harmonic content is removed by reflection from the sample. The reflectivity in second order is obtained by assuming, for example, that a measure of the reflectivity in the range E=700-800 eV, where second order is

negligible in the incident beam, approximates the second order
reflectivity at E/2 = 350 - 400 eV. Similar arguments can be made for
third order. After corrections for harmonic contamination, the
reflected intensity, I, is divided by the monitor intensity, I_o, to
remove the effects of fluctuations in the incident beam intensity.
Finally, geometric corrections are made for the specimen length and the
beam size.

For each sample, refl-EXAFS data were accumulated for a minimum of
five angles between ϕ = 0.5° and 3.5°. One method for deriving δ and β
as a function of incident photon energy is to fit each set of
reflectivity spectra. Alternatively, we used pairs of reflectivity
spectra to solve directly for the two optical constants. This type of
analysis permitted a careful examination of surface roughness and
related reflection angle effects. The energy dependence of the
absorption coefficient, μ, was obtained directly from the spectra for
β. From this point on, the analysis followed standard EXAFS
practice[8].

RESULTS AND DISCUSSION

Refl-EXAFS spectra from diamond, from HOPG with the photon
polarization within the basal plane, from HOPG with the photon
polarization perpendicular to the basal plane, and from glassy carbon
are shown in Fig. 1. Although the spectra shown do not include the
region of the oxygen K-edge, data was taken beyond that edge out to 800
eV. Whereas there was a small signal from oxygen from diamond and from
graphite, the oxygen signal from glassy carbon was quite pronounced.
Evidence of this is seen in the pre-edge region of the glassy carbon
spectrum, where the second order oxygen signal appears. For the angles
that we are using, the penetration depth is limited to 3 - 6 nm,
suggesting that the oxygen signal may be due to structure in the
surface layers. In a companion study[9] using x-ray Raman inelastic
scattering on the same sample, no evidence of an oxygen signal was
seen. Since x-ray Raman scattering is insensitive to surface
structures and contaminants in general, the lack of evidence of oxygen
here indicates that the oxide structure is indeed limited to surface
layers in the glassy carbon sample.

The possibility of nitrogen contamination in the samples was
considered because of the appearance of a feature near the N k-edge.
The feature actually shows up in only three of the four spectra and at
different energies around 425 ev which is above the edge. There is no
feature in the HOPG perpendicular spectrum which was obtained at the
same time and on the same sample. Thus, the oscillation near 60 nm^{-1}
in three of the X(k) spectra must be attributed to EXAFS from carbon.

The EXAFS signals, X(k) multiplied by k^3, for diamond, for HOPG in-
plane and out-of-plane, and for glassy carbon are shown in Fig. 2. The
amplitudes shown are all on the same scale, but displaced from one
another to permit comparison. The low coordination of the out-of-plane
HOPG result is evident, as is the similarity of the other three
results.

Fourier transforms of the data in Fig. 2 are shown in Fig. 3.
These results were studied by back-transforming the first neighbor
peaks from approximately r = 0.05 nm to 0.15 nm for HOPG parallel,
diamond and glassy carbon, and comparing the derived phases and
amplitudes to determine the relative amounts of sp^2 and sp^3 bonding in
carbon. Diamond and graphite were first compared with each other and
then with the glassy carbon sample. When the graphite in-plane (sp^2)

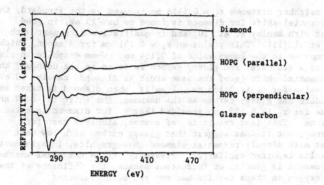

Figure 1. Refl-EXAFS spectra for diamond (top), HOPG in plane (2nd from top), HOPG out-of-plane (3rd from top), and glassy carbon (bottom).

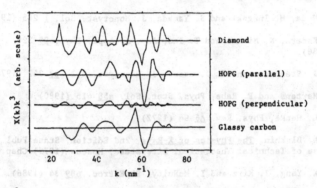

Figure 2. $k^3 X(k)$ for diamond, HOPG (in and out of plane) and glassy carbon. The amplitudes are all on the same scale, but vertically displaced.

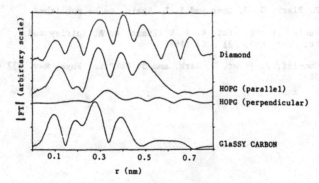

Figure 3. Magnitudes of the Fourier transforms of the data in Fig. 2.

nearest neighbor distance $r_1 = 0.1415$ nm is used as the standard, the inner potential shift for diamond is found to be -25 eV, in good agreement with Denley et al.[10] and in qualitative agreement with Comelli et al.[11]. This yields an $r_1 = 0.151$ nm for diamond, which is to be compared to the known value of 0.154 nm. When graphite in-plane is used as a standard, and glassy carbon is taken as the unknown, the inner potential shift is of the same order as diamond, namely -32 eV, and the value for r_1 is 0.155 nm. Finally, when diamond is taken as the standard and glassy carbon as the unknown, the shift is -8 eV and the value for r_1 is 0.155 nm, provided that r_1 for diamond is taken to be 0.151 nm. This last result is, of course, simply a test for consistency, but it does suggest that glassy carbon with low heat treatment more closely resembles diamond than graphite, in agreement with results reported earlier[2]. Presently, it is not clear whether this agreement is general or fortuitous because the influence of the surface oxygen on these results has not yet been evaluated.

REFERENCES

1. T. Noda, M. Inagaki and S. Yamada, J. Noncryst. Sol. 1 285 (1969).

2. S. Kodera, N. Minami and T. Ino, Jap. J. Appl. Phys. 25 328 (1986).

3. R. J. Stenhouse and P. J. Grost, J. Noncryst. Sol. 27 247 (1978).

4. G. Martens and P. Rabe, Phys. Stat. Sol. a58 415 (1980).

5. B. L. Henke, Phys. Rev. A6 94 (1972).

6. M. A. Blokhin, The Physics of X-Rays, 2nd Edition, State Publ. House of Technical-Theoretical Literature, Moscow (1975) Chap. 5.

7. B. X. Yang, J. Kirz and I. McNulty, SPIE Proc. 689 34 (1986).

8. P. A. Lee, P. H. Citrin, P. Eisenberger, B. M. Kincaid, Rev. Mod. Phys. 53 769 (1981).

9. D. R. Black, G. G. Long and I. L. Spain, to be published.

10. D. Denley, P. Perfetti, R. S. Williams, D. A. Shirley and J. Stohr, Phys. Rev. B21 2267 (1980).

11. G. Comelli, J. Stohr, W. Jark, and B. B. Pate, Phys, Rev. B37 4383 (1988).

OCCUPATION of DISTORTED Cu(1) SITES by Co and Fe in $Y_1Ba_2Cu_3O_{7-\delta}$

F. BRIDGES[1], J. B. BOYCE[2], T. CLAESON[3], T. H. GEBALLE[4] and J. M. TARASCON[5]

[1]Department of Physics, University of California, Santa Cruz, CA 95064
[2]Xerox Palo Alto Research Center, Palo Alto, CA 94304
[3]Physics Department, Chalmers Univ. of Techn., S–41296 Gothenburg, Sweden
[4]Department of Applied Physics, Stanford University, Stanford, CA 94305
[5]Bell Communications Research Laboratory Red Bank, New Jersey 07701

INTRODUCTION

Small perturbations of the structure and composition of the 90K superconductor $Y_1Ba_2Cu_3O_{7-\delta}$ (YBCO) have profound effects on its superconducting properties [1]. In an effort to understand the interaction between the structure of YBCO and its superconducting properties, we have made an extensive investigation of YBCO doped with Fe and Co, using x–ray absorption spectroscopy.

The structure of $Y_1Ba_2Cu_3O_{7-\delta}$ is a distorted, oxygen–deficient, trilayer perovskite [2,3], which as a 90K superconductor ($\delta < 0.2$) is orthorhombic with two distinct Cu sites: the Cu(2) site in the two Cu–O planes and the Cu(1) site in the one-dimensional Cu–O chains. The structure and superconducting properties are affected dramatically by the introduction of transition–metal substitutions. T_c is strongly suppressed, and the orthorhombic distortion vanishes at very low concentrations of the transition–metal dopant [4–6] — e.g. a tetragonal structure for ~2% of the Cu replaced by Co. T_c's in excess of 77K persist in some of these apparently tetragonal compounds. Determination of which Cu site(s) the dopants occupy is of importance for understanding the role of the two sites, and the importance of the orthorhombic distortion.

Another aspect of the substitutional doping of these materials is the assumption that the environment about a defect is essentially the same as the environment about a Cu atom on either the Cu(1) or Cu(2) sites. Mossbauer studies indicate that Fe dopants occupy mainly the Cu(1) sites in YBCO with at least three distinct Fe(1) sites within the chain layer [7]. Neutron diffraction studies of Co–doped YBCO indicate that Co dopants also occupy Cu(1) sites [8]. None of these investigations include the possibility of an off–center displacement of the substituted sites. The EXAFS experiments presented here indicate that some of the Co occupies strongly distorted Cu(1) sites. The results for the Fe substituted samples are similar, but the distortion is less and there appears to be a small occupation of the Cu(2) site as well.

EXPERIMENTAL DETAILS

Samples of transition–metal doped $Y_1Ba_2(Cu_{1-x}M_x)_3O_{7-\delta}$ for $0.017 < x < 0.3$ and M = Fe and Co, were prepared by a solid–state reaction.. Resistive and magnetic measurements of the superconducting critical temperatures are reported in Ref. 6. The samples were characterized using x–ray diffraction measurements which indicate that the samples are single–phase within a few percent. In addition, powders of the following EXAFS structural and near–edge standards were also prepared: FeO, Fe_2O_3, Fe_3O_4, CoO, Co_3O_4,Cu_2O, and CuO.The x–ray absorption measurements were performed in the vicinity of the Cu K–edge and dopant K–edge.

Co (and Cu) FIRST NEIGHBOR ENVIRONMENT

In Fig. 1 we compare the real space Co K–edge EXAFS data [FT $k\chi(k)$] for the $Co_{0.17}$ sample with the Cu K–edge data for normal YBCO. Both the amplitude and the real component of the complex fourier transform are plotted. The k–space transform range is 3.6–11.7 Å^{-1} with a gaussian broadening of 0.5 Å^{-1}. The main features to note are: 1) the first neighbor environment is similar for the two atoms but the main peak for Co is shifted to lower r by roughly 0.1Å (Note that this peak in the r–space EXAFS data includes a phase shift — the actual distance is about 1.8Å), 2) there is additional weight near 2.0Å for the Co edge, and 3) the second neighbor peaks, in the range 2.5 – 4.5Å, are considerably reduced in amplitude for the Co edge data.

Mat. Res. Soc. Symp. Proc. Vol. 143. ©1989 Materials Research Society

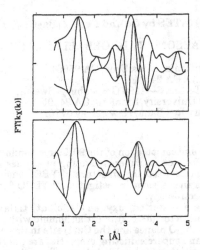

	Peak 1		Peak 2	
Cobalt Content x	# of O Nbrs.	r_{Cu-O} (Å)	# of O Nbrs.	r_{Cu-O} (Å)
0.017	3.2	1.75	1.4	2.39
0.03	3.6	1.80	0.8	2.42
0.07	3.4	1.81	1.2	2.38
0.17	3.8	1.84	0.9	2.33
0.30	3.8	1.87	1.0	2.39

Fig. 1.   Comparison of the Cu K–edge data for normal YBCO (top) with the Co K–edge data for YBCO(Co$_{0.17}$).

Table I.   Co K–edge results for the Co–O near–neighbor environment in tetragonal YBa$_2$(Cu$_{1-x}$Co$_x$)$_3$O$_{7-\delta}$.

The data for all the Co samples are quite similar. The first neighbor results are qualitatively inconsistent with the predictions obtained by assuming a simple replacement of the Cu(1) by Co. Additional weight in the 1.9–2.3Å region is quite unexpected since *no* long Cu–O bonds occur for the Cu(1) sites. A significantly shortened Co–O bond length along the chains compared to the Cu–O chain distance of 1.94Å appears inconsistent with the fact that the lattice constant increases very slightly when Co is introduced.

The first neighbor Co peak was fit to a sum of Co–O standards starting with the known distances and number of O neighbors around the Cu in YBCO. Fits assuming that the Co are all on the Cu(1) sites, all on the Cu(2), or uniformly distributed on both sites were carried out. In each case the fits converged to an unexpected result — the first neighbor shell about Co contains about 5 atoms, but with ~3.5 O at a short distance (~1.8Å) and about ~1.1 O at a much larger distance (2.4Å). For comparison, the weighted number of O neighbors in YBCO are 2/3 neighbor at 1.85Å, 10/3 neighbors at 1.94Å, and 2/3 neighbor at 2.3Å In Table I we summarize the results of the fits to the first neighbor peak for several samples. These results confirm the qualitative results discussed above. The quality of these two peak fits is very good — the weighted % error is less than 1%, except for the very low concentration sample.

To proceed with the analysis of the Co–O peak, we first anticipate a result from the next section:- that the Co substitutes primarily on the Cu(1) site (consistent with other investigations). If we further assume that the Co–O(4) distances along the c–axis are essentially unchanged, we must then accommodate both long and short Co–O distances within the Cu(1)–oxygen layer. In addition, because of the tetragonal structure, the nearest neighbor planar O are no longer constrained to the chains and it is possible that some Co have 3 or 4 nearest neighbor O within the plane containing the Cu(1) sites. We note at this point that the sum of one long plus three short Co–O distances is very close to four Cu–O distances in the tetragonal crystal. We return to possible models after a discussion of the second neighbor environment.

Co (and Cu) SECOND NEIGHBOR ENVIRONMENT

The second neighbor environment for a direct substitution of Co at the Cu(1) sites should be simpler than that for the usual mixture of Cu(1) and Cu(2) sites in normal YBCO. The important points to emphasize are 1) the total Cu–Ba amplitude for the Cu(1) site alone is *50% larger* than the total weighted Cu–Ba amplitude in YBCO, and 2) only two main peaks contribute to the undistorted Cu(1) second neighbor shell. Such a Cu(1) environment should yield a large and well defined second neighbor shell between 2.6 and 3.9Å with an amplitude that is larger than observed for the Cu K–edge data because less interference should occur with fewer peaks. The experimental reality is quite different for the Co K–edge data (see Fig. 1); the second neighbor peak is surprisingly small.

In our initial attempts to fit the second neighbor Co data we used two peaks, corresponding to the expected Co–Ba and Co–Cu(Co) bonds for the Cu(1) site. Even with significant broadening, the fit to the data is very poor and the number of Ba neighbors is far too low. Adding a Co(2)–Y peak at 3.2 Å and a small Co(2)–Ba peak near 3.4Å to include the possibility of a partial Cu(2) site occupancy for Co did not improve the fit. In such fits, the Co–Y amplitude is always reduced to values corresponding to less than 15% of the Co on Cu(2) sites and the goodness of fit parameter is not improved significantly. The reduced second shell amplitude is indicative of significant disorder for the Co second-neighbor–atom distances. Structure in the F.T. r–space data suggests a strong interference between several peaks, somewhat similar to the effect observed in O–depleted YBCO, in which the net Cu–Ba peak is greatly reduced [9].

Before discussing the Co second neighbors in more detail, it is instructive to first consider the question: is the Co–Ba second neighbor peak smeared out because the Ba atoms are highly disordered? The answer, even for the highly doped $Co_{0.3}$ sample, is no, based on the Cu K–edge second neighbor results for this sample. Unlike the Co second neighbor peak, the Cu second neighbor peak for the $Co_{0.3}$ sample is quite large and rather similar to normal YBCO. We fit the second neighbor multipeak structure in [FT $k\chi(k)$] from r = 2.5 to 3.8Å to one Cu–Y, two Cu–Ba and one Cu–Cu standard, and obtained a very good fit (see Fig. 2).Within a 13% uncertainty in the number of second nearest neighbors, the amplitudes of the two Ba peaks and the Y peak are consistent with a Co site distribution of 11% of the Co on Cu(2) sites and 89% on Cu(1) sites in this highly doped sample. We also note that the Cu–Ba bond lengths agree very well with those measured by neutron diffraction for a $Co_{0.27}$ sample and that the Cu–Ba peaks are only slightly broadened. Together these results indicate that the Ba atoms, as determined from the Cu(2) planes, are well ordered and at their expected positions.

We now consider the Co second neighbor environment. To achieve the large reduction in the observed Co–Ba peak amplitude without any disorder on the Ba sites, requires that some of the Co must be displaced a large distance (≥ 0.3Å) from the normal Cu(1) sites. To motivate and constrain possible models, we make use of the following: a) For all the samples studied, the crystal structure is tetragonal, with some oxygen atoms occupying the O(5) sites between the original Cu(1)–O(1) chains, consistent with the increased O content in the Co doped samples. b) Bordet et. al. [10] observed streaks in their diffraction patterns for low concentration Fe substituted samples, and interpreted this result as evidence for chains of Fe atoms along the <110> direction. Since the Fe and Co data are quite similar, we assume <110> chains for the Co atoms also. c) As discussed above, many of the Co(1)–O bonds are much shorter than the normal Cu(1)–O(1) distance, while a small but significant fraction of the Co–O bonds are much longer. A combination of both long and short Co–O bonds must be found in the Cu(1)–O plane such that the average lattice constant in the plane is essentially unchanged as observed in diffraction. d) We assume for the lower concentration samples that the amount of Co on the Cu(2) sites is negligible. e) Lastly we assume that the Co(1) atoms are not displaced along the c–axis (the peak for the Cu(2)–O(4) bond in the $Co_{0.3}$ sample has the expected position and amplitude).

A model that promotes the tetragonal structure and also generates the <110> chains suggested by the diffraction measurements for the low concentration samples, is a zigzag chain formed of 3–atom segments (Fig. 3a). Alternate Co atoms have a <110> off–center displacement. A more complicated double chain version is shown in Fig.3b. Assuming the c–axis O(4) sites are occupied, these segmented chains have either a 4–fold or 6–fold coordination of O about a Co atom. It is quite likely that vacancies at the O(4), O(1) or O(5) positions would provide a number of 5–fold coordinated Co as well. Zig-zag chain formed from 4–atom segments are also possible.

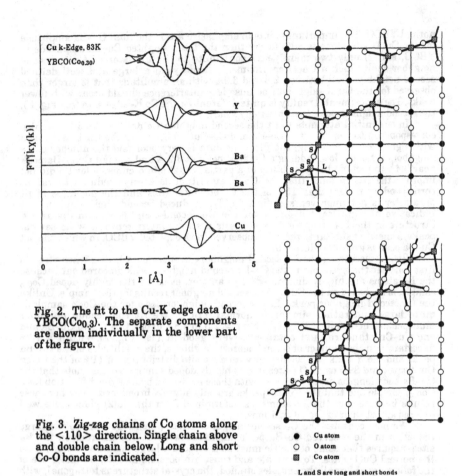

Fig. 2. The fit to the Cu-K edge data for YBCO(Co$_{0.3}$). The separate components are shown individually in the lower part of the figure.

Fig. 3. Zig-zag chains of Co atoms along the <110> direction. Single chain above and double chain below. Long and short Co-O bonds are indicated.

● Cu atom
○ O atom
▨ Co atom

L and S are long and short bonds

The zig-zag chain models result in specific predictions for the multi-peak second neighbor Co-Ba and Co-Cu(Co) environment. Assuming the short Co(1)-O bonds are 1.80 Å, the number of Ba and Co (or Cu) neighbors and the corresponding bond lengths are easily calculated. Several features emerge in these fits. First a reduced amplitude peak, very close to the expected Cu(1)-Ba distance, is always present indicating some undisplaced or <110> displaced Co atoms. <100> displaced atoms would not contribute to this peak. Second, small Co displacements, such that the Co-Ba distances differ by less than 0.2Å, are not consistent with the data; the fits always tend to increase the displacement. The most consistent fits require at least three Co-Ba peaks, at 3.57Å and 3.57 ± 0.32Å.

In Table II, we compare the predictions from the 3-Co-atom single and double chain models (Fig.3) to the corresponding fit results. The predictions of these models are very similar in that the positions of four of the peaks are identical. For the single chain model, in addition to the above constraints on the range of the parameters, we set the amplitudes of the long and short Co-Ba bond peaks equal. For the double chain model, only one Co-Co distance occurs within 4.0Å. In this case we constrained the number of Co-Ba neighbors in the ratio 1:4:1 and kept the number of Co-Co neighbors constant st 2.67 (see Table II). The quality of the fit is good in both cases; 0.3 and 0.6% respectively for the single chain fits to the Co$_{0.033}$ and Co$_{0.067}$ samples and 1.1 and 0.9% respectively for the highly constrained double chain fits. The resulting fit to the data for the double chain model is tabulated in Table II. Clearly a good fit can be obtained that is in reasonable agreement with the predictions. This constrained, double chain model fit, requires a broadening of the Co–Ba central peak ($\Delta\sigma^2$ =.017) and the Co–Co peak ($\Delta\sigma^2$= 0.01). The 4–Co–atom single chain is only slightly better, even though this model has 5 Co–Ba distances. The point here is not that we have found a specific model, but that displacements of the Co atoms can account for both the nearest neighbor O peak — the combination of long and short bonds — as well as the reduced amplitude of the multi–peak, second shell of neighbors for the low Co concentrations. It is quite likely that several types of chains could co–exist, making a unique assignment impossible.

Fe ENVIRONMENT

The EXAFS results for the Fe substituted samples are qualitatively similar to the Co results discussed above. Less data was collected for the Fe substituted samples and only the YBCO(Fe$_{0.10}$) and YBCO(Fe$_{0.17}$) samples were analyzed in detail. For both, the Fe first neighbor peak is well defined while the second neighbor multi–peak has a low amplitude and exhibits structure similar to the Co samples. This again suggests a distorted environment for the Fe atoms in YBCO. The detailed analysis is, however, not as simple. If we assume a substitution on an undistorted Cu(1) chain site we obtain a marginal fit with roughly two neighbors at r=1.83Å and two neighbors at 1.93Å. The major problem is that the phase in the real and imaginary parts of the fit to [FT kχ(k)] does not agree well with the data for either sample. If instead we assume that a distortion exists, similar to that found for the Co substituted samples discussed above, then better fits are obtained — the phase of the fit matches the data very well and for the YBCO(Fe$_{0.10}$) sample the goodness of fit parameter *decreases by a factor of 6*. These fits again require long and short bonds; 3.5 to 4 neighbors at 1.88Å and ~1.5 neighbors at 2.36Å.

CONCLUSIONS

Our detailed EXAFS study of Co–doped YBCO indicates that the Co primarily replaces the Cu(1)atoms, in agreement with earlier investigations. However the near neighbor O peak is composed of ~ 3.5 O atoms at 1.8Å and ~1 O atom at 2.4Å, which means that both long and short bonds must be present within the Cu(1)layer. The Co second neighbor peak is unexpectedly low in amplitude, but has considerable structure that is inconsistent with a simple gaussian broadening of the expected Co–Ba and Co–Cu(Co) bond distances. Measurements of the Cu environment in the highly doped Co$_{0.3}$ sample show a well defined second neighbor peak composed of a (larger than in normal YBCO) Co–Y peak, two Co–Ba peaks of unequal amplitude, plus the Co–Cu(Co) contribution. These results indicate that most of the remaining Cu in the YBCO(Co$_{0.3}$) sample is on the Cu(2) site, and that, viewed from the Cu(2) site, the Y, Ba, and Cu(Co) atoms are at their expected distances. The amplitudes are

Co-X Pair	Predicted Double Chain		Fit Co$_{0.033}$		Fit Co$_{0.067}$	
	N	r	N	r	N	r
	1.33	3.25	1.6	3.26	1.7	3.24
Co(1)-Ba	5.33	3.55	6.7	3.57	6.8	3.55
	1.33	3.89	1.6	3.87	1.7	3.84
Co(1)-Co(1)	2.67	3.55	2.67	3.52	2.67	3.56
Co(1)-Cu(1)	-	-	-	-	-	-

Table II. The second neighbor environment for the double chain shown in Fig. 3 for bond lengths between 3.0 and 4.0Å. N is the weighted number of neighbors. Fits to the low concentration data using the predicted values as starting parameters are shown.

consistent with a small amount of Co (~11%) on the Cu(2) site in this highly doped sample. The most important point of this analysis is that the Ba atom positions are *not* strongly *disordered*, and therefore cannot account for the small second neighbor amplitude observed for the Co K-edge data. Therefore, some of the Co atoms must be *displaced considerably* from the normal Cu(1) site, resulting in several different Co–Ba distances. The interference of these Co–Ba peaks produces the small overall second neighbor Co–peak amplitude. The analysis of the Fe EXAFS data suggests that Cu(1) sites occupied by Fe atoms are similarly distorted.

We propose a simple local structure model that can account for the different Co–O bond lengths in the Cu(1) plane, provides a good fit to the Co second neighbor peak for the low concentration data, and suggests an explanation for the apparent tetragonal structure of the Co doped samples. Zigzag chains of three–Co–atom segments can be formed along the <110> direction, that 1) have Co–Co distances that are shorter than the usual Cu–Cu planer distance, 2) provide a combination of long and short Co–O bonds as are observed in the first neighbor peak, 3) provide a good fit and a simple explanation for the reduced amplitude of the second neighbor Co multi–peak structure, and 4) fit easily into the square lattice obtained using the lattice constant from diffraction experiments. In this model, many of the O(5) sites near a Co atom are occupied; linear Cu–O chains can then leave the zigzag chain in either the x– or the y– direction. We think this promotes twinning on a microscopic scale and thus leads to the observed tetragonal structure.

Acknowledgment

The experiments were performed at SSRL, which is supported by the U. S. Department of Energy, Office of Basic Sciences, and the National Institutes of Health, Biotechnology Resource Division. We thank R. Howland for assistance in the early stages of the analysis. This research was supported in part by NSF Grant No. DMR 85–05549, AFOSR Grant No. F49620–82CW014 and the Swedish National Science Research Council.

References

1. See, for example, *High–Temperature Superconductors*, M Brodsky, R. Dynes, K. Kitazawa, and H. Tuller, eds. (Materials Research Society **99**, Pitt, 1988).
2. T. Siegrist, S. Sunshine, D. W. Murphy, R. J. Cava, and S. M. Zahurak, Phys. Rev. B **35**, 7137(1987); M. A. Beno, L. Soderholm, D. W. Capone II, D. G. Hinks, J. D. Jorgensen, J. D. Grace, I. K. Shuller, C. U. Segre, and K. Zhang, Appl. Phys. Lett. **51**, 57 (1987).
3. J. D. Jorgensen, M. A. Beno, D. G. Hinks, L. Soderholm, K. J. Volin, R. L. Hitterman, J. D. Grace, I. K. Shuller, C. U. Segre, and K. Zhang, and M. S. Kleefisch, Phys. Rev. B **36**, 3608 (1987).
4. Y. Maeno, T. Tomita, M. Kyogoku, S. Awaji, Y. Aoki, K. Hoshino, A. Minami, and T Fujita, Nature **328**, 512 (1987).
5. I.Sankawa, M. Sato and T. Konaka, Jap. J. Appl. Phys. **27**, L28 (1988).
6. J. M. Tarascon, P. Barboux, P. F. Miceli, L. H. Greene, G. W. Hull, M. Eibschutz and S. A. Sunshine, Phys. Rev B **37**, 7458 (1988).
7. M.Eibschutz and M. E. Lines J. M. Tarascon, and P. Barboux, Phys. Rev. B **36**, 2896 (1988); I. Nowik, M. Kowitt, I. Felner, and E. R. Bauminger, Phys. Rev. B **38**, 6677 (1988); C. Blue, K. Elgaid, I. Zitkovsky, and P. Boolchand, D. McDaniel, W. C. H. Joiner and J. Oostens, and W. Huff, Phys. Rev. B **37**, 5905, (1988).
8. P. F. Miceli, J. M. Tarascon, L. H. Greene, P. Barboux, F. J. Rotella and J. D. Jorgensen, Phys Rev. B **37**, 5932 (1988); P. Zolliker, D. E. Cox, J. M. Tranquada, and G. Shirane, Phys. Rev. B **38**, 6575 (1988).
9. J. B. Boyce, F. Bridges, T. Claeson, and M. Nygren (preprint).
10. P. Bordet, J. L. Hodeau, P. Strobel, M. Marezio, and A. Santoro, Solid State Comm. **66**, 435 (1988).

SOME INDUSTRIAL APPLICATIONS OF GRAZING INCIDENT
EXAFS TECHNIQUES

I. S. DRING, R. J. OLDMAN, A. STOCKS*, D. J. WALBRIDGE**, N. FALLA***,
K. J. ROBERTS and S. PIZZINI****.
*ICI Chemicals and Polymers Ltd., Runcorn, Cheshire, U.K.
**ICI Paints Division, Slough, U.K.
***The Paint Research Association, Middlesex, U.K.
****The University of Strathclyde, Department of Pure and Applied Chemistry,
Glasgow, U.K.

ABSTRACT

An instrument to undertake grazing incident EXAFS has been developed.
This is discussed along with some applications of its use on industrially
important systems.

INTRODUCTION

EXAFS has become an established technique for the study of industrially
important systems[1]. Its applicability has been enhanced by the development
of its grazing incident variant. This has been applied to the study of
surfaces[2], buried interfaces[3] and thin films[4].

When a monochromatic x-ray beam is incident to a surface below the
critical angle total external reflection occurs[5,6]. Under such conditions
for many materials, the penetration of the incident beam into the surface is
in the region of 20-30Å. Surface specific EXAFS is present both in the
reflected beam (REFLEXAFS) and in the fluorescence radiation[7]. A major
advantage over other surface sensitive probes such as LEED, ESCA, SIMS and
SEXAFS, is that is not restricted to investigations of materials under
vacuum.

Instrumentation for total external reflection measurements can also be
employed at low angles of incidence above the critical angle to increase the
applicability of standard fluorescence EXAFS (FLEXAFS). FLEXAFS is now
recognised as the most appropriate method for measurements on dilute
species[8]. The standard geometry has an incident angle of 45°. This is not
suitable for the study of thin films as the small volume swept out by the
beam within the film precludes good S/N statistics. Counting statistics can
be considerably improved using grazing angles of incidence. Above the
critical angle there is an increase of the penetration depth with angle.
Therefore by controlling the angle of incidence, depth specific information
can be obtained.

INSTRUMENTATION

The instrument used for grazing incident studies is described in detail
elsewhere[9], along with the procedure adopted for its alignment. It is
depicted in figure I. Briefly it consists of a sample holder mounted on a
two circle goniometer, which defines the incident angle and levels the
sample. The beam is collimated by two Slits S_1 and S_2 (typical slit widths
are 50 μm). During reflectivity measurements S_2 acts as a knife edge to
block any unreflected light. There are four ion chamber detectors on the
rig: a reference (Io) and reflectivity (Ir) chamber before and after the
sample holder, a large area fluorescence chamber (If) vertically above the
sample and a monitor chamber (Im) immediately behind the reflectivity
chamber. The insertion of foils between the latter two chambers allows
determination of edge shifts. The measurements described here were taken at

station 9.2 on the wiggler beam line at the Daresbury SRS.

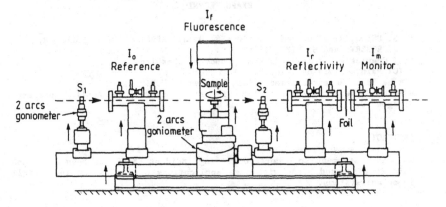

FIGURE I   Grazing Incident Instrumentation

REFLEXAFS OF COPPER SURFACES

An area where grazing incident EXAFS techniques can be usefully employed is in the study of corrosion inhibitors. A system that we are actively studying is the protection of copper surfaces by benzotriazole. The first stage in this work has been to characterise air exposed copper surfaces. Two films were prepared by vacuum evaporating CU onto glass substrates to produce a copper layer approximately 1000Å thick. One was left to develop a 'natural' oxide coating in the lab, whilst the degree of surface oxidation on the other was increased by heating it to 150°C. The angle at which EXAFS data was collected was defined from the reflectivity curves shown in figure II. The steep fall in the reflectivity at Φ_c and the well resolved fringe pattern for the natural oxide surface is indicative of a smooth homogeneous surface. The fringes are due to interference between light reflected from the CU/air interface and the CU/glass interface. The depth of the film can be calculated from their frequency[10]. This has been done and a value of 1095Å obtained. The shape of the film oxidised at 150°C can be explained due to the presence of a thick oxide layer. There are two critical angles one for the air/oxide interface and the other for the oxide/CU interface, a third critical angle for the CU/glass interface is not present because the critical angle for glass is below copper. The shallow drop off in reflectivity and lack of fringes can be explained by interfacial roughness. Any oxide present on the air exposed film at R.T. is probably not discernable in the reflectivity curve because it is very thin.

An XAS spectra for the air exposed surface at R.T. was recorded at the glancing angle Φ_1 in figure II. The XANES of this spectra is compared with CU_2O, CUO and CU in figure III. The spectra is typical of a macroscopic mixture of CU and CU_2O. At $\Phi/\Phi_c \ll 1$ the absorption coefficient can be deconvoluted from the reflectivity[11]. Least-square fits were obtained with the curved wave program EXCURV[12] using central atom and backscattering phaseshifts refined from CU and CU_2O reference compounds. The fits are shown in figure IV and the refined structural parameters for the first two shells summarised in table I. The Fourier transformed data is dominated by the first shell in FCC CU at 2.56Å (feature A). In the surface data this shell is reduced in amplitude by about half its bulk value. This is due to

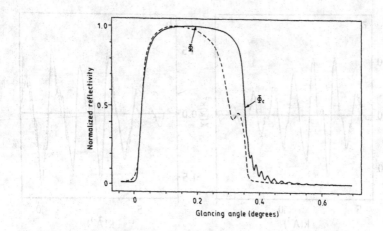

FIGURE II Reflectivity Profiles for air exposed copper film (————) and oxidised film at 150°C (– – – –)

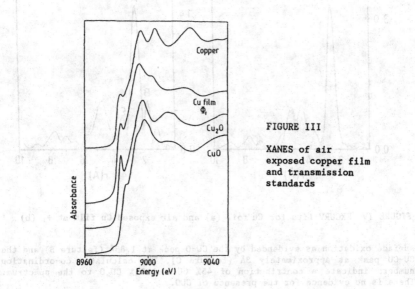

FIGURE III

XANES of air exposed copper film and transmission standards

TABLE I: Coordination numbers (N) and distances d(Å) for shells A and B obtained for the air exposed copper film compared with standard compounds.

	Cu		Cu₂O		CuO		Cu Film	
	N	d(Å)	N	d(Å)	N	d(Å)	N	d(Å)
			2	1.85	4	1.95	1.2	1.85
	12	2.55					5.6	2.55

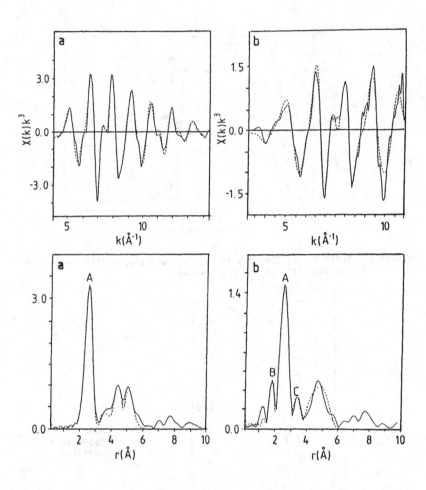

FIGURE IV   EXCURV fits for Cu foil (a) and air exposed Cu film at Φ_1 (b)

surface oxidation as evidenced by the CU-O peak at 1.85Å (feature B) and the CU-CU peak at approximately 3Å (feature C). The calculated co-ordination numbers indicate a contribution of 45% CU and 55% CU_2O to the spectrum. There is no evidence for the presence of CUO.

FLEXAFS OF THIN PAINT FILMS

Metallic soaps are added to paint resins to improve through film drying properties. The ability to carry out fluorescence measurements at glancing angles has made possible an EXAFS study of a zirconium drier (0.5 wt% Zr) in a 75 μm thick polyalkyde resin film on a glass substrate. The maximum in the fluorescence yield at approximately 2° was determined by scanning the incident angle. The penetration depth at this angle is that for a bulk measurement. EXAFS were obtained for wet and dry films. The transforms of these are compared in figure V to zirconium acetate taken in transmission.

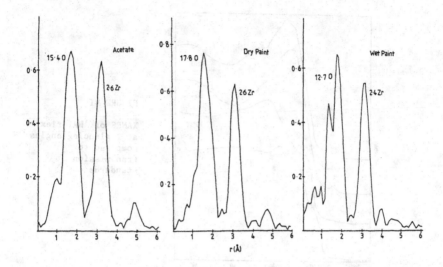

FIGURE V   Fourier transformed EXAFS of a zirconium acetate transmission
standard and zirconium dried paints. The co-ordination numbers
were derived in an EXCURV analysis

The co-ordination numbers in figure were obtained from an EXCURV
analysis with phase shifts refined from Zr foil, ZrO_2 and $Zr(CH_3COOH)_4$.
Zirconium co-ordination shells at 3.2Å and possibly 4.9Å are present in all
the transforms. An even dispersion of metallic drier in the paint would give
a zirconium-zirconium nearest neighbour distance of approximately 15Å. The
results are therefore indicative of extensive zirconium clustering in the
paint films in which the zirconium has a similar environment to that in the
acetate. The increase in both the oxygen co-ordination number and the
feature at 4.9Å suggests that the clusters become more extensive and more
ordered in the dry paint.

DEPTH SPECIFIC FLEXAFS OF 'HALOFLEX' COATINGS

'Haloflex'[13] is an ICI water based vinyl chloride/ vinylidene
dichloride polymer emulsion which has impressive properties as an
anti-corrosive coating. When applied to mild steel surfaces, microscopy
reveals that there is some leaching of substrate material into the polymer.
X-ray diffraction indicates the presence of various crystalline oxide phases
as well as a large amount of amorphous material. By carrying out a
fluorescence experiment at glancing angles the measurement can be made
specific to the polymer region and the contribution of a large substrate
signal avoided. Xanes spectra taken from a 2 μm thick 'Haloflex' coating on
mild steel at two glancing angles are shown in figure VI. They are compared
with an Fe and Fe_2O_3 spectrum taken in transmission.

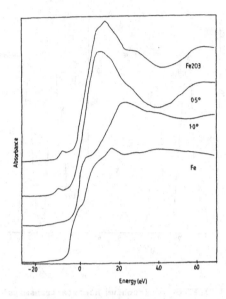

FIGURE VI

XANES of 'Haloflex' at 2 glancing angles compared to transmission standards

At 1° it can be seen that there is a large contribution to the signal from the iron substrate, which is not present at 0.5°. The xanes spectrum taken at the lower angle has strong similarities to an Fe_2O_3 spectrum, indicating that the amorphous material is 'Fe_2O_3 like'. This is supported by an EXCURV analysis which resulted in first shell parameters similar to that of the Fe_2O_3 standard.

REFERENCES

1. R. J. Oldman, J. de Phys. C8 47, 321 (1986).

2. N. T. Barrett, P. N. Gibson, G. N. Greaves, P. Mackle, K. J. Roberts, M. Sacchi, XAFS V 5th International Conference on X-ray Absorption Fine Structure, Seattle WA. (1988).

3. S. M. Heald, J. de Phys. C8 47 825 (1986).

4   T. W. Barbee Jr., J. Wong, XAFSV 5th International Conference on X-ray Absorption Fine Structure, Seattle WA (1988).

5. L. G. Parratt, Phys. Rev. 95, 359 (1954).

6. G. H. Vineyeard, Phys. Rev. B 26, 4146 (1982).

7. S. Affrossman, S. Doyle, G. M. Lamble, M. A. Morris, K. J. Roberts, D. B. Sheen, J. N. Sherwood, R. J. Oldman, D. Hall, R. J. Davey, G. N. Greaves, J de Phys. C8 47, 167 (1986).

8. S. M. Heald, in Cemical Analysis V.92 X-ray Absorption, edited by D. C. Koningsberger and R. Prins (John Wiley & Sons, New York 1988) pp 87-118.

9. G. N. Greaves, N. Harris, P. Moore, R. J. Oldman, E. Pantos, S. Pizzini, K. J. Roberts, 3rd International Conference on Synchrotron Radiation Instrumentation, Tsukuta Japan (1988).

10. A. Segmuller, Thin Solid Films 18, 287 (1973).

11. R. Fox, S. J. Gurman, J. Phys. C. 13, L249 (1980).

12. S. J. Gurman, N. Binstead, I. Ross, J. Phys. C. 17, 143 (1988).

13. R. G. Humphries, J. O. C. C. A. 1987 (6), 150.

PART IV

Crystallography and Small-Angle Scattering

ANOMALOUS SCATTERING OF POLARIZED X-RAYS

DAVID H. TEMPLETON AND LIESELOTTE K. TEMPLETON
University of California, Department of Chemistry, Berkeley, CA 94720

ABSTRACT

Some elements in some chemical states exhibit strong dichroism and birefringence near x-ray absorption edges. The atomic scattering factor is a complex tensor. This polarization anisotropy has profound effects on the transmission and scattering of x-rays even when the incident radiation is unpolarized. The linear polarization of synchrotron radiation makes it easier to study the effects and to use them for new methods of structure determination. Several of these anomalous scattering tensors have been measured by absorption spectroscopy and in diffraction experiments. New polarization terms enter the calculation of diffraction intensities, with interesting consequences. Reflections forbidden by a screw-axis rule are observed in sodium bromate near the Br K edge and permit direct observation of the structure factor phases of their second order reflections. This technique is a method of selective diffraction in which atoms of single element in a single chemical state contribute to the signal, and it can reveal their positions with precision. These effects can be a handicap for some applications of near-edge anomalous scattering in the study structures of crystals and amorphous materials.

INTRODUCTION

Scattering of x-rays is a powerful tool for studying the structures of materials at the atomic level. Some of the techniques make use of resonance effects, called anomalous scattering, which are most pronounced for photon energies close to electronic binding energies. These effects change both the magnitude and the phase of the x-ray wave scattered by the particular kind of atom, and thus can be used to isolate its contribution to a diffraction pattern. One reason for working with synchrotron radiation is that it gives a free choice of wavelengths for using these effects to best advantage. Synchrotron radiation is also strongly polarized. The subject of this paper is some interesting x-ray optical effects related to polarization which offer new ways to study structure and which can make traditional methods more complicated.

Calculations for x-ray scattering experiments usually contain the atomic scattering factor f:

$$f = f_0 + f' + if''$$

(1)

which is the amplitude of the wave scattered by a single atom. Here f_0 is the Fourier transform of the atomic electron density; it would be the only term if all electrons scattered like free electrons. The anomalous scattering terms f' and f'' include the effects of electronic binding. They are also called dispersion corrections because they describe how the complex index of refraction changes with wavelength. Complex numbers are used so that both the magnitude and phase are represented. The scattering factor, the refractive index, the absorption coefficient, and the dielectric constant are all related to each other for x-rays, as for light of any other wavelength. The fact that wavelength is similar to distances between atoms introduces some features which are absent from the familiar phenomena of crystal optics for visible light.

The scattering always depends on the polarization direction of the photon. In the traditional theory [1], where the dielectric constant is assumed to be isotropic for x-ray frequencies, the effect of polarization is the same for all atoms and can be allowed for as a simple correction factor which depends on the Bragg angle θ and the extent of polarization of the incident beam. However, when the dielectric constant is different for different directions of the electric vector, things are much more complicated. The optical properties, including the atomic scattering factor, need to be represented by tensors rather than as scalar quantities. Polarization effects are not the same for different components, and they can not be factored out of the intensity equation.

Anomalous scattering terms are often derived from absorption spectra. The imaginary term f'' at wavelength λ or angular frequency $\omega = 2\pi c/\lambda$ is proportional to the absorption cross section σ at that frequency [1]:

$$f''(\omega) = (mc\omega/4\pi e^2)\sigma(\omega). \qquad (2)$$

If f'' is known as a function of wavelength, f' can be calculated by a method called Kramers-Kronig inversion [2]:

$$f'(\omega) = \frac{2}{\pi} \int_0^\infty \frac{\omega' f''(\omega')}{\omega^2 - \omega'^2} \, d\omega' \qquad (3)$$

Because of these relationships, when f has tensor character the absorption also is different for different directions of polarization. This phenomenon is called dichroism when there are two independent spectra, and pleochroism for the most general case with three independent spectra.

X-RAY DICHROISM AND PLEOCHROISM

Dichroism has been observed for x-rays near absorption edges in $KClO_3$ [3], WSe_2 [4], Br_2 [5], vanadyl bisacetylacetonate [6], and many other materials where the electronic environment of the absorbing atom has low

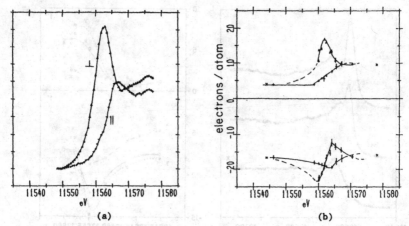

Fig. 1. (a) Absorption spectra of K_2PtCl_4 single crystal at Pt L_3 edge for two directions of polarization. (b) Principal values of anomalous scattering terms measured in diffraction experiments (f'' above, f' below) [7].

enough symmetry. The effects are particularly strong in the tetragonal crystals of K_2PtCl_4 near the L_3 absorption edge of platinum (Fig. 1) where they correspond to as much as 10 electrons/atom change in f'' with direction of polarization [7]. Here each Pt atom is at the center of a square-planar $PtCl_4^{2-}$ complex ion, and all of these ions have the same orientation, perpendicular to the crystallographic c axis. In this symmetry the two independent spectra are found with polarization parallel with or perpendicular to the c axis. Other directions of polarization give linear combinations of these two principal spectra.

Strong pleochroism was measured [8] near the selenium K edge in monoclinic crystals of a hydrate of selenolanthionine [9]:

$$\begin{array}{c} NH_3^+ \\ | \\ CH_2-CH-COO^- \\ Se \\ CH_2-CH-COO^- \\ | \\ NH_3^+ \end{array}$$

Here the polarization directions for principal spectra (Fig. 2) follow the C_{2v} local symmetry of the bonds to selenium. The spectrum for polarization perpendicular to both bonds lacks the resonance line at 12658 keV, as predicted by a simple molecular-orbital model. The orientation of the molecules in this specimen was determined with a precision of 3° by a search for minimum absorption at this wavelength.

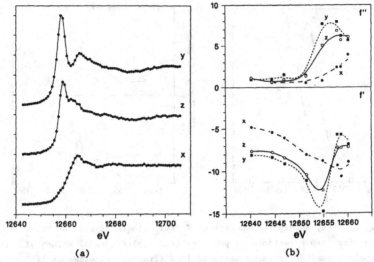

(a)   (b)

Fig. 2. (a) Absorption spectra of selenolanthionine hydrate at Se
K edge for three directions of polarization. (b) Principal values
of f'' and f' from diffraction intensities for the anhydrous
crystals [8].

POLARIZATION ANISOTROPY OF ANOMALOUS SCATTERING

When birefringence exists, f becomes a complex tensor, and several
combinations of linear polarization components need to be considered [10].
Polarization directions are unit vectors $\vec{s}$ and $\vec{p}$ for incident radiation and $\vec{s}$
and $\vec{p}'$ for scattered radiation. In matrix notation the f-tensor is a 3×3
symmetric matrix $\underline{f}$ with complex elements, and polarization directions may be
indicated by column vectors. Then the four atomic scattering factors are:

$$f_{ss} = \vec{s}^T \underline{f} \vec{s}, \qquad f_{sp'} = \vec{s}^T \underline{f} \vec{p}', \qquad f_{pp'} = \vec{p}^T \underline{f} \vec{p}', \qquad f_{ps} = \vec{p}^T \underline{f} \vec{s}, \qquad (4)$$

where T indicates transpose. The high degree of linear polarization that is
typical of synchrotron radiation beams (90 to 95%) permits two of these to be
neglected or treated as minor corrections.

The effects described by equations (4) can cause substantial changes in
diffraction intensities. This has been demonstrated by measuring these
intensities near absorption edges with crystals whose structures are known
from determinations at wavelengths where the birefringence is negligible.
The simplest method [7] is applicable to crystals like K_2PtCl_4 in which there
is a single orientation of the birefringent molecules. If the incident
radiation is considered to be perfectly polarized, an orientation of the

crystal can be found for some Bragg reflections such that only one principal value of $\underline{f}$ enters the intensity calculation. Then this principal value can be derived from measured intensities by least-squares methods the same as those used previously for cases where f is a scalar [11]. The values derived for platinum in this way are shown in Fig. 1(b). A comparison with Fig. 1(a) shows how the principal values of f'' are similar to the principal absorption spectra. Several experiments like this have indicated that eqs. (2) and (3) are valid for the tensor components as well as for the scalar quantities usually considered.

In many crystal structures there are several orientations of equivalent molecules, and these polarization phenomena are more complicated. Absorption cross section and index of refraction remain single-valued functions of polarization direction, and each may be represented by a single 3×3 matrix, just as for visible light. However, each molecular orientation for a birefringent atom requires its own matrix, and atoms which are equivalent in the usual crystallographic sense may have different atomic scattering amplitudes. Furthermore, the absorption coefficient for the scattered beam may not be the same as for the incident beam. This last complication is absent in cubic crystals, and may be minor in other symmetries if there is a suitable distribution of molecular orientations.

Anhydrous crystals of selenolanthionine, which are tetragonal, contain four orientations of the molecules such that there is relatively little anisotropy observed in absorption spectra. The anisotropy of the anomalous scattering, however, is quite evident in the diffraction intensities, yielding the principal values of the $\underline{f}$ tensor shown in Fig. 2(b) [8]. As in the absorption spectra, Fig. 2(a), there is clear evidence that the tensor is biaxial (has three independent principal values), and there is good correspondence of f'' with σ. The relation of these principal values to the molecular geometry is the same in both kinds of crystals.

PARTIAL RADIAL DISTRIBUTION FUNCTIONS

These polarization effects may also be present when anomalous scattering is used to study amorphous materials and can introduce a complication. A radial pair distribution $\rho(R)$ can be derived by Fourier inversion of intensity as a function of scattering angle. If the material contains two kinds of atoms, A and B, ρ is the sum of partial distributions ρ_{AA}, ρ_{AB}, and ρ_{BB}, weighted according to products of atomic scattering factors. A change in anomalous scattering, particularly in f', makes these weights different at another wavelength. Data measured at multiple wavelengths can be used in various procedures to determine these partial radial functions [12]. To get big changes in f' one must work close to an absorption edge, just where polarization anisotropy is most likely.

The usual analysis is based on the integration over a sphere of solid angle for directions of interatomic vectors, assuming that atomic scattering factors are equal for all directions. If f for the jth atom has tensor character, the orientation of that tensor will be correlated with the direction of whichever neighbor is most responsible for the electronic anisotropy. The integral over the sphere will contain additional terms. If the Fourier transformation is carried out in the usual way, fictitious distances are added to or subtracted from the distribution. The fact that the material is isotropic on a macroscopic scale does not cause these effects to cancel, nor does the use of unpolarized incident radiation. Further work is needed to show how serious this complication is in practical experiments.

THE CRYSTALLOGRAPHIC PHASE PROBLEM

The description of electron density in a crystal structure by a Fourier series requires the magnitudes and phases of structure factors. The magnitudes are easier to measure than the phases. For complicated structures the "phase problem" may be very difficult. The application of synchrotron radiation to this problem has been directed largely toward macromolecular crystals, but the same principles are valid for any material.

Most of the macromolecular crystal structures determined to date were solved by the method of isomorphous replacement. If a heavy atom can be added to the crystal in a specific site, with little effect on the locations of other atoms, the changes of Bragg intensities contain information both about phases and this site. A second source of this information, if the anomalous scattering f'' is large enough, is the intensity difference between a Bijvoet pair such as hkℓ and -h,-k,-ℓ. If several wavelengths are used, particularly those near absorption edges which are accessible with synchrotron radiation, the changes in both f' and f'' can modulate the intensities in a manner like that of isomorphous substitution. Several protein crystal structures have been solved recently using this multiple-wavelength method for phase determination [13].

The polarization anisotropy adds another dimension to these phasing methods. The same Bragg reflection can be measured at the same wavelength but at different azimuthal settings, thus changing the orientation of the f-tensor with respect to polarization direction. Then the changes in intensities are caused by atoms of a specific element in a specific chemical state. The combination of the anisotropy effects with the scattering from other atoms in the diagonal components (ss and pp') is similar to that in the multiple-wavelength method. The cross terms (sp and ps) arise only from the anisotropic atoms. When they can be isolated, they constitute a diffraction pattern for the simpler structure consisting only of those atoms. They can be observed directly, without any need for polarized radiation or polarization filters, in the case of some "forbidden reflections" to which normal atoms do not contribute. Dmitrienko has tabulated the peculiar polarization properties of these special reflections in many space groups [14].

DIRECT OBSERVATION OF STRUCTURE-FACTOR PHASES

Sometimes the polarization-dependent anomalous scattering gives phase information in a very direct way. Sodium bromate is cubic, with space group $P2_13$, with Br atoms in positions x,x,x. According to the usual rules for systematic absences, reflections of the type 00ℓ (ℓ odd) have zero intensity, but these rules assume that f is a scalar. Near the Br K edge, f_{Br} in the bromate ion depends strongly on polarization [15]. In general f is different for bromine atoms related by a screw axis, and the sp′ and ps components of scattering for these forbidden reflections can be observed with intensities proportional to [16]:

$$|F|^2 = (8/9)|f_\sigma - f_\pi|^2 \cos^2\theta \ (1 - \cos 4\pi\ell x \cos 2\psi), \quad\quad (5)$$

where f_σ and f_π are principal values of f′ + if′′, and ψ is an azimuthal angle for the orientation of the specimen. According to this equation, the intensity varies as ψ is changed (Fig. 3). This modulation gives both the magnitude and phase of the factor $\cos 4\pi\ell x$; it is positive or negative according to the position of the maximum intensity at $\psi = 90°$ or $0°$. The contribution of bromine to the structure factor for the $0,0,2\ell$ reflection is

$$F_{Br}(0,0,2\ell) = 4f_0(Br) \cos 4\pi\ell x, \quad\quad (6)$$

which has the same sign as the factor $\cos 4\pi\ell x$ because $f_0(Br)$ is a positive real number. That is, the azimuthal variation of the intensity of the forbidden reflection gives the structure-factor phase of its second order. Data like these yielded the phases: +, -, +, -, -, + for 00ℓ, $\ell = 2, 6, 10, 14,$ 18, 22.

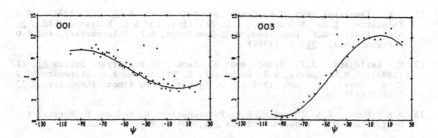

Fig. 3. Plots of intensity of forbidden reflections vs azimuth of polarization vector. The positions of the maxima and minima for 001 and 003 show the phases of 002 and 006 to be 0° and 180° [16].

CONCLUSIONS

The polarization-anisotropy of anomalous scattering can have large effects in experiments with x-rays near absorption edges. The tensor character of the optical properties adds complications to calculations of intensities of scattered radiation. These effects offer new ways to study the structures of materials. They may need attention if error is to be avoided in traditional methods.

ACKNOWLEDGEMENT

Preparation of this report was supported by National Science Foundation Grant CHE-8515298.

REFERENCES

1. R.W. James, The Optical Principles of the Diffraction of X-rays, (Ox Bow Press, Woodbridge, 1982).

2. H. Wagenfeld, in Anomalous Scattering, edited by S. Ramaseshan and S.C. Abrahams (Munksgaard, Copenhagen, 1975), p. 13-24.

3. H.W. Schnopper, in Roentgenspektren Chem. Bildung. Vortr. Int. Symp. (Leipzig, 1965), p. 303 [Chem. Abstr. 69, 6410e (1968)].

4. S.M. Heald and E.A. Stern, Phys. Rev. B 16, 5549 (1977).

5. S.M. Heald and E.A. Stern, Phys. Rev. B 17, 4069 (1978).

6. D.H. Templeton and L.K. Templeton, Acta Cryst. A 36, 237 (1980).

7. D.H. Templeton and L.K. Templeton, Acta Cryst. A 41, 365 (1985).

8. L.K. Templeton and D.H. Templeton, Acta Cryst. A (in press).

9. G. Zdansky, Arkiv Kemi 29, 443 (1968); W.A. Hendrickson, Trans. Amer. Cryst. Assoc. 21, 11 (1985).

10. D.H. Templeton and L.K. Templeton, Acta Cryst. A 38, 62 (1982).

11. L.K. Templeton and D.H. Templeton, Acta Cryst. A 34, 368 (1978); D.H. Templeton, L.K. Templeton, J.C. Phillips, and K.O. Hodgson, ibid., 36, 436 (1980); L.K. Templeton, D.H. Templeton, R.P. Phizackerley, and K.O. Hodgson, ibid., 38, 74 (1982).

12. M. Laridjani, J.F. Sadoc and D. Raoux, J. Non-Cryst. Solids 91, 217 (1987); K.F. Ludwig, W.K. Warburton, L. Wilson and A.I. Bienenstock, J. Chem. Phys. 87, 604 (1987); O. Lyon and J.P. Simon, Phys. Rev. B 35, 5164 (1987).

13. J.M. Guss, E.A. Merritt, R.P. Phizackerley, B. Hedman, M. Murata, K.O. Hodgson and H.C. Freeman, Science 241, 806 (1988).

14. V.E. Dmitrienko, Acta Cryst. A 39, 29 (1983); 40, 89 (1984).

15. D.H. Templeton and L.K. Templeton, Acta Cryst. A 41, 133 (1985).

16. D.H. Templeton and L.K. Templeton, Acta Cryst. A 42, 478 (1986); 43, 573 (1987).

POLYCRYSTALLINE DIFFRACTION AND SYNCHROTRON RADIATION

MICHAEL HART* AND WILLIAM PARRISH**

* Department of Physics, Schuster Laboratory, The University, Manchester, M13 9PL, U.K.

** IBM Research, Almaden Research Center, 650, Harry Road, San Jose, California 95120-6099, U.S.A.

ABSTRACT

Synchrotron radiation sources allow a new generation of parallel beam instruments to be developed for polycrystalline xray diffraction. Spectroscopy can be combined with diffraction in either angle or energy scanning modes. The high intensity narrow instrument profile and freedom from aberrations make possible a wide range of measurements; of anomalous dispersion, of peak–shape analysis to determine particle size and strain, of preferred orientation and structure as a function of depth and of lattice parameters with a precision approaching 1 part per million. The parallel beam instrument which we have developed is now to be installed on a regular basis at SSRL and at Daresbury.

INTRODUCTION

A very large fraction of the polycrystalline diffraction research done in the last few years has concentrated upon crystal structure solution and refinement wherein the high quality of the instrument function is exceptionally important. Recent work has been summarised in two conference reports[1,2]. In this article we will concentrate on aspects of synchrotron radiation parallel beam polycrystalline diffraction other than structural studies. Three attributes of the parallel beam arrangement are important; the ease with which either angle or energy scanning can be selected, the complete freedom of choice so far as specimen geometry is concerned and the fact that the instrumental resolution function is independent of xray wavelength and quite symmetrical.

DIFFRACTOMETER DESIGN

The main features of diffractometer design are illustrated in Figure 1 in which all three arrangements have been used by us and others at synchrotron radiation sources around the world. We have concentrated on the scheme shown in Figure 1(a) because it is the most versatile and provides higher intensities than the other arrangements. Two diffractometers are necessary; one provides control of the monochromator angle (D1) while the second (D2) determines the sample angle θ, and the detector angle 2θ

The Monochromator

The energy resolution is set at not better than 133 ppm[3] by the choice of a doubly reflecting channel–cut silicon 111 crystal CM as the monochromator. We have found no need for higher resolution but, if the need arose, higher order reflections could be used. In practice, on bending magnet beamlines, the monochromator is quite stable and requires no cooling. Combined with a scintillation detector SC2 and a pulse height analy-

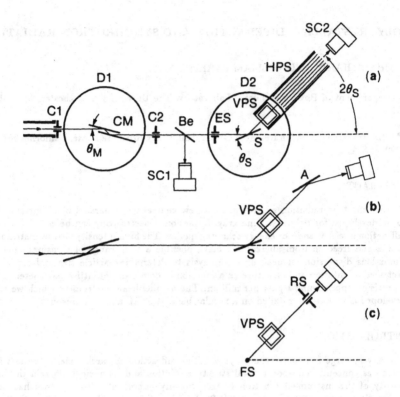

Figure 1: Experimental arrangements for parallel beam polycrystalline diffraction

ser there is no detectable harmonic contamination in the diffracted beam. For high power applications or where extreme stability is essential we have developed an internally water-cooled offset harmonic–free channel–cut monochromator. The xray optical relationship between monochromator CM and sample S can be dispersive as in Figure 1(a) or non-dispersive as in Figure 1(b). In principle, the narrowest resolution function is obtained in the non–dispersive position when the monochromator and sample Bragg angles (θ_M and θ_s respectively) are the same. This "focussing" is important in neutron powder diffraction but in the xray case, as systematic measurements show[4,5], there is little practical advantage to be obtained.

Sample two–theta resolution

We sought to maximise the intensity–resolution product by using a (large volume) flat sample geometry but with a long set of parallel slits HPS which collimate the diffracted beam to a full width at half maximum of 0.05° 2θ. In recent experiments at Daresbury[5] we have found that the resulting powder diffraction peaks are symmetric to the limit set by counting statistics and by the specification of the diffractometer two–theta angle transducer (which is $\frac{1}{3}$sec.arc or 0.0001°). We know of no powder samples which require superior resolution and have therefore not sacrificed intensity to gain resolution beyond 0.05°. Higher resolution can be obtained with an analyser crystal A (Figure 1(b)) or with a standard receiving slit RS (Figure 1(c)). In the latter case specimen displacement and transparency errors arise, even with a fibre specimen FS.

Beam Definition

The physical dimensions of the beam striking the sample and zero angle of the diffractometer are set by the entrance slit ES and not by the secondary shielding apertures C1 and C2. Axial divergence control is required, as in laboratory methods, and that is achieved with the vertical parallel Soller slits VPS. Primary beam intensity is monitored by SC1 which receives scattered monochromatic radiation from a kapton or beryllium foil Be.

OPERATING MODES

The key advantage of the setup in Figure 1(a) stems from the fact that the monochromator CM can select different energies with no change in the angle of the primary xray beam while the angle transducer HPS is achromatic. Thus two parameters in the diffraction equation (1) can be controlled separately or together

$$2d \, \sin\theta = \lambda = hc/\mathcal{E} \qquad (1)$$

At a fixed detector angle powder patterns can be obtained in the energy dispersive mode by scanning the monochromator angle θ_M. At any chosen energy powder patterns can be obtained by scanning 2θ and the specimen angle is a free parameter, unlike the situation in laboratory experiments when 2θ and θ_s must be coupled together. It is important to realise that the two modes can be chosen under software control since no reconfiguration of the diffractometers is required; 2θ, $\mathcal{E}$ and θ_s are truly independent variables, as the following examples illustrate.

Angle Dispersive Diffraction

Figure 2 shows a set of theta:two–theta powder patterns from a mixture of three rare earth oxides on the left hand side, and from the individual pure oxides on the right. These are chosen to illustrate several features of the technique which permits independent control of $\mathcal{E}$ and 2θ. In many materials fluorescence background can be a problem; the laboratory example of ferrous materials studied with copper $K\alpha$ radiation is well known. Each of the pure oxide patterns was obtained with an xray energy just lower than that of the rare earth's L_{III} absorption edge. For example, the Dy L_{III} edge is at 1.5916Å and the pattern was taken 20eV below the edge energy at 1.5957Å wavelength, which provides the highest signal to background. The triple mixture is chosen to illustrate in the left hand half of the figure how anomalous dispersion can be used to separate the patterns. The three 440 peaks, for example, consist of an unresolved doublet and a singlet. Dy_2O_3 has the largest and Yb_2O_3 the smallest unit cell. In the left hand graphs we also took patterns 20eV in energy below the respective L_{III} absorption edges. The quantitative details are complicated, but the main effect of working at an energy 20eV below the Dy L_{III} edge is to reduce substantially the Dy scattering factor (by about 20 electrons) and so the Dy_2O_3 Bragg peak is relatively suppressed. The low angle part of the doublet is suppressed. Similarly 20eV below the Yb L_{III} absorption edge the right hand peak in the triplet (from Yb_2O_3) is relatively suppressed. Thus, anomalous dispersion can be used to separate diffraction patterns from mixtures into their component patterns and this will aid indexing and analysis. Conversely, using structure refinement methods we are able to determine the anomalous dispersion f' [6].

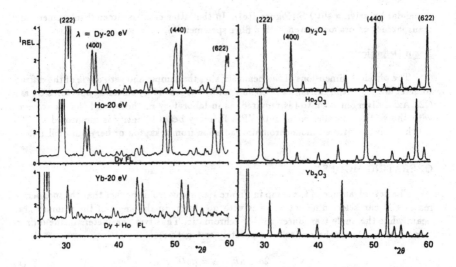

Figure 2: Angle scanned diffraction patterns for three rare–earth oxides and their admixture obtained at 20eV below the respective L_{III} absorption edge energy. $\theta : 2\theta$ mode.

Variable Specimen Geometry

In conventional powder diffraction techniques which use divergent beams, the sample geometry is fixed. With the synchrotron radiation parallel beam geometries in Figure 1 no special relationship between sample angle θ_s and scattering angle 2θ is necessary unless absolute lattice parameters are required. Figure 3 illustrates this point.

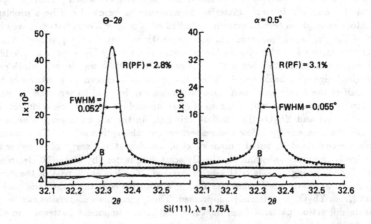

Figure 3: Bragg angle peaks from silicon in symmetric and very oblique geometries

The left hand peak was obtained in $\theta:2\theta$ geometry ($\theta_s \sim 16°$ and equal to $\frac{1}{2}(2\theta)$) whereas the right hand curve was obtained with a stationary sample and $\alpha = 0.5°$. Within the limits set by counting statistics the two peaks are indentical in shape. The Bragg angles are slightly different because of the effect of xray refraction[7] and the heights are quite different because the effective beam size is reduced at low sample angles.

Energy Dispersive Diffraction

Figure 4 shows energy dispersive diffraction patterns from quartz at fixed scattering angles of $2\theta = 10°$, $45°$, $90°$ and $135°$. Two octaves of energy are accessible on bending magnet beamlines such as the one used at SSRL in these experiments, and it would be straightforward on a higher energy source to extend the range of d's beyond 0.3Å. The rather high background arises because we used a scintillation detector with only a low level discriminator; it would be suppressed if we had used a solid state detector with a computer controlled single channel analyser locked in energy to the monochromator. Note that the peak intensity exceeds 10^5 s^{-1}, quite enough to saturate most detectors!

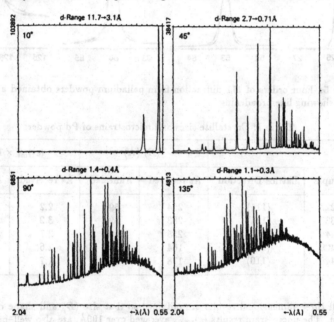

Figure 4: Energy dispersive diffraction patterns from quartz extending to very small d–spacings, well beyond the limiting sphere for copper $K\alpha$ radiation.

APPLICATIONS OF PARALLEL BEAM DIFFRACTION

The three unique features of parallel beam diffraction can be exploited in a variety of ways. We will concentrate here on problems of interest in materials science, where well established laboratory techniques can now be fully developed at synchrotron radiation sources.

Line Shape Analysis: Warren–Averbach Method

The Warren–Averbach method depends on determining peak shape variations between the hkl and nh nk nl Bragg reflections from a sample. For three dimensional information three non–coplanar couples are required; a situation which cannot be achieved in most materials with copper $K\alpha$ radiation. With synchrotron radiation sources one can work at sufficiently short wavelengths that three non–coplanar couples can be obtained and an extended series of reflections, for example n = 1 to 4 in figure 5, can be obtained[8]. Because the instrument function is constant and symmetrical, peak shape analysis is a straightforward and reliable procedure. Table 1 shows the results of analysis for data on palladium powder before and after annealing.

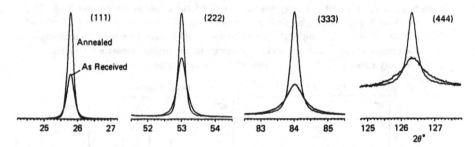

Figure 5: Four orders of 111 diffraction from palladium powders obtained at 1Å wavelength, showing line broadening.

Table I: Crystallite sizes and microstrains of Pd powders

		$\langle D \rangle$ (Å)		$\langle \varepsilon \rangle_{rms} \times 10^{-3}$	
hkl couple	Lattice Direction	As received	Annealed	As received	Annealed
111–222	$\langle 111 \rangle$	264	598	2.2	1.0
111–333	$\langle 111 \rangle$	283	597	3.2	1.3
111–444	$\langle 111 \rangle$	258	600	2.1	0.8
200–400	$\langle 100 \rangle$	164	406	2.6	1.4
220–440	$\langle 110 \rangle$	178	369	1.7	1.0

Full three–dimensional results are obtained so that size $\langle D \rangle$ and shape changes are detected. The microstrain results $\langle \varepsilon \rangle_{rms}$, averaged over 100Å, are also well–behaved and the data probably represent the first complete test of the Warren–Averbach method. It should be remembered that the 511 Bragg reflection is superimposed on 333 in the powder method so that the 111–333 couple is contaminated by 511 information.

Preferred Orientation and High Resolution Pole Figures

Another well–established laboratory method determines texture in materials. Two problems arise in situations where the sample is a quasi–crystalline thin film. Strong preferred orientation requires higher angular resolution than conventional pole figure methods can provide and the intensities diffracted are rather low. This is a very common case where small strains, which determine specimen properties, must be characterised.

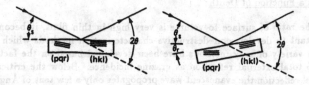

Figure 6: Geometry of Bragg reflections which are oblique to the sample surface.

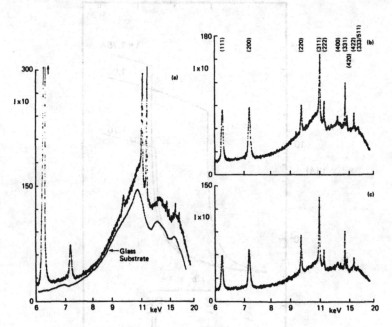

Figure 7: Energy dispersive Bragg reflection patterns from a Pd/Xe film obtained with various (fixed) angles of incidence.

The particular samples which we studied were 4100Å thick films of palladium containing xenon which were deposited on quartz substrates. The ⟨111⟩ direction is strongly preferred and normal to the film. By obtaining patterns in the energy dispersive mode both the angle of incidence, $(\theta_r + \theta_s)$ in Figure 6 and the scattering angle 2θ are free parameters. In near symmetric reflection (Figure 7(a)) the 111, 222 and 333 reflections are strong. One of the angles between ⟨111⟩ and ⟨331⟩ is 22° so that with $\theta_r = 22°$, in Figure 7(c), a strong 311 reflection is seen. The intensities of other Bragg peaks vary in a systematic way as the relative orientation of surface and incident angle are changed. At present we have had insufficient beam time to make a full exploration of the potential of this high intensity, high resolution pole–figure technique, but this demonstration shows clearly some new possibilities.

Structure as a Function of Depth

Since the ratio of surface to volume is very high in thin films, it becomes increasingly important to develop non–destructive characterization methods which are surface sensitive. A variety of xray methods have been developed based on the fact that xrays can undergo total external reflection at grazing incidence. Below the critical angle for total external reflection the evanescent wave propogates only a few tens of Ångström units into the material. Thus all the diffracted intensity relates only to that part of the sample[9].

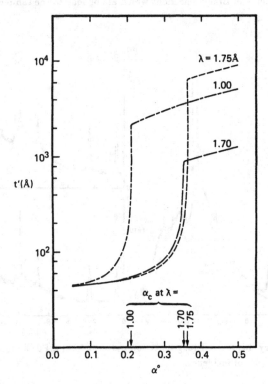

Figure 8: Calculated penetration depth as a function of angle for $\gamma-Fe_2O_3$ at several wavelengths.

Provided that the film is optically continuous, the penetration depth t' can be calculated from the composition (if known!) as Figure 8 shows for $\gamma-Fe_2O_3$. Above the critical angle the penetration depth can be reduced to about 1000Å by working just below the iron K-edge (1.74346Å) at, say, 1.70Å or may be increased to include the whole film by working at longer (1.75Å) or shorter (1.00Å) wavelengths. By choosing α to be less than the critical angle the topmost layer of the thin film can be studied. With $\lambda = 1.75$Å and $\alpha = 0.25°$ only the top 60Å layer contributes, while at $\alpha = 0.33°$ twice that depth is probed.

Figure 9 shows two patterns from a 5000Å thick iron oxide film on a glass substrate. The sample has been heated in air at 325°C for 6.5h to induce the formation of $\gamma-$ and

$\alpha-Fe_2O_3$. The preparation method and subsequent thermal treatment are critical in determining the iron valence and the lattice site occupancy. While no trace of the Fe_2O_3 Bragg reflections can be seen in the $\theta : 2\theta$ scan (Figure 9(a)), which includes contributions from the whole film, the grazing incidence pattern Figure 9(b)), which arises from only to top 60Å of material, shows clear evidence of the $\gamma-Fe_2O_3$ superstructure 211 peak (S.S.) and of the $\gamma-Fe_2O_3$ 422 and $\alpha-Fe_2O_3$ 116 peaks.

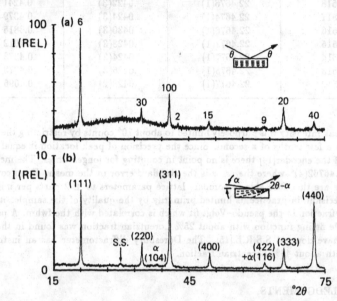

Figure 9: (a) Symmetric reflection $\theta : 2\theta$ scan from iron oxide film on glass substrate. The numbers indicate the intensities expected from a randomly oriented sample. (b) Grazing incidence scan with $\alpha = 0.25°$.

Determination of Absolute Lattice Parameters

The first extensive survey[10] of precision lattice parameters showed conclusively that lattice parameters were not consistently measured to better than part in 10^4. Even now the National Bureau of Standards Standard Reference Material *silicon* is not certified to much higher precision[11,12] and the recommended lattice parameter differs substantially from that of pure single crystal silicon. In view of the very widespread use of the Bond single crystal method, it is clear that it would be valuable if the powder method could be developed to achieve accuracies in the one part per million range.

In our earlier measurements at SSRL[13] we were able to reproduce measurements to within 0 to 0.0004° two theta but the difference between observed and least–squares calculated Bragg angles was still 0.001° to 0.004°. This year we were able to repeat some of these measurements on the diffractometer which is installed at Daresbury[5]. Table II shows results for repeated scans of the silicon 111 Bragg peak.

Table II: Repeatability and precision of the Daresbury diffractometer

Run No.	2θ	fwhm	n
R1521	22.4676(1)	.0426(3)	0.4141
R1520	22.4678(1)	.0422(2)	0.4279
R1519	22.4676(1)	.0418(2)	0.4468
R1518	22.4676(1)	.0422(3)	0.4241
R1517	22.4674(1)	.0424(3)	0.4379
R1516	22.4676(1)	· .0430(3)	0.3815
R1515	22.4677(1)	.0428(3)	0.4012
R1514	22.4677(1)	.0424(2)	0.4373
R1513	22.4675(1)	.0426(3)	0.4073
R1512	22.4677(1)	.0428(2)	0.4096

The peak intensity was purposely limited to about 10^4 counts by restricting the counting time to a few tenths of a second. Since the precision of peak location is equal to the precision of the encoder[14] there is no point in counting for longer times. The mean 2θ-value is 22.46762(4)° where the error is the standard error in the mean. From complete patterns we are therefore able to compare lattice parameters to ± 2 parts per million[5] and the precision, in practice, is limited primarily by the quality of the sample. n is the Lorentzian fraction in the pseudo–Voigt fit which is correlated with the fwhm. A pseudo–Voigt profile fitting function with about 25% Lorentzian fraction was found in the work which we have done at S.S.R.L.[15]. The Daresbury diffractometer has an instrument function with about 40% Lorentzian fraction.

ACKNOWLEDGEMENTS

We are grateful to the staff of the Stanford Synchrotron Radiation Facility and of Daresbury Laboratory for providing the facilities for this research. Instrumentation and computer programmes at S.S.R.L. were prepared by C.G. Erickson and G.L. Ayers of I.B.M. The success of this programme was strongly dependent on the help we received from our collaborators, G. Lim from I.B.M., M. Bellotto and N. Masciocchi (visiting scientists from Milan) and R.J. Cernick from Daresbury Laboratory.

REFERENCES

[1] Materials Science Forum, Vol. 9, edited by C.R.A. Catlow (Trans. Techn. Publications, Switzerland, 1986)

[2] Xray Powder Diffractometry (Aust. J. Phys. 41 (2), (1988))

[3] J.H. Beaumont and M. Hart, J. Phys. E. 7, 823–829 (1974)

[4] D.E. Cox, B.H. Toby and M.M. Eddy, Aust. J. Phys. 41 (2), 117–131 (1988)

[5] R.J. Cernick, M. Hart and W. Parrish, to be published

[6] G. Will, N. Masciocchi, M. Hart and W. Parrish, Acta Cryst. A43, 677–683 (1987)

[7] M. Hart, W. Parrish, M. Bellotto and G.S. Lim, Acta Cryst., A44, 193–197 (1988)

[8] T.C. Huang, M. Hart, W. Parrish and N. Masciocchi, J. Appl. Phys., 61, 2813–2816 (1987)

[9] G. Lim, W. Parrish, C. Ortiz, M. Bellotto and M. Hart, J. Mater. Res. 2 (4), 471–477 (1987)

[10] W. Parrish, Acta Cryst., 13, 838–50 (1960)

[11] C.R. Hubbard, H.E. Swanson and F.A. Mauer, J. Appl. Cryst. 8, 45–48 (1975)

[12] C.R. Hubbard, Adv. Xray Anal., 26, 35–44 (1983)

[13] W. Parrish, M. Hart, T.C. Huang and M. Bellotto, Adv. Xray Anal., 30, 373–381 (1987)

[14] The Heidenhain ROD800 transducer gives an incremental precision of $\frac{1}{3}$ sec.arc.

[15] G. Will, N. Masciocchi, W. Parrish and M. Hart, J. Appl. Cryst, 20, 394–401 (1987)

APPLICATION OF MULTILAYER STRUCTURES TO THE DETERMINATION OF OPTICAL CONSTANTS IN THE X-RAY, SOFT X-RAY AND EXTREME ULTRA VIOLET SPECTRAL RANGES

TROY W. BARBEE, JR.
Lawrence Livermore National Laboratory, P.O. Box 808, Livermore, CA 94550

ABSTRACT

The dispersion of x-rays (XR), soft x-rays (SXR) and extreme ultra-violet (EUV) light by multilayer structures is dependent on the scattering and absorption cross-sections of the elements used to synthesize the multilayer. In this paper it will be shown that this dependence provides a means for the accurate experimental determination of the optical constants of the multilayer constituents. Two specific approaches will be presented and discussed. First, it will be shown that detailed analysis of the energy dependence of the reflectivity of a simple depth periodic multilayer allows the unfolding of the optical constants. Secondly a new optic structure, the multilayer diffraction grating, will be described and it will be demonstrated that such combined microstructure optics allow the scattering cross-sections of the multilayer constituents to be accurately determined over broad spectral ranges.

INTRODUCTION

In this paper the use of multilayer [1,2,3] Bragg reflecting structures in the determination of the optical constants of their component elements is considered. Multilayers are manufactured structures of alternating layers of two materials, A and B, that have been vapor deposited to obtain uniform layer thicknesses t_A and t_B. Therefore multilayers Bragg diffract XR's, SXR's and EUV light as if they had a characteristic superlattice parameter $d = t_A + t_B$. These structures are also directly analogous to quarterwave stacks used at longer wavelengths. Their properties, including reflectivity, resolution and the angular dependence of the Bragg peak positions on photon energy, are determined by the optical constants of the component layers and the multilayer period. Therefore, accurate measurement of these multilayer properties as a function of wavelength provides data that may be analyzed for the optical constants of the elements contained in the multilayer structure.

A short discussion of the optical constants in the XR, SXR and EUV is presented first as background. The dependence of multilayer reflectivity on the optical constants of their component materials is described and modeling procedures outlined. The refraction correction to the multilayer period is then presented and its use with simple multilayers and multilayer gratings described. Experimental results obtained by these techniques are then presented. An assessment of the potential of these techniques is given as a summary.

Optical Constants

Traditionally, the interaction of XR, SXR and EUV light with condensed matter has been treated from two points of view [4]. The first, an extension of the classical electromagnetic description of reflection/refraction to the appropriate frequency regime, describes the motion of photons across interfaces between media that are described by complex indices of refraction $n = 1-\delta-i\beta$ where δ and β are small. The second treatment is scattering theory, which describes x-rays as being scattered from the incident beam by the electrons of the atoms making up the sample. Here, the important quantity

Mat. Res. Soc. Symp. Proc. Vol. 143. ©1989 Materials Research Society

is the atomic scattering factor $f^2 = [f_0 + f']^2 + [f'']^2$, where f_0 is the number of electrons associated with a given atom active in the scattering process. f' and f" correct for deviations reflecting the energy states of the electrons associated with the atom and the interaction of these electrons with the environment of the atom. These two terms are typically significant at low photon energies and in the vicinity of absorption edges and are referred to as anomalous dispersion coefficients. In both treatments the real portion changes the magnitude of the scattering and the imaginary term describes absorption. Since the same physical phenomena are being described in both approaches relationships between δ and f' and β and f" may be established. It is important to note that given either f' (δ) or f" (β) is accurately known over an extended energy range the other may be calculated by use of the Kramers-Kronig [4,5] dispersion relationship between the real and imaginary parts of the atomic scattering factors of a given element. Therefore, accurate measuremelt of one component [f'(or δ) or f" (or β)] defines both components.

The emphasis in the following will be on determination of the real or scattering component of the refractive index by means of multilayer Bragg diffraction. Thin multilayer transmission samples [6,7] may also be used to measure the imaginary or absorption component [8] [f"(β)]. The layering provides the ability to passivate the ambient to sample interfaces and improves the mechanical properties of such thin fragile samples.

Measurement Approaches

Modeling of multilayer reflectivity is performed using a standard multi-layer code [9] based on the Fresnel construction. Briefly, the Fresnel reflection and transmission coefficients are calculated at each interface in a multilayer and a recursion relation is developed for the amplitude of the reflected beam in the jth layer in terms of the amplitude in the (j+1)th layer. The reflectivity is calculated by iterative application of the recursion relation; working from the substrate to the multilayer surface. The Fresnel coefficients are functions of the angle of incidence of the light and the optical constants of the layers forming the interfaces. These calculations are typically performed for a fixed angle of incidence as a function of the energy of the incident light or at a fixed light energy as a function of the angle of incidence.

These model calculations require eight input data parameters: t_A, t_B – thicknesses of the A and B layers; δ_A, β_A, δ_B, β_B - optical constants of the A and B layers; θ - the angle of incidence of the light; E(λ) - the energy or wavelength of the incident light. Four of these are independently determined: t_A, t_B, θ, E(λ). At this time this full reflectivity modeling approach has only been used in the XR energy domain where three of the needed optical constants could be reasonably estimated. The fourth optical constant (δ) was then determined by fitting experimental data using this optical constant as the fitting parameter. This approach is particularly effective in the vicinity of absorption edges where the anomalous dispersion effects are large and strongly energy dependent.

As previously stated multilayers Bragg diffract light at angles determined by their superlattice parameters ($d=t_A+t_B$). There is a significant refraction correction [4] for small θ to the angle of incidence at which the Bragg peaks appear. Application of Snell's Law gives the refraction corrected Bragg equation [4]:

$$n\,\lambda = 2d\,\mathrm{Sin}\theta \left[1 - \frac{2\,\bar{\delta} + \bar{\delta}^2}{\mathrm{Sin}^2\,\theta}\right]^{1/2} . \tag{1}$$

λ is the wavelength of the light, d the multilayer lattice parameter, θ the angle of grazing incidence and $\bar{\delta}$ the spatially averaged scattering component of the refractive index. $\bar{\delta}$ is given as

$$\bar{\delta} = \delta_A \frac{t_A}{d} + \delta_B \frac{t_B}{d} \quad . \tag{2}$$

Therefore, measurement of the angular positions of the Bragg peaks as a function of energy allows $\bar{\delta}$ to be determined given d is known. If two multilayers having the same or known periods (d_1, d_2) and differing known compositions $(t_{A1}/t_B = 2\ t_{A2}/t_{B2})$ the optical constants δ_A and δ_B of the component materials A and B may be determined.

A new multilayer optic structure, the multilayer diffraction grating [10], has recently been developed that may also be used to accurately determine, $\bar{\delta}$, the spatially averaged scattering component of the refractive index. In this approach the grating dispersion angles, ϕ_m, relative to the zeroeth order (m=o) simple Bragg diffracted light at angle θ relative to the multilayer surface are measured. An analysis [10] based on scalar grating theory defines the relationship between the dispersion angle ϕ_m, the grating order m, the multilayer period d, the grating period d_g and the refraction term of Eq. 1 and gives:

$$\sin \phi_m = \frac{2md}{nd_g} \left[1 - \frac{2\bar{\delta} + \bar{\delta}^2}{\sin^2 \theta} \right]^{1/2} \tag{3}$$

If $2\delta >> \delta^2$ (as is often the case) Eq. 3 may be rearranged to give

$$\bar{\delta} = \frac{\sin^2 \theta}{2} \left[1 - \frac{\sin^2 \phi_m}{\sin^2 \phi_m^o} \right] \tag{4}$$

where $\sin \phi_m^o = 2md/nd_g$. Therefore, given ϕ_m^o (m,d,n,d_g) δ may be measured as a function of energy by determining the dependence of ϕ_m on θ. Note that there is no energy or wavelength dependence in Eq. 4 so that data must be energy calibrated.

Reported Results

The real part of the anomalous dispersion coefficient of titanium was determined [11] at the titanium K edge (4965 eV) using a titanium-carbon multilayer (d = 56.4 Å, t_{Ti} = 26.4 Å, N = 70). These data were analyzed as previously described and δ_{Ti} determined. The results are shown in Fig. 1 and compared to calculated an ideal estimation of δ_{Ti} over this energy range. The well defined minima seen is in general agreement with the predictions of fixed oscillator strength models though the detailed structure is dependent on the electronic and atomic structures of the titanium. The anomalous dispersion coefficients, f'_{Ti}, calculated from these data were compared to values obtained by integration through the Kramers-Kronig dispersion relationship [12] of absorption data on pure titanium. The agreement shown was good though EXAFS structure at energies above 4965 eV is not seen in the multilayer data as too coarse an energy scale was used in the multilayer measurements. These results demonstrate that applicability of the Fresnel optical multilayer dispersion formalism to calculation of the properties of multilayer structures and the general effectiveness of this approach to the determination of XR optical constants.

Refraction changes the propagation angle of light entering a multilayer and therefore changes the angular position of the Bragg peak maxima. As discussed earlier this refraction effect can be used [3] to measure refractive index of multilayer structures. Also, given the refractive

200

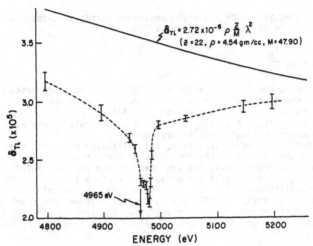

$$\delta_{Ti} = 2.72 \times 10^{-6} \, \rho \, \frac{Z}{M} \, \lambda^2$$
$$(\bar{z} = 22, \, \rho = 4.54 \, gm/cc, \, M = 47.90)$$

4965 eV

ENERGY (eV)

Fig. 1 δ_{Ti} obtained from the analysis [11] of the energy dependence
of multilayer reflectivity.

indices of the multilayer components is known it can be used [4,14] to
determine the composition of the multilayer. I note here that this
technique was initially used with atomically ordered solids [4] (crystals),
its application to multilayers extending its useful range from the XR to
the SXR and EUV.

This method has been applied to the determination of the dispersion
coefficient δ of a single multilayer component element for vanadaium
[3,15] at its K absorption edge, for carbon [16] at its K absorption edge,
for nickel [17] at its L absorption edges and is inherent in the titanium K
edge results previously described. In all cases it was necessary to
include a correction [18] for absorption effects as absorption introduces
an asymmetry into the angular dependence of the Bragg diffracted intensity.
This is typically not a large correction but should be included if absolute
values are sought. Experimental results recently reported by Spiller [16]
for δ of carbon over the energy range 215 eV $<E<$ 295 eV are shown in Fig.
2. Note, the carbon K absorption edge is at 283 eV. These results demonstrate
that the δ of carbon is negative over the energy range 275 eV $<E<$ 288 eV so
that the refractive index of carbon, n, is greater than 1.

δ for rhodium/carbon multilayer has also been measured [10] using a
symmetric amplitude diffraction grating. The multilayer had a period of 80
Å, contained 40 layer pairs of rhodium/carbon with $t_{Rh}=t_C$. The grating
period was 2 microns. $\bar{\delta}$ values derived using Eq. 4 and measured values of
ϕ_m are compared to values calculated using Eq. 2 and values of (f_0+f')
for carbon and rhodium given by Henke et al. [19] in Fig. 3.

These results for δ (C, Rh) may be unfolded to give the energy dependence of δ for carbon by using the literature values for the refractive index of rhodium [19]. This analysis demonstrates that the δ of
carbon is strongly negative at its K absorption edge. I note that for δ
carbon to be non-negative the value of f_0+f' for rhodium would be required
to decrease from the literature value [19] of 11.53 to approximately 3.4.

CONCLUSIONS

The results described here demonstrate that multilayers provide an
effective means for determination of optical constants in the SR, SXR and

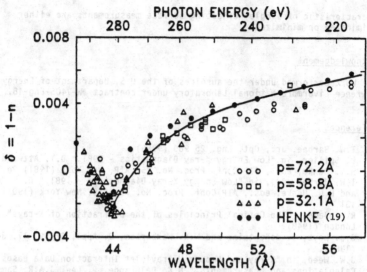

Fig. 2 Refractive index ($\delta=1-n$) of amorphous carbon in cobalt-carbon multi-layers (period p) is compared to literature [19] values.

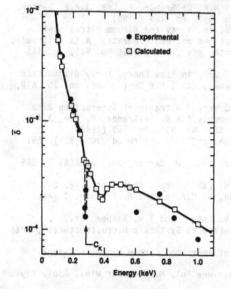

Fig. 3 $\bar{\delta}$ for a Rh/C multilayer on a grating ($d_g=2000$ nm, $d_o=8.0$ nm, $N=40$, $t_{Rh}=t_c$) is plotted as a function of photon energy.

EUV. The absolute accuracy attainable is at this time undemonstrated. It is dependent on knowledge of the period of the multilayer, its composition, energy calibration of the optical constant data and, for the case of multi-layer diffraction gratings, on an accurate value for the grating period. Error analysis indicates that well characterized samples will allow determination of δ(or f_o+f') to an accuracy of ± 3%. Also, note that such multilayer based results are not strongly sensitive to the multilayer-ambient interface so that experimental difficulties characteristic of single reflecting surface measurements are either eliminated or minimized.

Acknowledgement

Work performed under the auspices of the U.S. Department of Energy by Lawrence Livermore National Laboratory under contract #W-7405-Eng-48.

References

1. T.W. Barbee, Jr., Opt. Eng. <u>25</u> 989 (1986).
2. E. Spiller, in "Low Energy X-ray Diagnostics - 1981," D.T. Attwood and B.L. Henke, eds., AIP Conf. Proc. No. 75, AIP, New York (1981) Pg. 124.
3. T.W. Barbee, Jr., in "Low Energy X-ray Diagnostics - 1981," D.T. Attwood and B.L. Henke, eds., AIP Conf. Proc. No. 75, AIP, New York (1981) Pg. 131.
4. R.W. James, "The Optical Principles of the Diffraction of X-rays" Bell, London (1948).
5. J.J. Hoyt, D. deFontaine and W.K. Warburton, J. Appl. Cryst. <u>17</u>, 344 (1984).
6. J.W. Weed, in "X-ray and Vacuum Ultraviolet Interaction Data Bases, Calculations and Measurements," N.K. DelGrande, P. Lee, J.A.R. Samson, D.Y. Smith, eds., Proc. SPIE No. 911, Pg. 166 (1988).
7. R.G. Musket, in "X-ray and Vacuum Ultraviolet Interaction Data Bases, Calculations and Measurements," N.K. DelGrande, P. Lee, J.A.R. Samson, D.Y. Smith, eds., Proc. SPIE No. 911, Pg. 169 (1988).
8. K.G. Tirsell and N.K. DelGrande, in "X-ray and Vacuum Ultraviolet Interaction Data Bases, Calculations and Measurements," N.K. DelGrande, P. Lee, J.A.R. Samson, D.Y. Smith, eds., Proc. SPIE No. 911, Pg. 146 (1988).
9. J.H. Underwood and T.W. Barbee, Jr., in "Low Energy X-ray Diagnostics 1981," D.T. Attwood and B.L. Henke, eds., AIP Conf. Proc. No. 75, AIP, New York, (1981) Pg. 179.
10. T.W. Barbee, Jr., in "X-ray and Vacuum Ultraviolet Interaction Data Bases, Calculations and Measurements," N.K. DelGrande, P. Lee, J.A.R. Samson, D.Y. Smith eds., Proc. SPIE No. 911, Pg. 169 (1988).
11. T.W. Barbee, Jr., W.K. Warburton, and J.H. Underwood, JOSA(B) <u>1</u>, 691 (1984).
12. W.K. Warburton, K.F. Ludwig, Jr. and T.W. Barbee, Jr., JOSA(B) <u>2</u>, 565 (1985).
13. P.F. Miceli, D.A. Neumann and H. Zabel, Appl. Phys. Lett. <u>48</u>, 24 (1986).
14. J.P. Simon, O. Lyon, A. Brunson, G. Marchal and M. Piecuch, J. Appl. Cryst. <u>21</u>, 317 (1988).
15. O.J. Petersen, J.M. Thorne, L.V. Knight and T.W. Barbee, Jr., "Reflectivity and Roughness of Layered Synthetic Microstructures", SPIE Proc. No. 448, Pg. 27 (1984).
16. E. Spiller, "Refractive Index of Amorphous Carbon Near its K-edge," (submitted for publication).
17. H. van Brug, M.P. Brujin, R. van der Pol, M. Jvan der Wiel, Appl. Phys. Lett. 49, 914 (1986).
18. A.E. Rosenbluth and P. Lee, Appl. Phys. Lett. <u>40</u>, 466 (1982).
19. B.L. Henke, P. Lee, T.J. Tanaka, R.L. Shimabukuro, and B.K. Fujikawa, "Low energy x-ray interaction coefficients: photoabsorption, scattering and reflection. E = 100-2000 eV, Z = 1-94," in <u>Atomic and Nuclear Data Table 27</u>, Academic Press, New York (1982).

TIME-RESOLVED SMALL ANGLE X-RAY SCATTERING AND DYNAMIC LIGHT SCATTERING STUDIES OF SOL-GEL TRANSITION IN GELATIN

DAN Q. WU and BENJAMIN CHU[*]
Department of Chemistry, State University of New York at Stony Brook, Long Island, NY 11794
* author to whom all correspondence should be addressed.

ABSTRACT

Structural and dynamical properties of an aqueous gelatin solution (5 wt%, 0.1M NaCl, pH≈7) in a sol-gel transition were studied by time-resolved small angle x-ray scattering (SAXS) and dynamic light scattering (DLS) after quenching the gelatin sol at ~45°C to 11°C. SAXS intensity measurements suggested the presence of gel fibrils which grew initially in cross-section. The average cross-section of the gel fibrils reached a constant value after an initial growth period of ~800 sec. Further increase in SAXS intensity could be attributed to the increase in the length of the gel fibrils. Photon correlation, on the other hand, clearly showed two relaxation modes in both the sol and the gel (~1 hr after the quenching process) states: a fast cooperative diffusion mode which remained constant from the sol to the gel state after correction for the temperature dependence of solvent viscosity; and a slow mode that could be attributed to the self-diffusion of the "free" gelatin chains and aggregates. The slow mode contribution to the time correlation function was reduced from ~40% in sol to ~20% in gel signaling a decrease but not the elimination of "free" particles in the gel network. The decrease in the intensity contribution by the slow mode is, however, accompanied by a large increase in the characteristic line-width distribution.

INTRODUCTION

Gelatin is a partially denatured collagen which is formed by three left-handed α-helix polypeptide chains. An aqueous gelatin solution (higher than ~1%) forms a thermodynamically reversible gel upon cooling below the melting temperature of ~35°C [1]. Such a sol-gel transition is a renaturation of gelatin molecules through a coil-helix conversion and aggregation. Several molecular models were proposed to explain the kinetics of such a transition [2-4]. Godard *et.al.* [5] showed, by using differential scanning colorimetry (DSC), that upon quenching the semidilute gelatin solution, a two step gelation process could be derived: the primary step took about 5-10 min and involved the nucleation of gelatin chains. The nucleation process could be fitted by the Avrami crystallization equation with $n = 1$ suggesting a unidirectional growth of the gelatin gel. They also suggested that the fringe-micelle model, as proposed by Lauritzen and Hoffman [6], could explain the observed gelatin gelation kinetics which was supported by Djabourov *et.al.* [7,8] using polarimetry. However the two-step model seemed to work only for gelling temperatures <20°C [8]. Electron micrographs showed gelatin networks with broad distributions of mesh sizes and different cross-sections of gelatin fibrils in the gel state [7,10].

In this paper, we report measurements of time-resolved small angle x-ray scattering (SAXS) and dynamic light scattering (DLS) on a gelatin solution (5 wt%) undergoing the sol-gel transition.

EXPERIMENTAL SECTION

Sample Preparations:

Gelatin samples were obtained from ossein by the lime process. Gel permeation chromatography (GPC) showed a very broad molecular weight distribution with molecular weight ranging from 10^4 to 10^6 g/mole and M (peak value) ~1×10^5 g/mole [8]. Aqueous gelatin solutions were prepared by swelling the "dry" gelatin in 0.1M NaCl and a small amount of NaN_3 (to prevent bacterial contamination), at ~5°C, for ~2 hrs, then stirring the sample at 50°C for ~2 hrs. pH (≈7) was controlled by addition of NaOH solution. The gelatin solution was filtered through a 0.22 μm Millipore Millex-GV filter at 50°C before DLS experiments.

Experimental Methods:

Synchrotron Small Angle X-ray Scattering:

SAXS experiments were carried out at the SUNY Beam Line [11], National Synchrotron Light Source, Brookhaven National Laboratory, using a modified Kratky collimation system [12] and a one-dimensional EG&G photo-diode-array detector (Model 1463). A 5 wt% gelatin solution (0.1M NaCl, pH≈7) at ~45°C was quenched to a gelling temperature of 11°C. The sample could reach 12°C in ~200 sec. SAXS profiles were recorded in 100-second time intervals up to 3000 sec corresponding to ~40% coil-helix conversion [9]. All SAXS profiles were corrected for the incident x-ray intensity fluctuations, sample attenuation, detector dark counts and non-linearity; and normalized to 1 sec. The x-ray wavelength was set at 0.154 nm.

The measured SAXS profiles contained the scattered intensities of the gelatin gel and the remaining sol. As an approximation, the excess scattered intensity of gel: $I_{ex}(q,t) \approx I(q,t) - f(t) \cdot I(q,0)$, with $f(t)$ and $I(q,0)$ being, respectively, the mole fraction of sol content and the scattered intensity of the sol at ~45°C ($t=0$). $f(t)$ can be measured by DSC [5], polarimetry [7,8,9], or fluorescence recovery after photobleaching [18]. In this work, we used the results from polarimetry [9], $f(t) = 1 - \chi(t)$, where $\chi(t)$ is the measured coil-helix conversion.

The aims of our SAXS measurements are: (1) to verify if the gelatin molecules aggregate to form fibrils; and (2) to estimate an average fibrillar diameter if aggregation does occur.

For a very long rod particle of length L with $L \gg R$ and R being the cross-sectional radius of the rod, the excess scattered intensity due to the solute, $I_{ex}(q)$, can be written as a product of two terms: a term from the rod length ($L\pi/q$) and a term from the cross-section $I_c(q)$ [13]:

$$I_{ex}(q) = (L\pi/q) \cdot I_c(q) \tag{1}$$

where $q = (4\pi/\lambda)sin(\theta/2)$, with λ and θ being, respectively, the x-ray wavelength and the scattering angle. If $qL \gg 1$ and $qR \lesssim 1$, we get

$$q \cdot I_{ex}(q) \sim L \cdot I_c(q) \sim L \cdot exp(-q^2 R_c^2/2) \tag{2}$$

The radius of gyration of the cross-section ($R_c = R/\sqrt{2}$ for a cylindrical rod) can then be determined from a plot of $log [q \cdot I_{ex}(q)]$ vs. q^2. Eq.(2) has been applied successfully to study the the cross-section dimension of protein molecules in dilute solution [14]. In our case, we have assumed that short lengths in the broad rod size distribution do not interfere with the quantity $I_c(q)$, or that the cross section dimension is relatively uniform in size. Further more, the interparticle interference effect has been neglected. Thus, the use of Eq. (2) is qualitative at best, and we consider only the gross feature of the SAXS profiles.

Dynamic Light Scattering:

DLS was carried out on a standard light scattering spectrometer using a Brookhaven BI-2030AT 136 channel correlator. The time correlation functions of the same gelatin sample were measured at (1) 46°C before quenching, (2) 5 min after quenching to 11°C and (3) ~1 hr after quenching, when gel was formed but not yet at equilibrium. A multiple delay-time increment mode was used to cover the delay time from about 10 μs up to a few seconds in order to cover the entire decay range of the intensity-intensity time correlation function decay curves.

The intensity-intensity time correlation function $G^{(2)}(q,\tau)$ in the self-beat mode has the form [15]:

$$G^{(2)}(q,\tau) = B(1 + \beta | g^{(1)}(q,\tau)|^2) \tag{3}$$

where B is the base-line, β is a coherent factor and $|g^{(1)}(q,\tau)|$ is the normalized electric field correlation function which is equal to $exp(-\Gamma\tau)$ for a monodispersed system, and is related to the normalized linewidth distribution function $G(q,\Gamma)$ for a polydispersed system by:

$$|g^{(1)}(q,\tau)| = \int_0^\infty G(q,\Gamma) \, exp(-\Gamma\tau) \, d\Gamma \tag{4}$$

where Γ is the characteristic linewidth. At small q, $\Gamma = Dq^2$ with D being the translational diffusion coefficient. Many numerical Laplace inversion methods have been developed to obtain $G(q,\Gamma)$, one of which, called CONTIN, was used in the present work. [16]

For the gelatin solution before and after the gel point, the measured β value agreed to within a few percent of the instrumental β value. The difference between the calculated baseline and the measured baseline was less than 0.1% for the sol and less than 0.2% after quenching. Generally the measurements were made at three scattering angles: 30°, 60°, and 90°.

RESULTS AND DISCUSSION

Time-resolved SAXS measurements on a 5 wt% gelatin solution (0.1M NaCl, pH≈7) after quenching the sample at ~45°C to 11°C revealed that the scattered intensity increased rapidly within the first few hundred seconds. Then the rate of increase in the scattered intensity slowed down. To demonstrate this,

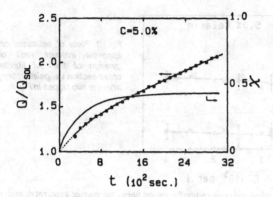

Fig. 1. Normalized integrated scattered intensity of a gelatin solution (5 wt%, 0.1M NaCl, pH≈7) in a sol-gel transition (quenched from ~45°C to 11°C) (-□-). The solid line represents coil-to-helix conversion measured by polarimetry [9].

we plotted an integrated scattered intensity $Q(t)$ normalized by $Q(0)=Q_{sol}$ vs. the gelation time as shown in Figure 1. The integrated scattered intensity is defined by

$$Q(t) = \int_0^{q^{max}} q^2 \cdot I(q,t) dq \tag{5}$$

where q_{max} is the maximum q value as determined by the experimental setup. Together with the coil-helix conversion $\chi(t)$, measured by Djabourov et. al. [9], both Q/Q_{sol} and χ showed the similar increases in the initial time period of ~1000 sec. However at $t \geq 1000$ sec, χ became relatively constant while Q/Q_{sol} continued to increase, suggesting further aggregation mainly among the triple helices, which could not be measured by polarimetry.

According to Eq. 2, plots of $log\ q \cdot [I(q,t) - f(t) \cdot I(q,0)]$ vs. q^2, as shown in Figure 2, at selected gelation times, could reveal qualitative information on the cross-section of gel fibrils. Two distinct slopes were observed. The straight line at $\sim 0.05 \leq q \leq \sim 0.4$ nm^{-1}, extended itself toward smaller angles as more gel was formed. Another one at $\sim 0.6 \leq q \leq \sim 2$ nm^{-1} might suggest the presence of fibrils with a smaller cross-section.

The apparent radii of gyration of the cross-section, $R_{c,app}$, of the fibrillar gelatin gel were computed using Eq. 2 and plotted in Figure 3. The solid triangles represent the $R_{c,app}$ (1) obtained from the slope at smaller q values; the solid squares represent $R_{c,app}$ (2) from the slope at larger q values. The former grew gradually from 1 ± 0.2 nm at $t = 300$ sec, to a plateau value 1.4 ± 0.1 nm at $t \sim 800$ sec; the latter fluctuated

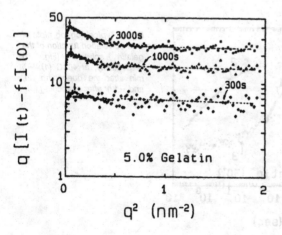

Fig. 2. Plots of $log\ [q \cdot I(q,t) - f(t) \cdot I(q,0)]$ vs. q^2 for the gelatin solution (Fig. 1) at selected gelation time. $f(t) = 1-\chi$. χ is the coil-helix conversion measured by polarimetry [9]. The curves show two distinct slopes.

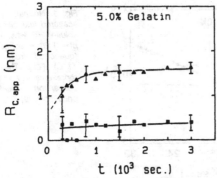

Fig. 3. Plots of estimates on apparent average radii of gyration of the gel fibrillar cross-section vs. gelation time from the two slopes in Fig. 2.

about 0.4±0.2 nm. If we assumed a cylindrical cross-section for the gel fibrils, the average apparent diameter $d = 2/2 \cdot R_{c}^{app}$ of the fibrils, would vary from $d(1) = 2.8\pm0.6$ nm at $t = 300$ sec. to 4.0±0.3 nm at $t \sim 800$ sec; and $d(2) \approx 1.1\pm0.6$ nm, in agreement with the diameter of a triple-helix collagen. Thus $d(1)$ represents an aggregate of a few three-strand helices. This observation is consistent with Fig. 1 showing further increase in the normalized integrated scattered intensity with little change in coil-helix conversion after 1000 sec. Fig. 3 might also suggests a two-step gelation process: a growth in diameter of the gelatin fibrils up to $t \sim 800$ sec and then a growth in the length of these fibrils as supported by a shift of the maximum toward smaller q and an increase in $q \cdot I_{ex}(q,t) \sim L \cdot I_{c}(q,t)$ in Fig. 2.

Electron micrographs, though strongly dependent on the sample preparation, showed a gel network of fibrils, with ~ 2-~ 20 nm in cross-sectional diameter and ~ 100-~ 1000 nm in length for 2% gelatin [7], and 3-7 nm and 8-10 nm in cross-sectional diameter, respectively, for 1% and 10% gelatin [10]. Dodard $et.al.$ [5] calculated that $d \approx 5.8$ nm for a 5.1% gelatin quenched at 10°C and the fibrillar diameter decreased with increasing concentration or decreasing gelling temperature. Our qualitative estimate on the diameters of the fibrils seems to agree reasonably well with these findings.

Figure 4 shows typical log-log plots of the intensity-intensity time correlation function of the gelatin (5 wt%, 0.1M NaCl, pH$\approx$7) in a sol-gel transition (quenched from 46°C to 11°C), measured at a scattering angle of 30°, where the curves 1, 2, and 3 are, respectively, for the gelatin solution at 46°C, for the gelatin solution at ~ 5 min after quenching (the sample could still flow, though very viscous), and for a gelatin gel at ~ 1 hr after quenching (the gel has not yet reached an equilibrium state). We could give the following observations:

(1) At least two relaxation modes are present;
(2) The initial fast decay is less sensitive to gel formation;
(3) Contribution of the slow mode decreases in the gel state;

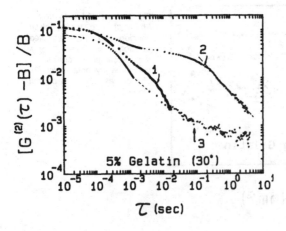

Fig. 4. Typical log-log plots of time correlation function of the gelatin solution (Fig. 1) respectively, at 46°C (1); ~5 min after the quenching (2), and ~1 hr after quenching (3).

(4) A power law region *seems* to appear at long delay times: $G^{(2)}(q,t) \propto t^{-2\alpha}$, e.g. $\alpha = 0.38 \pm 0.02$ for curve 2.

Before quenching, the gelatin solution is in the semidilute regime. The fast initial decay of the time-correlation function could be attributed to the cooperative diffusion caused by local concentration fluctuations, while the slow decay would come from the self-diffusion of gelatin chains due to the broad size distributions of the gelatin used. We used CONTIN, a method of Laplace inversion, to obtain the linewidth distribution $G(\Gamma)$ from the normalized electric field time correlation function $g^{(1)}(q,\tau)$ (Eq. 4).

Figure 5 plots a typical $G(\Gamma)$ for curves 1 and 3 in Fig. 4. It is clear that when a gelatin solution undergoes a sol-gel transition, the fast mode (higher Γ) remains relatively constant, but the slow mode is changed from an already broad distribution to an even broader one. The possible presence of a power law would further extend the size domain as for example suggested by the tail section of curve 2 in Fig. 4. The ratio of the slow mode to the fast mode changes from ~36/53 in the gelatin solution at 46°C (curve 1) to ~17/75 after gelation (curve 3). Both the slow mode and the fast mode follows the q^2-dependence implying a translative diffusive nature. On closer examination, an additional "intermediate" peak in $G(\Gamma)$ located near the fast-mode peak could be observed. This intermediate peak, which was often difficult to separate from the fast peak, contributed ~10% intensity to the time correlation function before and after gelation and might be attributed to the diffusion of the short "free" gelatin chains. The separation of $G(\Gamma)$ into three (instead of two) modes by means of CONTIN could be an artifact of the fitting procedure. It does suggest that at least up to ~1 hr after quenching there is a very broad distribution of particle motions in addition to the cooperative diffusive motion.

Recently, Adam *et.al.* [17] demonstrated that a polydispersed polymer cluster solution behaved dynamically like a glass as the time-correlation function changed from a stretched-exponential function to a power law function and the exponent decreased with increasing concentration where the largest clusters were partly penetrated by smaller clusters. As the gelatin gelation has been modelled by a nucleation and growth process of the fibrils, eventually forming a three-dimensional fibrillar gel network, our preliminary observation of the possible power law behavior on the slow mode near the gelation threshold would suggest the presence of the diffusion of the gelatin sol molecules and the small aggregates within the gel network during gel formation. Further study on the form of $G^{(2)}(q,\tau)$ at different gelation temperature and on different gelatin concentration is in progress.

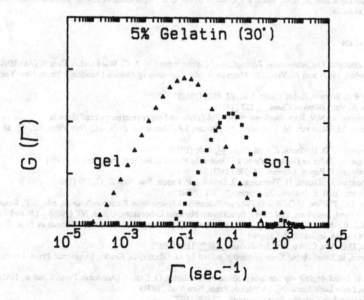

Fig. 5. Typical plots of characteristic linewidth distributions of the gelatin solution based on curves 1 (sol) and 3 (gel) in Fig. 4. The results are listed in Table 1.

Table 1. CONTIN Results of Fig. 5

	D_{slow}	A_{slow}	V_{slow}	D_{fast}	A_{fast}	V_{fast}	D_3	A_3	V_3	$\eta(cP)$
sol	1.7 ± 0.2	0.36	0.6	50 ± 3	0.53	0.06	12 ± 4	0.10	0.07	0.58
gel	0.5 ± 0.3	0.17	1.0	25 ± 5	0.75	0.05	0.1 ± 0.04	0.08	0.09	1.27

$D: 10^{-8} \text{cm}^2/\text{sec}; A_{slow} + A_{fast} + A_3 = 1$, typically $\Delta A = \pm0.03; V$: variance of the peak in $G(\Gamma); \eta$: viscosity of water at $46°C$ and $11°C$.

CONCLUSIONS

Time-resolved SAXS measurements on a 5 wt% gelatin solution in a sol-gel transition, showed that gelatin molecules tended to form a fibrillar gel network. The gelatin gel fibrils initially grew in cross-section, then in length and our results are consistent with the findings of electron microscopy, of DSC and of polarimetry.

Dynamic light scattering measurements showed, before and after gelation, at least two relaxation modes: the fast cooperative diffusion remained constant after correction due to solvent viscosity, while the slow mode(s) could represent the self-diffusion of "free" particles in the system.

ACKNOWLEDGEMENT

We are indebted to Dr. J. Leblond for the gelatin samples, to Drs. M. Adam and J. Leblond for simulating discussions, to Dr. J.C. Phillips for his assistance in setting up the X21 SUNY Beamline for SAXS measurements, and to Dr. S.W. Provencher for offering us his CONTIN program. B.C. gratefully acknowledges support of this research by the U.S. Department of Energy (DEFG0286-ER45237A002), the NSF Polymers Program (DMR8617820), and NATO (850389). The work was carried out at the SUNY Beamline supported by the U.S. Department of Energy (DEFG0286-ER45231A001) at the National Synchrotron Light Source, BNL, which is sponsored by the U.S. Department of Energy under contract (DE-AC02-76CH00016).

REFERENCES

[1] See, for example, *The Science and Technology of Gelatin*, edited by A. G. Ward and A. Courts, (Academic Press, London, 1977) and A. Veis, *The Macromolecular Chemistry of Gelatin* (Academic Press, New York, 1964)

[2] P. Flory, E.S. Weaver, *J. Am. Chem. Soc.*, 82, 4518 (1960)

[3] L. Yuan, A. Veis, *Biophys. Chem.*, 1, 117 (1973)

[4] W.F. Harrington, N.V. Rao, *Biochemistry*, 9, 3714 (1970) and other references cited there in.

[5] P. Godard, J.J. Biebuyck, M. Daumerie, H. Naveau, J.P. Mercier, *J. Poly. Sci., Poly. Phys. Ed.*, 16, 1817 (1978)

[6] J.I. Lauritzen, J.D. Hoffman, *J. Appl. Phys.*, 44, 4340 (1973)

[7] M. Djabourov, *Thése de Doctorat d'etat*, a l'Universite Pierre et Marie Curie (Paris 6) (1986)

[8] M. Djabourov, P. Papon, *Polymer*, 24, 539 (1983)

[9] M. Djabourov, J. Maquet, H. Theveneau, J. Leblond, P. Papon, *Brit. Poly. J.*, 17, 169 (1986)

[10] Y.F. Titova, Y.M. Belavtseva, *Biophysics*, 29, 372 (1984)

[11] B. Chu, J.C. Phillips, D.Q. Wu, in *Polymer Research at Synchrotron Radiation Sources*, eds., T.P. Russel and A.N. Goland, report no. BNL51847, Brookhaven National Laboratory, Upton, NY (1986) p.126 and J.C. Phillips, K.J. Baldwin, W.F. Lehnert, A.D. LeGrand, C.T. Prewitt, *Nucl. Instr. and Methods in Phys. Res.*, A246, 182 (1986)A. leGrand, B. Lanarnd

[12] B. Chu, D.Q. Wu, C. Wu, *Rev. Sci. Instrum.*, 58(7), 1158 (1987)

[13] G. Porod, in *Small Angle X-ray Scattering*, edited by O. Glatter, O. Kratky, (Academic Press, London, 1982)

[14] I. Pilz, in *Small Angle X-ray Scattering*, edited by O. Glatter, O. Kratky, (Academic Press, London, 1982)

[15] B. Chu, *Laser Light Scattering*, (Academic Press, New York, 1974)

[16] S.W. Provencher, *Comput. Phys. Commun.*, 27, 229 (1972)

[17] M. Adam, M. Delsanti, J.P. Munch, D. Durand, Phys. Rev. Lett., 61, 706 (1988)

[18] P.S. Russo, in *Reversible Gels and Related Systems*, ACS Symp. Ser. #350, edited by P.S. Russo, (American Chemical Society, 1987), and P.S. Russo, M. Mustafa, D. Tipton, J. Nelson, D. Fontenot, *Proceedings of the ACS Division of PMSE*, 59, 605 (1988)

DISPERSIVE X-RAY SYNCHROTRON STUDIES
of Pt-C MULTILAYERS

B. RODRICKS*, F. LAMELAS*, D. MEDJAHED*, W. DOS PASSOS*,
R. SMITHER**, E. ZIEGLER†, A. FONTAINE‡, and R. CLARKE*
*Department of Physics, The University of Michigan, Ann Arbor, MI, 48109
**Argonne National Laboratory, Argonne, IL 60439
† ESRF, 38043 Grenoble, France
‡ LURE, Bat. 209D, 91405 Orsay, France

ABSTRACT

We demonstrate the simultaneous acquisition of high-resolution x-ray absorption spectra and scattering data, using a combination of energy-dispersive optics and a two-dimensional CCD detector. Results are presented on the optical constants of Pt and on the reflectivity of a platinum-carbon multilayer at the L_{III} absorption edge of Pt.

INTRODUCTION

Unique characteristics of synchrotron radiation, such as tunability and excellent vertical collimation, offer many advantages for the study of surfaces and interfaces[1]. Two of the most powerful structural probes which exploit these characteristics are the EXAFS technique,[2,3] and x-ray diffraction carried out close to the glancing angle for total external reflection.[4] While the former methods address basically short-range structural correlations, the latter provide information principally on the long-range atomic ordering. The two types of measurement provide complementary information but they require quite different experimental conditions and are not usually performed together on the same sample.

In this paper we describe a new method which permits simultaneous recording of x-ray absorption spectra and x-ray scattering data. As a demonstration of the technique we measured the optical constants of Pt near its L_{III} edge ($\sim$11.564 keV) where the anomalous corrections become important.[5] Using this information we compare the measured glancing-angle reflectivity with values calculated using an iterative Fresnel method.

The use of energy dispersive optics[6] and a high-resolution charge coupled device (CCD) detector[7] makes for fast, efficient, data acquisition. The methods thus have much to offer for a wide range of measurements including near-edge spectroscopy, EXAFS, and small-angle scattering. The parallel nature of data acquisition also presents some interesting possibilities for time-resolved measurements.[8]

EXPERIMENTAL DETAILS

The measurements were performed on the DCI storage ring at LURE. The heart of the dispersive optics is a triangular-shaped Si(311) bent crystal with an energy band-pass given by[6] $\Delta E = E_o L(R^{-1} - \sin \theta_B/p) \cot \theta_B$, where L is the length of the crystal illuminated, p is the source to crystal distance and R is the radius of curvature. The nominal energy, E_o, is determined by the choice of Bragg angle, θ_B.

Fig. 1 shows the experimental layout with a white beam incident on the bent crystal and an outgoing beam which is energy-resolved in space. The sample is placed at the polychromatic focus point and, if a position sensitive detector is located on the Rowland circle, the energy resolution is Darwin-width limited (typically ~2eV for $E_o \approx$ 11 keV). The dispersive method is discussed in more detail by Fontaine et al.[9] in this Proceedings. The bent crystal was adjusted to have $\Delta E \approx 500eV$ centered on the L_{III} edge of platinum.

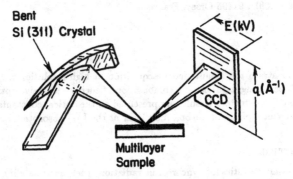

**Bent
Si (311) Crystal**

E(kV)

CCD

q(Å⁻¹)

**Multilayer
Sample**

Fig. 1: Schematic arrangment for energy-dispersive measurements
using a CCD position-sensitive detector.

One axis of the CCD detector (parallel to the pixel rows) is oriented accurately along the energy fan of the bent crystal so that the positional coordinate along this axis corresponds to a particular energy. In the perpendicular direction, the detector registers the scattering angle (momentum transfer). In this way we can obtain spectroscopic data and scattering information simultaneously.

The CCD detector has 390x584 pixels each of area 22 x 22 $(\mu m)^2$. In order to maximize the effective collection area the detector chip is coupled to a Trimax phosphor[10] screen 40 mm in diameter, using a pair of photographic lenses. This resulted in a 4:1 image reduction at the CCD chip. Dark current was suppressed by cooling the chip to -50°C with a thermoelectric element. The Pt-C multilayers were prepared by sputtering and consisted of ~20 bilayers, each layer (Pt or C) being 200Å thick. The top layer of the sample is Pt.

RESULTS AND DISCUSSION

In order to investigate the near-edge reflectivity our first task was to measure the optical constants for Pt, these being the dominant factors compared to the values for carbon. The anomalous scattering corrections, $\Delta f'$ and $\Delta f''$, are related to the absorption cross section, $\sigma(\omega)$, in the usual way:[5]

$$\Delta f''(\sigma) = \frac{mcw\sigma(\omega)}{4\pi e^2} \tag{1}$$

and

$$\Delta f'(\omega) = \frac{2}{\pi} \int_0^\infty \frac{\omega' \Delta f''(\omega) d\omega'}{\omega^2 - \omega'^2} \tag{2}$$

The imaginary part of the anomalous structure factor, $\Delta f''$, was obtained directly from absorption cross-section measurements on a thin foil of Pt placed at the polychromatic focus (see Fig. 1); no self-absorption corrections were applied. The real part, $\Delta f'$, was then calculated using the Cauchy principal value of the Kramers-Kronig integral (Eq. 2) spanning $\hbar\omega_{LIII}$ to 100keV.

The energy dependences of $\Delta f'$ and $\Delta f''$ around the L_{III} edge obtained in this way are shown in Figs. 2(a) and 2(b), respectively. The data were averaged over 44 frames each with an exposure time of 100 msec.

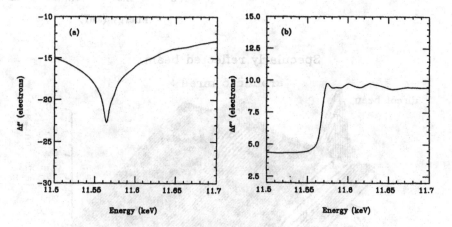

Fig. 2: Anomalous corrections to the structure factor of Pt.
(a) real part; (b) imaginary part.

Reflectivity Measurements

The arrangement shown in Fig. 1 was used to measure the reflectivity of the multi-layer as a function of incidence angle. The two-dimensional CCD readout permits one to measure reflectivity values simultaneously over a range of photon energies (Fig. 3). For simplicity we present the behavior at just one energy (11.56 keV) slightly below the absorption edge. Data were accumulated over a range of glancing angles from 0.2 mrad to 7 mrad, the sample tilted with respect to the incoming beam in increments of 0.1 mrad.

Fig. 4 is an example of the data obtained in this way; the absorption edge is clearly visible on the specularly reflected beam. Note that at small angles a portion of the direct beam escapes over the top of the sample; this is useful for calculating values of the absolute reflectivity and as a reference "I_o" for absorption measurements.

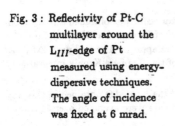

Fig. 3 : Reflectivity of Pt-C
multilayer around the
L_{III}-edge of Pt
measured using energy-
dispersive techniques.
The angle of incidence
was fixed at 6 mrad.

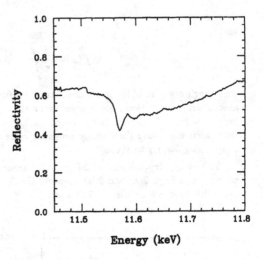

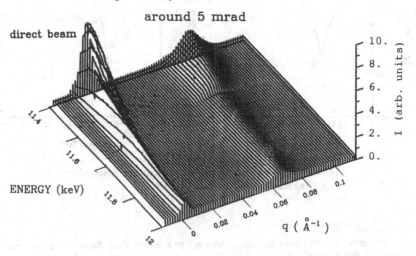

Fig. 4 : Two-dimensional CCD readout of Pt-C multilayer glancing-angle reflectivity.

In Fig. 5(a) we show the measured reflectivity curve at 11.56keV. The reflectivity decreases rapidly, as expected, towards the critical angle, θ_c, at ~6.6 mrad. This is in good agreement with the calculated value from $\theta_c = \sqrt{2\delta}$ (= 6.56 mrad) using the complex refractive index ($\tilde{n} = 1 - \delta$) determined from the optical constants of the multilayer. As a first approximation the contribution of the carbon layers can be neglected. More accurate calculations, as described below, include the carbon optical constants explicitly.

Dynamical Calculations of the Reflectivity

The fine structure on the reflectivity curve is a result of the multilayering geometry. In order to prove this, it is necessary to calculate the reflectivity allowing reflection and transmission at each interface. Note that well below θ_c the evanescent wave penetrates only the first ~ 50 Å below the surface.[11] At higher angles, the multiple scattering at the various interfaces will become important. The reflection coefficient at the interface between the jth and (j+1)th layers given by:[12]

$$R_{j,j+1} = a_j^4 \left(\frac{R_{j+1,j+2} + T_{j,j+1}}{R_{j+1,j+2}T_{j,j+1} + 1} \right) \tag{3}$$

where

$$T_{j,j+1}^\sigma = \frac{g_j - g_{j+1}}{g_j + g_{j+1}} \tag{4}$$

and

$$g_j = (\tilde{n}_j^2 - \cos^2\theta)^{1/2} \tag{5}$$

$T_{j,j+1}^\sigma$ is a transmission coefficient (for σ polarization) which depends on the complex refractive index, $\tilde{n}_j$, and the angle of incidence θ; a_j is an absorption factor. The computational scheme starts at the substrate and works iteratively back to the surface. The resulting reflectivity is related to the intensity by $I(\theta)/I_o = (R_{12})^2$.

The calculated reflectivity is compared with the measured curve in Fig. 5. The feature at ~ 5.5 mrad is clearly reproduced in the calculated reflectivity curve. The other prominent feature, a dip at ~ 4 mrad is found to be very sensitive to the carbon optical constants. Better values for the latter are required in order to improve the agreement with the measured curve. Even so, the overall fit is reasonable indicating that the interfaces are abrupt and that there is very little intermixing of the constituents. This is also confirmed in our conventional diffraction measurements of the 00ℓ profile.

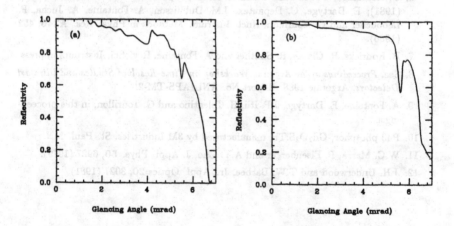

Fig. 5 : Glancing-angle reflectivity curves for Pt-C multilayer at 11.56 KeV:
(a) measured; (b) calculated.

CONCLUSIONS

We have demonstrated that a two-dimensional CCD detector, in conjunction with energy dispersive optics, can be used to obtain simultaneous scattering data and photon energy dependence. The method is useful for studies of the complex optical constants and for reflectivity determination close to absorption edges. These techniques are also promising for time-resolved surface EXAFS and glancing angle diffraction measurements.

ACKNOWLEDGEMENTS

The work was supported in part by DoE contract 31-109-Eng-38.

One of us (RC) was supported by NSF Low Temperature Program Grant DMR8805156 and is grateful for a Visiting Scientist grant from the French Ministry of Research. BR is supported by an Argonne Studentship and FL by ARO Fellowship DAAL-03-86-G-0053.

REFERENCES

1. See, for example, I.T. McGovern, D. Norman and R.H. Williams in *Handbook on Synchrotron Radiation*, Vol. 2, edited by G.V. Marr (North Holland, Amsterdam, 1987) p. 467.

2. E.A. Stern and S.M. Heald, *Ibid*, Vol. 1B, edited by E.E. Koch (North Holland, Amsterdam, 1983) p. 955.

3. S.M. Heald and G.M. Lamble, in this proceedings.

4. I.K. Robinson, W. Waskiewicz, P.H. Fuoss, J.B. Stark and P.A. Bennett, Phys. Rev. B33, 7013 (1986).

5. R.W. James, *The Optical Principles of the Diffraction of X-rays*, (Ox Bow, Woodbridge, CT, 1982) p. 135.

6. U. Kaminaga, T. Matsushita and K. Kohra, Japan, Jnl. Appl. Phys. 20, L355 (1981); E. Dartyge, C. Depautex, J.M. Dubuisson, A. Fontaine, A. Jucha, P. Leboucher and G. Tourillon, Nucl. Instrum. Methods in Phys. Res. A246, 452 (1986).

7. B. Rodricks, R. Clarke, R. Smither and A. Fontaine, Rev. Sci. Instrum., in press.

8. See, *Proceedings of the Argonne Workshop on Time-Resolved Studies and Ultrafast Detectors*, Argonne 1988. Report No. ANL/APS-TM-2.

9. A. Fontaine, E. Dartyge, J.P. Itii, H. Tolentino and G. Tourillon, in this proceedings.

10. P43 phosphor, Gd_2O_2S:Tb, manufactured by 3M Industries, St. Paul. MN.

11. W.C. Marra, P. Eisenberger and A.Y. Cho, J. Appl. Phys. 50, 6927 (1979).

12. J.H. Underwood and T.W. Barbee, Jr., Appl. Optics 20, 3027 (1981).

SURFACE ROUGHNESS AND CORRELATION LENGTH DETERMINED FROM X-RAY DIFFRACTION LINE SHAPE ANALYSIS ON GERMANIUM (111)

Q. SHEN, J.M. BLAKELY[+], M.J. BEDZYK[*] AND K.D. FINKELSTEIN[*]
School of Applied and Engineering Physics
Cornell University, Ithaca, NY 14853
+ Also at Department of Materials Science and Engineering
* Also at Cornell High Energy Synchrotron Source (CHESS)

ABSTRACT

In an x-ray diffraction experiment performed on a germanium (111) crystal, both the rod-like and the diffuse-like scattering from the surface have been observed on a nonspecular crystal truncation rod. These scattering contributions can be explained using existing theory on surface roughness. Two treatments to the Ge (111) surface have been used to provide examples with different roughness characteristics for this study. Quantitative analysis results in a surface roughness of 2.5±0.3Å for a clean surface passivated with iodine and 4.3±0.5Å for a Syton polished surface covered with a naturally grown oxide layer. A typical lateral scale of flat surface regions has also been obtained from the transverse width of the diffuse-like scattering peak, and found to be 200 Å and 400 Å respectively.

INTRODUCTION

X-ray diffraction from crystal surfaces and interfaces can be used not only to determine reconstructed surface or overlayer structures, by analyzing superlattice peak positions and intensities [1,2], but also to study surface roughness and morphology with monolayer sensitivity. Measurements of the second type are made on the so-called crystal truncation rods (CTR); the intensity in these rods arises from the abrupt termination of bulk materials at the surface [3,4]. For a perfectly truncated flat crystal surface, the truncation rod is an exact 2-D δ-function (rod-like) with a structure factor $F_{CTR} = 1/\sin(Q_z c)$, where Q_z is the perpendicular momentum transfer and c is the lattice constant normal to the surface [4]. As pointed out by several authors [3,5], when the surface is rough, the diffracted intensity consists of two parts:

$$I(q) = |F_c F_{CTR}|^2 \left\{ g(q_z) \, \delta(q_r) + h(q_r, q_z) \, [1 - g(q_z)] \right\}, \quad (1)$$

($h(q_r, q_z)$ is peaked at $q_r = 0$ and $\int h(q_r, q_z) \, dq_r = 1$, where q_r and q_z are the momentum transfers parallel and normal to the surface respectively, measured from a nearest bulk Bragg point, F_c is the unit cell structure factor.) The true truncation rod intensity, namely the δ-function term, is multiplied by a Debye-Waller-like factor $g(q_z)$, which is the Fourier transform of the surface height probability distribution function $P(z)$; this causes the intensity to decay faster with increasing q_z than for a perfectly sharp interface. In addition, because of the decrease in long range order brought about by random surface steps, there is a broad diffuse scattering peak around the truncation rod, represented by $h(q_r, q_z)$, which is basically the Fourier transform of the surface height-height correlation function $C(r) \equiv \langle z(r)z(0) \rangle$, when the rms roughness is small compared to $1/q_z$.

For the specular rod near the origin (Fresnel reflectivity region), extensive studies on both the rod and the diffuse scattering on x-ray mirrors and liquid crystal surfaces can be found in the literature [5-11]. For nonspecular truncation rods, however, very few experiments on the diffuse scattering part have been reported [12]. In this article, we describe an x-ray diffraction experiment on a Ge (111) single crystal, in which we have observed both the rod and the diffuse contributions to the scattered intensity near a nonspecular crystal truncation rod. We will demonstrate the importance of the diffuse scattering and show how it helps to determine the surface roughness in both the normal and lateral directions.

EXPERIMENTAL

The experiment was performed at the C2 Station of the Cornell High Energy Synchrotron Source (CHESS). As shown in Figure 1, a flat and a saggital focusing Si (111) crystal were used to provide monochromatic radiation of wavelength 1.459 Å, with higher order harmonics eliminated by a flat quartz mirror. The incident beam was reduced to 2 mm horizontally by 0.7 mm vertically by slits and monitored with an ionization chamber. The diffracted beam was measured by a NaI detector, with its in-plane resolution $\Delta q_{2\theta}=0.022$ Å^{-1} for the full width at half maximum (FWHM), as determined by a 5 mrad Soller slit. The resolution perpendicular to the vertical diffraction plane was 0.15 Å^{-1} (FWHM), due to the horizontal divergence of the incident beam and a 1.2cm horizontal detector slit. The transverse resolution of $\Delta q_{\theta}=0.00025$ Å^{-1} was dominated by the vertical divergence of the incident beam, which was measured by a θ-scan on the weak Ge (222) forbidden bulk reflection and found to be 15 arc-seconds (FWHM). It was this high resolution that allowed our line shape analysis, while the relatively low resolutions in the other two directions actually helped the observation of the diffuse-like scattering since the arrangement effectively integrated the diffuse peak perpendicular to the rocking scan direction.

The Ge (111) sample with a surface area of 2.0cm x 1.4cm was mounted at the center of a standard four circle diffractometer, operated in its symmetric mode ($\omega=0$). The sample was cut with its surface parallel (<0.1°) to the (111) atomic planes and was aligned in such a way that the (111) reciprocal vector coincides with the ϕ-axis of the diffractometer. The sample had been pretreated by Syton polishing, sputtering and annealing under ultra-high vacuum conditions and then exposed to air for about two months. Two sets of data were obtained on the same sample: one with the sample as described (Case A), and the other after the sample was treated by a HF etch, followed by an iodine-methanol nonabrasive pad polishing procedure [13], and then put into a helium gas environment (Case B). The wet chemistry treatment causes the Ge surface to be passivated by iodine.

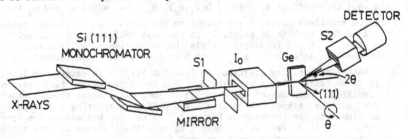

Fig. 1. Illustration of the experimental set-up and the scattering geometry.

For convenience, a hexagonal reciprocal unit cell is adopted, defined by $(300)_h=(22\bar{4})_c$, $(030)_h=(24\bar{2})_c$, $(003)_h=(111)_c$, so that a perpendicular momentum transfer is represented by a single Miller index ℓ. The measurements were done mainly on the $(10)_h$ truncation rod, the one connecting the $(11\bar{1})_c$ and $(220)_c$ (cubic) bulk reflections. At each ℓ along the rod, a transverse rocking curve was measured with a fixed 2θ detector position. A typical counting time at each θ point was 2 seconds, with an incident flux of 4×10^9 photons per second.

RESULTS AND ANALYSIS

A series of typical rocking curve scans are shown in Figure 2, for both Case A and Case B. After subtracting a constant background, which is mainly due to thermal diffuse scattering, we found that the line shape of each rocking curve could be described by a sum of two Lorentzians, one sharp and the other broad. The sharp Lorentzian is from the true rod-like scattering,

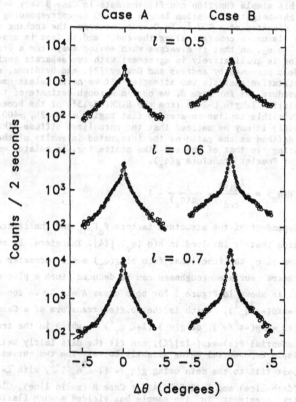

Fig. 2. Rocking scans transverse to the crystal truncation rod $(10\ell)_h$, for Case A and Case B, at ℓ = 0.5, 0.6 and 0.7. The open circles are experimental data points and the solid lines are the χ^2 fits using two Lorentzians. The thermal diffuse scattering background has been subtracted out on each plot.

whose finite width is due to the combined effects of instrumental resolution, surface domain size and domain misorientation. The averaged values of the χ^2 fitted FWHM's are 0.028° for Case A and 0.070° for Case B, which yield minimum surface domain sizes of ~1100 Å and ~450 Å respectively. In both cases the instrumental resolution is sufficiently good (~0.004°) that it need not by considered. We believe that the smaller domain size or the greater misorientation in Case B is due to the fact that the sample was not annealed after the wet chemistry treatment.

The broad Lorentzian is due to the diffuse-like scattering represented by $h(q_r,q_z)$ in eq.(1). It can be shown that an isotropic exponential correlation function $C(r)= \exp(-r/L)$ has a Fourier transform which is a Lorentzian in reciprocal space, with a FWHM of $\Delta q_x = 2/L$, if one integrates over the q_y direction, i.e.

$$h(q_r,q_z) = \int dq_y \int \exp\left(-\frac{r}{L} + iq_r \cdot r\right) dr = \frac{4\pi/L}{q_x^2 + 1/L^2} . \tag{2}$$

The integration over q_y is effectively done in the experiment because of the broad resolution function perpendicular to the q_x (θ-scan) direction. We found that this simple function can fit the data in Case B very well, with a surface height-height correlation length $L = 100$ Å, corresponding to a mean flat region diameter of ~ 200 Å. In Case A, however, the rocking curve data show that Δq_x is not a constant along the rod, and in fact is approximately proportional to q_z, so that it diverges when moving away from a Bragg point. This behavior is qualitatively in agreement with two separate surface statistical models proposed by Andrews and Cowley [3], and by Sinha, et al [5], but no quantitative analysis was attempted. If we nevertheless use the same fitting procedure as for Case B, we obtain a rough estimate of the height-height correlation length L~200Å from the HWHM(=0.15°) of the broad Lorentzian at $\ell = 0.5$; this implies an average flat region size, 2L, ~400 Å.

From eq.(1) it may be noticed that the normalized diffuse-like intensity, $\beta(q_z)$, defined as the ratio of the integrated intensity of the diffuse-like scattering to that of the rod-like scattering, is related only to the surface height Fourier transform $g(q_z)$,

$$\beta(q_z) = \frac{I_{diff}}{I_{rod}} = \frac{1}{g(q_z)} - 1 , \tag{3}$$

which is independent of the structure factors $F_c F_{CTR}$ and the lateral correlation function that is involved in $h(q_r,q_z)$ [14]. Therefore, by analyzing β as a function of q_z the functional form of $g(q_z)$ and a parameter related to the mean square surface roughness can be deduced. Such a plot around the $(101)_h$ point is shown in Figure 3 for both cases A and B. We found that either $g(q_z) = \exp(-\sigma^2 q_z^2)$, which is the Fourier transform of a Gaussian distribution $P(z) = \exp(-z^2/\sigma^2)$, or $g(q_z)=(1+\zeta^2 q_z^2)^{-2}$, which is the transform of a simple exponential $P(z)=\exp(-|z|/\zeta)$, can fit the data fairly well, the fit with the simple exponential being slightly better. The two curves in Figure 3 are the best fits to the data using $g(q_z)= (1+\zeta^2 q_z^2)^{-2}$, with $\zeta= 4.3\pm0.5$ Å for Case A (dash line) and $\zeta= 2.5\pm0.3$Å for Case B (solid line). Clearly, the wet chemistry treatment to the sample has yielded a much flatter crystal surface. This type of atomic-scale flatness on semiconductor surfaces after this treatment has been previously inferred from spectroscopic ellipsometry observations [13]. It should be noted that because the measurements involve large momentum transfers on a nonspecular rod, the surface that the x-rays

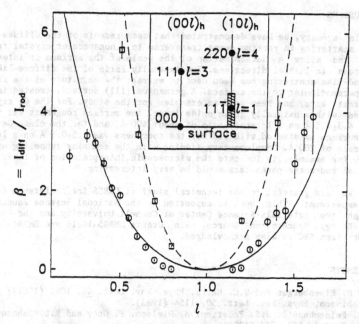

Fig. 3. Intensity ratio $\beta = I_{diff}/I_{rod}$ as a function of perpendicular momentum transfer q_z on the nonspecular rod $(10\ell)_h$. The experimental data are plotted as open circles for Case B and squares for Case A. The curves are the calculations using $\beta(q_z) = (1 + \zeta^2 q_z^2)^2 - 1$, with $\zeta = 4.3$ Å (dash line) for Case A and $\zeta = 2.5$ Å (solid line) for Case B. Note that no corrections for surface area and other scattering geometry related factors are necessary because of the normalized ratio β. The shaded region in the inset shows the locations of the experimental points in reciprocal space.

are probing is actually the interface between the amorphous oxide layer and the underneath crystal structure.

One could also describe the surface roughness from the total integrated intensity $I_{tot}(q_z) = I_{rod} + I_{diff} = |F_c F_{CTR}|^2$ by inclusion of incomplete layers in F_{CTR}. However, in that description one loses the information on the lateral scale of the roughness. In our analysis the magnitude of surface roughness and the lateral scale of flat regions are treated as two independent aspects of surface morphology. The surface in Case A has a larger roughness magnitude ζ than in Case B, but it is actually "smoother" in the lateral scale because of its greater height-height correlation length L. This kind of distinction may be important in applying measurements of the type reported here to the study of thermal roughening transitions on crystal surfaces [12,15].

CONCLUSIONS

In summary, we have demonstrated that measurements of the diffuse-like x-ray scattering in rocking scans transverse to a nonspecular crystal truncation rod allow a determination of the scale of the surface or interface roughness in lateral directions. The intensity ratio of the diffuse-like to the rod-like scattering has been used to extract the magnitude of the roughness perpendicular to the surface. A germanium (111) surface, treated in two different ways, has been used as examples for the study. For the Ge crystal covered with a naturally grown oxide layer, the surface roughness was found to be 4.3±0.5 Å on a lateral scale of about 400 Å, and for the clean Ge crystal surface passivated with iodine, the roughness was 2.5±0.3 Å on a lateral scale of 200 Å. Complementary studies with the scanning tunneling microscope, for example, to indicate the microscopic interpretation of these statistical roughness parameters would be very interesting.

We are grateful to the technical staff at CHESS for assistance during our experiment. This work is supported by the National Science Foundation through the Materials Science Center at Cornell University and the Cornell High Energy Synchrotron Source, via Grants DMR85-16616 and DMR84-12465. Support from SRC is also acknowledged.

REFERENCES

1. P.M. Eisenberger and W.C. Marra, Phys. Rev. Lett. 46, 1081 (1981); I.K. Robinson, Phys. Rev. Lett. 50, 1154 (1983).
2. R. Feidenhans'l, J.S. Pedersen, M. Nielson, F. Grey and R.L. Johnson, Surf. Sci. 178, 927 (1986).
3. S.R. Andrews and R.A. Cowley, J. Phys. C 18, 6427 (1985).
4. I.K. Robinson, Phys. Rev. B 33, 3830 (1986).
5. S.K. Sinha, E.B. Sirota, S. Garoff and H.B. Stanley, Phys. Rev. B 38, 2297 (1988).
6. P.A.J., de Korte and R. Laine, Appl. Opt. 18, 236 (1979).
7. E.L. Church, Proc. Soc. Photo-opt. Instr. Eng. 184, 196 (1979).
8. D.H. Bilderback, Proc. Soc. Photo-Opt. Instr. Eng. 315, 90 (1981).
9. P.S. Pershan and J. Als-Nielsen, Phys. Rev. Lett. 52, 759 (1984).
10. T. Matsushita, I. Ishikawa and K. Kohra, J. Appl. Cryst. 17, 257 (1984).
11. H. Hogrefe and C. Kunz, Appl. Opt. 26, 2851 (1987).
12. G.A. Held, J.L. Jordan-Sweet, P.M. Horn, A. Mak and R.J. Birgeneau, Phys. Rev. Lett. 59, 2075 (1987).
13. D.E. Aspnes and A.A. Studna, Proc. Soc. Photo-Opt. Instr. Eng. 276, 227 (1981).
14. Strictly speaking, this separation between the structure factors and the statistical roughness is true for Case B only if the iodine coverage is independent of surface height and, therefore, of the lateral position r. We believe this is a reasonable assumption for relatively flat surfaces.
15. K.S. Liang, E.B. Sirota, K.L. D'Amico, G.J. Hughes and S.K. Sinha, Phys. Rev. Lett. 59, 2447 (1987).

PART V

X-Ray Microprobe, Lithography
and VUV Techniques

X-RAY MICROPROBE FOR THE MICROCHARACTERIZATION OF MATERIALS

C. J. SPARKS and G. E. ICE
Metals and Ceramics Division, Oak Ridge National Laboratory, P.O. Box 2008,
Oak Ridge, TN 37831-6117

ABSTRACT

The unique properties of X rays offer many advantages over those of
electrons and other charged particles for the microcharacterization of
materials. X rays are more efficient in exciting characteristic X-ray
fluorescence and produce higher fluorescent signal-to-background ratios than
obtained with electrons. Detectable limits for X rays are a few parts per
billion which are 10^{-3} to 10^{-5} lower than for electrons. Energy deposition
in the sample from X rays is 10^{-3} to 10^{-4} less than for electrons for the
same detectable concentration. High-brightness storage rings, especially in
the 7 GeV class with undulators, will have sources as brilliant as the most
advanced electron probes. The highly collimated X-ray beams from undulators
simplify the X-ray optics required to produce submicron X-ray probes with
fluxes comparable to electron sources. Such X-ray microprobes will also
produce unprecedentedly low levels of detection in diffraction, EXAFS, Auger,
and photoelectron spectroscopies for structural and chemical characterization and elemental identification. These major improvements in microcharacterization capabilities will have wide-ranging ramifications not only in
materials science but also in physics, chemistry, geochemistry, biology, and
medicine.

INTRODUCTION

Efforts to obtain micro-diffraction and micro-elemental analysis in
order to understand the properties of matter have been made since the early
part of this century when the first X-ray probes were constructed. Conferences have been held on the design and application of X-ray microprobe
sources [1]. In the early 1950s, there was an upsurge in interest in
microanalytical methods. Electron microprobes quickly became the instrument
of choice. Electrons could easily be focused to spot sizes less than
1 μm-diam with enormous intensities compared to the weak X-ray sources then
available. Though X-ray excitations were known to give a much better signal-
to-background ratio, their generation required electron bombardment of a
metal target with X rays emitted into 4π sr which then had to be gathered
and focused to a small spot with inefficient X-ray optics. With the commercial availability of high-intensity and high-resolution electron microprobes, the unique properties of photons could not overcome their lack of
intensity from weak sources. Interest in X rays as an excitation source for
microprobe analysis faded, leaving electrons as the dominant microprobe
source. It is a conservative estimate that 2500 electron probes and
microscopes currently are in use for microcharacterization of matter in the
United States. This represents an investment of $1 billion or more in
instrumentation alone. The wide usage of analytical microprobes is well
documented in the several yearly conferences and journals of the electron
microscopy and microprobe societies and attests to the great emphasis
being placed by the scientific community on the need for microstructural
characterization of matter.

With the advent of electron storage rings, an intense source of X rays
has become available. The energy spectrum in the hard X-ray region is 10^4
to 10^5 times more brilliant than our conventional X-ray sources (2 kW to
60 kW dissipated by electrons impinging on metal targets). In fact, with
magnetic devices especially suited for extracting the radiation from proposed low emittance storage rings, the spectral brilliance in units of

Mat. Res. Soc. Symp. Proc. Vol. 143. ©1989 Materials Research Society

photons (or electrons) s^{-1} mm^{-2} $mrad^{-2}$ is near 10^{19} and approaches the brilliance of 3×10^{19} from the most advanced electron probes having field-emission electron guns [2–4]. This brightness is for the energy range from 1 keV to 35 keV from undulators on low-emittance electron storage rings in the 7 GeV energy range. This X-ray energy range covers the electron energy levels of the K- and/or L-shells for all the atoms and is most useful for EXAFS, fluorescent and diffraction analysis. Existing X-ray storage rings are about 10^{-3} to 10^{-6} less bright in this energy range.

As electron microprobes have clearly dominated the field of micro-characterization, the merits of the use of X rays for excitation of the sample will be compared to electrons. There are too many possible applications where microcharacterization is important to advancing our understanding of materials properties to be covered here. For more extensive information the reader should refer to the electron microscopy and microprobe literature and to Ref. 5. Actual applications of X-ray microprobes analysis are just beginning to emerge.

DETECTION LIMITS WITH X RAYS VERSUS ELECTRONS

A standard definition of the minimum detectable mass fraction [6] based on Poisson counting statistics for 95% confidence in detection is

$$\text{Minimum Detectable Mass Fraction (MDMF)} = 3.29 \, C_Z(N_b)^{1/2}/N_S , \qquad (1)$$

where C_Z is the mass fraction of element Z, N_b the background counts, and N_S the counts in the signal. To evaluate Eq. (1), we need to know the fluorescent cross sections to calculate the number of events in the signal and to evaluate the background. These data are presented in Figs. 1–3 [7].

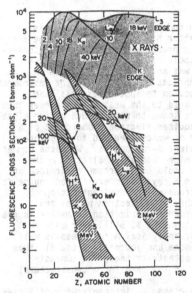 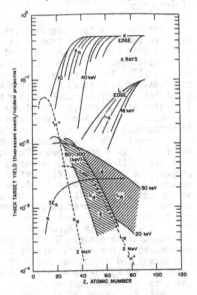

Fig. 1. Fluorescence cross sections for the elements when excited by X rays, protons ($^1H^+$), and electrons (e) of varying energies (after Ref. 7).

Fig. 2. Thick-target fluorescence yields of pure elements for X rays, protons, and electrons of varying energies (after Ref. 7).

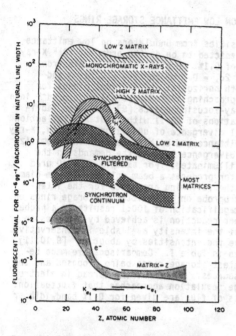

Fig. 3. Comparison of the fluorescent signal-to-background ratio for various excitation radiations at a concentration of 10^{-6} gg^{-1} for an X-ray fluorescent radiation detection system with an energy resolution of the natural linewidth (after Ref. 7).

As shown in Figs. 1 and 2, the number of characteristic fluorescent events is typically 10 to 200 times larger for X-ray excitation than for the same number of electrons. As shown in Fig. 3 the fluorescent signal-to-background ratio is approximately 10^4 times larger for X-ray excitation than for electron excitation. This extremely favorable property of X rays derives from the fact that about 90% or more of the incident X-ray energy is dissipated by ionization of the inner shells which gives rise to the fluorescence of interest. In contrast, only 0.1% of the energy dissipated by electrons gives rise to the fluorescent radiation of interest. Much of their energy is consumed by interactions with the least-bound outer-shell electrons. The deceleration of the incident electrons in the target produces a broad spectrum Bremsstrahlung radiation which is responsible for most of the background beneath the fluorescent signals of interest.

With these greater signals and lower backgrounds we find that 10^{-3} fewer X rays than electrons are required for the same MDMF. In terms of energy deposited, electron energies from 20 keV to 100 keV usually exceed by three to ten times the ionization energy of the bound electron. X-ray energies between 2 keV to 33 keV can be chosen to lie just above the ionization energy. As 10^3 more electrons with energies from three to ten times those of X rays are required for the same MDMF, the energy deposited by electrons is 3×10^3 to 10^4 times that deposited by X rays in thick targets. For very thin targets where most of the incident electron energy is transmitted, even more of the incident X-ray energy would be transmitted with still less damage to the sample. In air-dried blood cells and 8 μm-thick tissue sections [8] exposed to synchrotron radiation and charged particles, X-ray fluences 10^2 to 10^3 times greater were required to produce similar damage. Therefore, X-radiation damage to the samples may be even orders of magnitude less than the comparison made here on the basis of energy deposition alone. Thus with X-ray excitation we have the important choice of being able to either lower the detectable limits for the same fluences or to reduce the radiation damage and heat deposited in the sample for the same MDMF. Since heating of metal and ceramic samples and both heating and radiation damage of organic samples is of primary concern in modern electron probes, X rays have a major advantage. Before we can make a direct comparison between modern electron microprobes and the proposed X-ray microprobe, we must determine how many X rays can be put into very small beams.

X-RAY INTENSITIES FROM UNDULATORS ON LOW EMITTANCE STORAGE RINGS

With the proposed X-ray intensities from undulators on low emittance storage rings [2,3], fluxes are predicted to be on the order of 10^{18} X-rays s^{-1} in an energy bandwidth, $\Delta E/E$, of 0.1%. Projections of an electron source size in the storage ring of $2\sigma_x = 0.8$ mm and $2\sigma_y = 0.14$ mm and divergences of about 40 µrad in both horizontal and vertical planes for undulators produce a brilliance approaching 10^{18} X rays s^{-1} mm^{-2} mrad^{-2} in a 0.1% energy bandwidth. Actual X-ray-focusing optics [9] have been fabricated and shown to have demagnifications of 287:1 with about 25% reflection efficiencies. Because of the small divergence of undulator radiation, X-ray optics can intercept the entire emittance and both focus and monochromatize with high efficiency. This small divergence lessens the demands on the optics since only a small area is illuminated. For example, the 40 µrad divergence from a practical undulator produces a beam only about 1 mm-diam at 25 m from the source. We base the following arguments for the intensities obtainable for an X-ray microprobe on low emittance storage rings on a demagnification of 100:1. A demagnification of 100:1 results in a probe size of about 2×8 µm², and further reduction is achieved by pinholes in heavy metal foils. The estimate of the intensity available is conservative. Other optical systems could improve the intensities by about 10^2 [9,10,11]. The predicted intensities are given in Table I. Comparisons are made between photon fluxes from the brightest storage rings being proposed and the most intense electron microprobe sources. Note that the spread in electron energy is listed as 1 eV as voltage regulation approaches that fluctuation. For X-ray undulators, typical units of flux are given for 0.1% bandwidth or 10 eV spread for 10 keV X rays.

Table I. Comparison of X-ray microprobe intensities from undulators on a 7 GeV storage with a modern high brightness electron probe source.

	10 keV X-rays 7 GeV, 100 mA undulator	100 keV e⁻ Electron microprobe
Brightness photons or e⁻/s mm² mrad²	1×10^{18} [a,b,c]	3×10^{19} [d]
Intensity P or e⁻/area s	$\dfrac{1 \times 10^{18}}{(3.2\ \mu m)^2\ s\ 10\ eV}$	
	$\dfrac{1 \times 10^{14}}{\mu m^2\ s\ 10\ eV}$	$\dfrac{6 \times 10^{13}\ [e]}{\mu m^2\ s\ eV}$
	$\dfrac{3 \times 10^{11}}{(500\ \text{Å})^2\ s\ 10\ eV}$	$\dfrac{6 \times 10^{11}\ [e,f]}{(500\ \text{Å})^2\ s\ eV}$
		$\dfrac{6 \times 10^{9}\ [f]}{(30\ \text{Å})^2\ s\ eV}$
		$\dfrac{10^{7}\ [f]}{(4\ \text{Å})^2\ s\ eV}$

[a]National Synchrotron Light Source, Planned Evolution of NSLS, October 1983.
[b]H. Wiedemann, Nucl. Instr. and Methods in Phys. Res. A266, 24 (1988).
[c]G. K. Shenoy, P. J. Viccara, and D. M. Mills, Argonne National Laboratory Report ANL-88-19 (October 1988).
[d]N. J. Zaluzec, Quantitative X-ray Microanalysis: Instrumental Considerations and Applications to Materials Science, Chapter 4, pp. 121–67 in *Introduction to Analytical Electron Microscopy*, J. J. Hern, J. I. Goldstein, and D. C. Joy, eds., New York: Plenum Press, 1979.
[e]LaB₆ electron source.
[f]Field emission source.

At 0.1% bandwidth (10 eV at 10 keV), the brilliance of undulators will match that available from the most advanced electron probes. Though undulators peak the intensity at harmonic energy intervals, energies ranging from 2 keV to 40 keV would be available from the first through third harmonic. For fluorescent excitation, large energy spreads of $\Delta E/E = 1$ are acceptable [7]. For Bragg diffraction to identify the compounds present, a $\Delta E/E = 10^{-2}$ to 10^{-3} is useful. For microprobe EXAFS analysis and photoelectron spectroscopy, $\Delta E/E$ should be 10^{-4} to 10^{-5} to match the energy width of the electron energy levels. These various energy resolutions can be achieved with presently available optics from multilayers to nearly perfect crystals, and the microprobe energy resolution tailored to meet the experimental needs of a variety of applications.

SPATIAL RESOLUTION

Inherently, X-ray excitation for microprobe analysis offers the highest spatial resolution in thick samples of any radiation because the low-scattering cross section for X rays limits the lateral spreading of the beam in the sample. However, X rays are more difficult to focus, and their source size is larger. Electron field-emission source sizes approach 10 Å-diam and can be focused to diameters of atomic dimensions. However, lateral spreading of the electrons in matter produce interaction regions of about 1 μm-diam in thick samples [12]. Electron-probe analysis of ferrous materials having specimen thicknesses of 1000 Å to 2000 Å produce an interaction region of approximately 500 Å-diam even though the incident probe diameter is much less. Diffraction limits and beam penetration may keep the useful X-ray probe diameter to 500 Å and greater [11]. An X-ray microprobe with a beam diameter of 500 Å to 1 μm would compete favorably with the spatial resolution of electron microprobes in a great majority of the samples of interest. Because X rays have greater penetrating power than electrons, the convergence angle of the X rays at a demagnification of 100:1 places a limit of about 10 μm to the depth of beam penetration in the sample before divergence of the X-ray beam in the sample exceeds the intended probe size of 1 μm diam.

FLUORESCENT DETECTION LIMITS

To determine the fluorescent detection limits from Eq. (1), we use the data presented in Figs. 1–3 and Table I. We assume a fluorescent detection system equipped with a crystal analyzer intercepting 10^{-3} of the total solid angle and with an energy resolution matching the K_α or L_α fluorescent energy widths. In this case MDMF is about 10^{-4} times less than for the best electrons probes with similar wavelength dispersive optics.

As the detection of elements at interfaces is important to many materials problems, we calculate the detectable levels for a planar distribution of elements. Assume that one monolayer of an impurity element replaces one 2 Å-wide atomic plane. For an X-ray probe diameter of 1 μm², 5000 atomic planes end-on would be irradiated by the beam. Since one of the atomic planes out of 5000 consists of impurity atoms, the concentration of the impurity is 2×10^{-4} gg^{-1} in the volume irradiated by the probe. This assumes that none of the impurity exists outside the boundary. Since the MDMF is 10.5×10^{-9} gg^{-1} s^{-1}, then the detection of 5×10^{-5} of a monolayer s^{-1} of impurity is feasible. However, in most materials we expect some of the impurity to be distributed in the matrix. Typical elemental concentrations of 2×10^{-3} to 10^{-4} gg^{-1} are used to affect grain-boundary behavior. Plotted in Fig. 4 are the calculated profiles of X-ray-probe scans for an iron sample containing 0.1 wt % titanium uniformly distributed in the matrix and with one monolayer of titanium in the grain boundary. The shape and size of the microprobe beam can be defined by pinhole apertures to

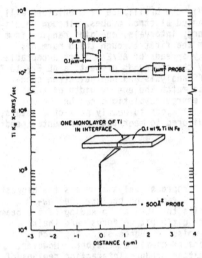

Fig. 4. Calculated fluorescent intensity profile for an X-ray microprobe scan over a grain boundary containing a monolayer of titanium when the iron matrix contains 0.1 wt % titanium (after Ref. 5).

better match the geometry of the interface to improve the contrast and lower the detectable limit. For this case of titanium in iron where 0.1 wt % of the same impurity is in the matrix, the minimum detectable impurity with the plane of the grain boundary parallel to the direction of the probe (end on) is 5×10^{-3} of a monolayer for both the 1 μm^2 and 500 $\AA^2$ probes. As predicted by Eq. (1) a decrease in the probe diameter does not change the detectable limits for a line distribution since the signal decreases linearly and the background decreases as the square of the probe size. If the region next to the boundary is denuded of the impurity, then the smaller probe size has the advantage of better providing the spatial resolution to determine that information. The rectangular-shaped probe has a detectable limit of 6×10^{-4} of a monolayer in the presence of 0.1 wt % in the matrix. Typical experience with advanced analytical electron probes is that 0.1 monolayer of impurity at a boundary is at the detection limit [13]. For typical surface sensitive electron excited Auger spectroscopy the detection limit is about 0.01 of a monolayer [5].

DIFFRACTION AND EXAFS ANALYSIS

X rays also have some advantages over electrons when used for diffraction. Diffraction measurements provide important information such as the crystal structure, compound identification, and how the geometrical arrangements of the atoms deviate from perfect periodicity. For the same number of 10 keV X rays or electrons impinging on a metal sample, the X rays are approximately 200 times more likely than electrons to undergo a useful elastic-scattering event. Electrons are most likely to lose energy by straggling energy-loss processes adding to the unwanted background unless removed by energy-analysis spectrometers. Electrons are also more likely to undergo multiple-scattering events in thick foils which complicate the interpretation of the measured diffraction pattern [14]. X-ray microprobes permit the use of thicker samples reducing the problem of defect migration to interfaces and strain relief which can be a problem in thin samples.

With the criteria expressed in Eq. (1), the MDMF by diffraction with a μm^2 X-ray probe in 1 s is:

10^{-2} of a monolayer, 1.6×10^3 atoms in a particle, and 28-$\AA$-diam particle.

This is a conservative estimate, since an experiment with less flux but for the favorable case of a monolayer of lead deposited on the surface of a copper single crystal produced a minimum detectable coverage of approximately 10^{-3} of a monolayer from the observed 5×10^4 signal counts s^{-1} with a signal to background of 500:1 [15]. For amorphous materials, the diffuse scattering from only six monolayers of matter could be measured with a 1 µm-diam probe. Recent X-ray diffuse-scattering measurements from thin amorphous layers convinced those authors to predict that analysis of 100 Å films is feasible even at intensities 10^{-3} of those proposed for the microprobe [16].

Among the most prominent applications of synchrotron radiation in the X-ray-energy region is the measurement of the extended X-ray absorption fine structure (EXAFS) [17]. Such measurements permit the determination of the average number of near-neighboring atoms and average bond distances about a central atom whose absorption edge is scanned by changing the X-ray energy. The ability to determine the chemical environment of a particular element at low concentrations has resulted in major contributions to our understanding of the role of minor elements in matter [18]. The projected 3×10^{10} photons s^{-1} (eV)$^{-1}$ for a 500 Å-diam probe would extend the ability to make EXAFS measurements on extremely small quantities of matter containing minor elements at concentrations as low as 100 ppb.

MICROPROBE RESULTS

The earliest application of synchrotron radiation to microprobe measurements was by Horowitz and Howell in about 1971 [19]. They achieved 2 µm resolution and rastered samples in front of the beam but reported no detectable limits. This work was followed in 1977 with a quantitative analysis of the detection limits for a probe area of 0.45 mm^2 containing 2×10^{11} X rays s^{-1} of 37 keV energy from the SPEAR storage ring at the Stanford Synchrotron Radiation Laboratory [20]. The detectable limit in one second was found to be 10^{10} atoms of cadmium though the solid state detector energy resolution was sacrificed (450 eV at 39.2 keV) to achieve high counting rates.

More recent developments in X-ray optics and improved storage ring brightness have resulted in beam dimensions of 10 µm × 10 µm containing fluxes of 3×10^8 X rays s^{-1} of 10 keV energy [9,21]. Fluorescent detection limits of 2×10^{-15} g in 60-s counting times were reported. An example of this work is shown in Fig. 5 where the elemental variation along a blue-green algae strand is plotted.

In a microprobe experiment where the beam was formed by a 25 µm-diam pinhole placed at the focus of an optical system with a magnification of unity, a flux of 10^7 X rays s^{-1} at 8 keV were obtained [22]. This 25 µm-diam collimated X-ray beam was used to measure the Bragg diffraction from a

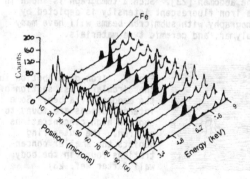

Fig. 5. Energy spectra as a function of position obtained at 10-µm intervals along a filament of blue-green algae. The variation of iron along the filament is highlighted (after Ref. 21).

niobium single crystal in the vicinity of a niobium hydride precipitate. X-ray rocking angle measurements were made to measure the tilt of the niobium crystal planes caused by the growth of the niobium hydride particle. The data are shown in Fig. 6 where the rocking curves are plotted as a function of position, Z, as the beam is stepped over the embedded particle. From data such as these and measurements of the Bragg positions, strain distributions can be determined. Sufficient X-ray flux is available from this simple optical arrangement for Bragg diffraction with beams as small as 0.5 µm in diameter since single crystal diffraction is an efficient process reflecting between 0.1 and 10% of the incident flux. With undulator sources on low emittance storage rings, detection by Bragg diffraction will be possible for single particles of matter no larger than about 30 Å in diameter.

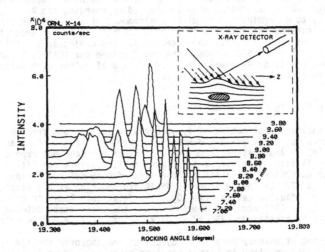

Fig. 6. X-ray diffraction rocking curves taken at 0.2-mm intervals show how the niobium crystal lattice is tilted in the vicinity of a NbH particle. Note also that the rocking curves broaden as the particle is approached (after Ref. 22).

With similar pinhole X-ray optics, X-ray tomography was performed on the abdomen of a bee. Iron fluorescence excited by 9 keV X rays collimated to a 150 µm-diam beam was used to map the iron distribution throughout a transverse cross section of the abdomen [23]. Such a tomograph is shown in Fig. 7 where the change in the iron fluorescent intensity is depicted by the vertical scale. X-ray tomography with submicron beams will have many applications in biological, polymer, and ceramic type materials.

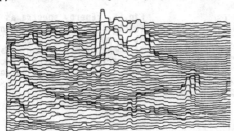

Fig. 7. Iron distribution in the circular cross section of the abdomen of a bee. Some internal organs are found to contain high concentrations of iron. The outer ring depicts the smaller concentrations found in the body wall (after Ref. 23).

CONCLUSION

Though the application of X-ray microprobe analysis is still in its infancy, the scientific promise is so appealing that high brilliance X-ray sources will be forthcoming. Table II compares some of the properties of X rays with those from electrons for the case of fluorescence. It takes little imagination to recognize the revolutionary impact on materials science that will arise when detection limits at surfaces, interfaces, and in bulk are reduced by 10^3 or more. As segregation of elements to defects is the rule, mapping their segregation behavior is essential to understanding their role on the physical and chemical properties of matter.

Though fluorescence detection has been emphasized here, compound identification by Bragg scattering, strain mapping about crack fronts, determining the structure of grain boundaries, and a host of other applications are obvious [5]. An X-ray fluorescent, diffraction, and EXAFS microprobe that can operate in environments such as air, moisture, or through millimeter thick water films opens the door to much new scientific progress. A scanning photoelectron microscope is being proposed in which the spot size on the sample will be diffraction limited to about 250 Å in diameter [24].

Table II. Comparison of the characteristics of an X-ray fluorescent microprobe on proposed low emittance storage rings with those of the most advanced electron microprobes for fluorescent chemical analysis on thick samples (after Ref. 5).

Characteristics	X rays	Electrons
Minimum detectable mass fraction s^{-1} for 1 μm-diam probe	0.01 ppm	50 ppm
Minimum detectable mass s^{-1} for 500 Å probe	250 atoms	10,000 atoms
Minimum spatial resolution (samples ≥ 1 μm thick)	~500 Å	10^3 to 10^4 Å
Minimum spatial resolution (samples 100 Å to 2000 Å thick)	~500 Å	10 to 500 Å
Number of electrons and X rays for the same MDMF	1	10^3
Number of energy units deposited in thick targets for same MDMF	1	10^3 to 10^4
Operating atmosphere	air, gas, water, vapors	vacuum
Relative signal to background (contrast)	10^4	1
Accuracy for quantitative analysis {similar standards / pure element standards	~1% / ~5%	~5% / ~10%
Relative fluorescent cross section	10 to 200	1
Relative thick-target fluorescent yields	10 to 150	1
Charge collection on electrically insulating samples	negligible	must be coated with conducting film

ACKNOWLEDGMENTS

This research was performed in part at the Oak Ridge National Laboratory Beamline X-14 at the National Synchrotron Light Source, Brookhaven National Laboratory, sponsored by the Division of Materials Sciences and Division of Chemical Sciences, U.S. Department of Energy, under contract DE-AC05-84OR21400 with Martin Marietta Energy Systems, Inc.

REFERENCES

1. V. E. Cosslett, A. Engström, and H. H. Pattee, Jr., eds., *X-ray Microscopy and Microradiography* (Academic Press, Inc., 1957).

2. H. Wiedemann, Nucl. Instr. and Methods in Phys. Res. A266, 24 (1988).

3. G. K. Shenoy, P. J. Viccaro, and D. M. Mills, Argonne National Laboratory Report ANL-88-9, February 1988.

4. N. J. Zaluzec, in Introduction to Analytical Electron Microscopy, edited by J. J. Hren, J. I. Goldstein, and D. C. Joy (Plenum Press, 1979), pp. 121–167.

5. C. J. Sparks, in *Major Materials Facilities Committee*, National Research Council, National Academy Press, Washington, D.C., p. 92 (1984).

6. L. A. Currie, Anal. Chem. 40, 586–593 (1968).

7. C. J. Sparks, Jr., in *Synchrotron Radiation Research*, edited by H. Winick and S. Doniach (Plenum Press, 1980), pp. 459–512.

8. D. N. Slatkin, A. L. Hanson, K. W. Jones, H. W. Kramer, and J. B. Warren, Brookhaven National Laboratory Report BNL-34555, 1984.

9. J. H. Underwood, A. C. Thompson, Y. Wu, R. D. Giauque, Nucl. Instr. and Methods in Phys. Res. A266, 296 (1988).

10. M. R. Howells and J. B. Hastings, Nucl. Instr. and Methods in Phys. Res. 208, 379 (1983).

11. G. E. Ice and C. J. Sparks, Nucl. Instr. and Methods in Phys. Res. 222, 121 (1984).

12. J. I. Goldstein, in *Introduction to Analytical Electron Microscopy*, edited by J. J. Hren, J. I. Goldstein, and D. C. Joy (Plenum Press, 1979), pp. 83–120.

13. E. A. Kenik, Scripta Metall. 21, 811 (1987).

14. J. W. Cowley, *Diffraction Physics* (American Elsevier), 1975.

15. W. C. Marra, P. H. Fuoss, and P. E. Eisenberger, Phys. Rev. Lett. 49(16), 1169–1172 (1982).

16. A. Fischer-Colbrie, P. H. Fuoss, M. Marcus, and A. Bienenstock, in Stanford Synchrotron Radiation Laboratory Report 83/01, edited by K. Cantwell, 1983.

17. H. Winick and S. Doniach, eds., Chapters 10–13 in *Synchrotron Radiation Research* (Plenum Press, 1980).

18. A. L. Bianconi, L. Incoccia, and S. Stupcich, eds., *Proceedings of Second International Conference on EXAFS and Near-Edge Structure* (Springer-Verlag, 1983).

19. P. Horowitz and J. Howell, Science 178, 608 (1972).

20. C. J. Sparks, S. Raman, E. Ricci, R. V. Gentry, and M. O. Krause, Phys. Rev. Lett. 40, 507 (1978).

21. A. C. Thompson, J. H. Underwood, Y. Wu, R. D. Giaugue, K. W. Jones, and M. L. Rivers, Nucl. Instr. and Methods in Phys. Res. A266, 318 (1988).

22. G. E. Ice, Nucl. Instr. and Methods in Phys. Res. B24, 397 (1987); and S. Stock, private communication.

23. P. Bouisseau, L. Grodzins, C. J. Sparks, G. E. Ice, and T. Habenschuss, National Synchrotron Light Source Annual Report, BNL-51947, 1985, p. 231.

24. F. Cerrina, G. Margaritondo, J. H. Underwood, M. Hettrick, M. A. Green, L. J. Brillson, A. Franciosi, H. Höchst, P. M. Deluca, Jr., and M. N. Gould, Nucl. Instr. and Methods in Phys. Res. A266, 303–307 (1988).

IMAGING WITH SOFT X-RAYS

D.M. SHINOZAKI
Department of Materials Engineering, The University of Western Ontario,
London, Ontario, N6A 5B9, Canada

ABSTRACT

A review of recent advances in soft X-ray imaging using synchrotron radiation is given.

INTRODUCTION

Interest in utilizing X-rays to study the microstructure of materials has been spurred by recent developments in a variety of areas, which include access to extremely high brightness, high intensity X-ray sources; the refinement of microfabrication methods to manufacture focussing elements usable in the X-ray range of wavelengths; and the increased use of very high resolution recording media. Much of the effort in modern X-ray imaging research has been devoted to instrumental improvements and been largely the domain of the physicist.

The principal reason for the continued interest in imaging with soft X-rays has been the recognition that biological specimens are made of materials with a small absorption coefficient at shorter wavelengths. Thin sections of unstained materials produce little visible contrast in the image in the hard X-ray range. Long wavelength photons, in the visible spectrum, can not be used to resolve extremely small structures. Between these two extremes, in the soft X-ray range, the spectrum between the oxygen and carbon absorption edges (2.33 and 4.36nm.) is particularly suitable for producing contrast in hydrated cellular material since the absorption for water is low in this region, while that for carbon is relatively high (Figure 1).

The prospect of revealing unmodified biological structures and examining structures which have the same dimensions as the wavelength of the imaging radiation has been a motivating force in modern soft X-ray imaging physics [1]. The best examples of images reveal good contrast, and reasonable resolution over very large areas when compared to transmission electron microscope images of similar specimens [2] [3] [4].

A quantitative comparison of the minimum radiation dosage for transmission electron microscopy and soft X-ray microscopy for a given resolution and thickness of specimen has shown that over much of the thickness-resolution ranges used in microscopy, photon imaging affects the specimen less [1]. This is clearly a significant advantage in studying radiation sensitive materials such as crystalline polymers (polyethylene, polypropylene, polyoxymethylene etc.)

The relatively small attenuation of soft X-rays compared to 100 keV electrons also means that specimens thicker than those normally usable for electron microscopy can be used. A comparison of the relative imaging characteristics of electrons and soft X-rays suggests that 2.4 nm X-rays can reveal microstructure in specimens about an order of magnitude thicker than those used in a typical 100 keV transmission of electron microscope (2 to 3 micrometers compared to 0.2 to 0.3 micrometers).

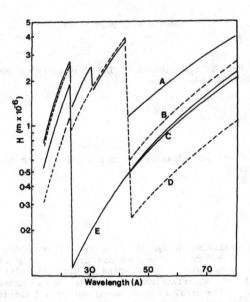

FIGURE 1:   Absorption coefficient as a function wavelength for (A) nucleic acid, (B) carbohydrate, (C) protein, (D) lipid, (E) water

Material	Illuminating Radiation Near Absorption Edge nm	Micrometers $t_{1/2}$
Be	Be-K (11.2)	0.61
BN	B-K (6.6)	0.42
C	C-K (4.37)	1.4
Polypropylene	C-K (4.37)	3.6
Formvar	C-K (4.37)	2.2
Mylar	C-K (4.37)	2.1
BN	N-K (3.1)	0.41
Al_2O_3	O-K (2.33)	0.64
SiO_2	O-K (2.33)	0.69
Fe	Fe-L_3 (1.75)	0.41
Ni	Ni-L_3 (1.45)	0.44
Cu	Cu-L_3 (1.33)	0.49
Mg	Mg-K_3 (0.95)	9.1
Al	Al-K (0.795)	7.3
Si	Si-K (0.674)	10.0
$(Ch_2\ CCl_2)$	Cl-K (0.44)	43.7

TABLE 1:   Thickness of material which reduces the incident intensity by 1/2. The incident radiation has a wavelength just above the relevant absorption edge

The thickness through which significant imaging information can be obtained depends on the material examined and the wavelength of radiation. The absorption coefficient for materials increases as the wavelength of the incident X-ray increases, with edges corresponding to the maximum wavelength (minimum photon energy) which will expel an electron from a given level in the atom. At wavelengths slightly greater than the relevant absorption edge, the transmission of X-rays is relatively high. The thickness of a variety of materials which will reduce the incident intensity to one-half ($t_{1/2}$) is shown in Table 1).

The flux of photons as a fraction of the incident intensity is shown as a function of depth in a specimen for a variety of typical wavelengths (Figure 2). These data indicate that a variety of important synthetic materials will be transparent to soft X-rays even in relatively thick sections.

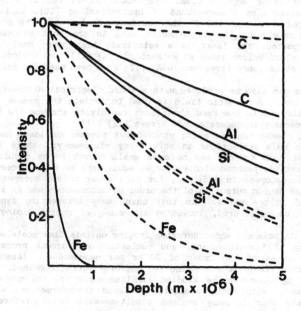

FIGURE 2: Attenuation of soft X-rays in materials using 5A (dotted lines) and 10A (solid lines) radiation. For Al and Si, more absorption is observed for 5A radiation because the absorption edge lies between 5A and 10A.

The contrast in X-ray images can be controlled and optimised for a given material by changing the wavelength. Changing the photon energy involves selecting a portion of the available spectrum using a grating or filter. Since this monochromatization of the X-rays involves removing a large part of the available photons, the overall intensity of the original X-ray source is a crucial limiting aspect of X-ray microscopy.

High Intensity Soft X-ray Sources

In the attempt to achieve diffraction limited resolution using X-rays, a serious restriction until recently has been the low intensity and low brightness of X-ray sources. In order to distinguish between neighboring resolution elements (pixels) which represent different contrast levels, the incident flux must be large enough that the measured difference between the pixels exceeds some detector sensitivity criterion [1] [5] [6]. In the soft X-ray region, the most intense sources are synchrotron radiation and plasmas.

Plasma sources for soft X-rays are those which produce very hot, short duration plasmas. High power lasers focussed on solid targets can be used to ionize a solid target and if the plasma temperature is high enough, significant quantities of UV and soft X-ray photons are produced [7] [8] [9]. Low power laser sources (less than about 10 J) can be used to produce line emission on a background of bremsstrahlung [10], while at higher powers (greater than 100 J) the continuum dominates. The source size is approximately 0.1 mm in diameter, and in the most successful applications reported, the laser is a relatively high power one. For example, the Nd-glass Vulcan laser at Rutherford (UK) [11], and the sources being developed at Lawrence Livermore (USA) [12] have been used for soft X-ray microscopy.

Hot plasmas can also be produced with a short electrical discharge in an appropriate gas. A magnetic field is used to isolate the plasma from the chamber walls and to compress the plasma rapidly to produce the high temperatures necessary to generate soft X-rays [13] [14].

Pulsed hot plasma sources using pinched gas plasmas or laser heated solids have two main attractions in soft X-ray microscopy: there is a possibility that such sources may be made small enough to be usable in standard laboratories, perhaps with resources equivalent to those necessary for electron microscopes; in addition, for high power pulsed sources the time scale of the photon output is of the order of nanoseconds and an image can be recorded in one pulse. In this case, many interesting dynamic experiments can be considered, including the study of motile biological (wet) cells.

For materials science, short duration imaging on this time scale would allow the study of transformation, and nucleation and growth processes which occur at rates of the order of 20 nm per nanosecond. (Assuming optimal achievable resolution). Using the microradiographic method, very large areas of specimen can be studied simultaneously (of the order of several mm square) with high resolution. Rapid transformations which nucleate randomly over in many regions simultaneously can therefore be examined with pulsed sources.

However, the best soft X-ray images produced by pulsed sources have been made using the larger laser sources, and small intense soft X-ray sources have yet to produce soft X-ray images which compare with those produced from synchrotron radiation sources or from laboratory electron impact sources. Attempts to image wet biological materials at high resolution using pulsed sources have not been successful, and many problems related to the specimen preparation and handling have yet to be solved.

There are a number of problems with pulsed soft X-ray sources that must be addressed before they can be used for microscopy. One is the problem of debris from the evaporated solid target in laser generated sources. The output from such sources includes ions, electrons, and particulate matter. The photon output can include UV as well as soft X-rays. In a typical microradiographic experiment, the distance between the source and the specimen to be imagined is kept as small as possible (perhaps of the order of cm) in order to keep the intensity high (the

divergence from small dimension sources is large). The undesirable
radiation must be filtered out using thin windows which also attenuate the
soft X-ray flux.

Zone Plates for Soft X-ray Focussing

Scanning X-ray microscopy involves focussing an X-ray beam to a small
spot on a specimen and recording the transmitted intensity while the beam
is moved in a raster pattern relative to the specimen (either the specimen
or the beam can actually be moved Figure 3 [15] [16] [17] [18].

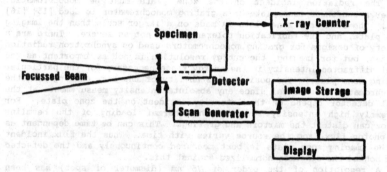

FIGURE 3: Schematic of a scanning soft X-ray microscope

The resolution of such a system depends on the spot size, the flux of
photons at the specimen, and the detector efficiency. In the relevant
wavelength range, photon detectors can be very efficient, and the major
limitation is in the poor optical efficiency of the focussing system and in
the low brightness of the source. The major advantage of scanned systems
is that the information is stored directly in digital form, and image
processing is possible with a minimum of difficulty.

In the system, shown in Figure 4, the zone plate has a focal length
about (diameter)^2/4nλ , where n is the number of zones. Being essentially
a circular diffraction grating (Figure 4) with progressively decreasing
zone width from inside to outside, different diffraction orders focus at
different points. The apodization of the zone plate and the pinhole are
arranged to eliminate unwanted orders. In an ideal zone plate
approximately 1/10 or the incident photon flux is focussed in the first
order spot, but in practice about 1% to 3% efficiencies have been measured
[18]. The diameter of the spot at focus is approximately equal to the
width of the narrowest (outermost) zone.

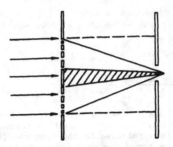

FIGURE 4: Focussing with an apodized zone plate. The aperture removes the extraneous diffracted orders

The radiation incident on the zone plate must be monochromatic. Either a condenser zone plate or a grating monochromator is used [15] [16] [17]. The condenser zone plate is made on a larger scale than the imaging zone plate, and the fabrication tolerances are not as severe. There are a variety of designs for grating monochromators used on synchrotron radiation sources, but for imaging, the energy resolution is not as important as the total diffracted intensity in the energy range necessary for the particular imaging zone plate. The positional stability of the beam exiting from the monochromator is critical since any absolute intensity measurements at the final detector depend on the intensity incident on the zone plate. For extremely high intensity sources, the thermal loading of the beamline optics can distort the mirrors and gratings. This can be time dependent as the photon flux from the source varies with time. Thus the flux incident on the imaging zone plate is best measured continuously and the detected flux behind the specimen normalized against this.

A resolution of the order of 75 nm (diameter of spot) has been demonstrated recently [18] and zone plates with outer zone widths of 55 nm have been reported, with suggestions of usable zone widths of 20 nm [19]. These results were obtained with electron beam written zone plates. Other methods which have been attempted include contamination writing with subsequent X-ray lithography [20], and UV holography [21]. The fabrication difficulties are considerable, since the outside diameter of the zone plate is of the order of 50 to 70 micrometers, and the accuracy of fabrication (circularity, zone boundary positions) determines the efficiency and minimum spot size achievable. In essence, the tolerances acceptable in the making of a good, high resolution zone plate are much smaller than in fabricating microelectronic circuits. A detailed discussion of the optical requirements for zone plates is given by Michette [22]. At the present time, the most advanced scanning soft X-ray microscopes use zone plate focussing (for example; Kirz's group at SUNY Stony Brook; the King's College group; and Schmahl's group at Göttingen).

A typical single image is recorded digitally pixel by pixel, and is acquired in times from 1 to 30 minutes. The recording speed depends on the frame size of the image and the dwell time at each pixel. The dwell time can be shortened by using higher brightness sources. Kirz and co-workers presently are installing their scanning microscope on the 37 period soft X-ray undulator at the X-ray ring at Brookhaven (NSLS).

Optical elements which focus soft X-rays can also be used directly to magnify images, and to act as objective lenses. The Gottingen group has been active in applying their zone plate fabrication methods to such a direct imaging microscope. A coarse condenser zone plate is used to monochromatize the radiation, and a micro-zone plate is used for imaging. (Figure 5) Resolutions of the order of 50 nm to 70 nm have been reported [23] [24], making the performance of these instruments comparable to scanning microscopes.

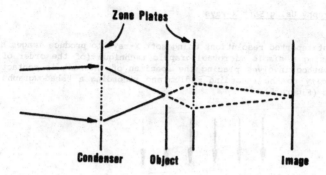

FIGURE 5:   Direct imaging soft X-ray microscope using two zone plates; a condenser and an objective lens

For radiation sensitive materials, the total dose to record an image can be larger with a direct imaging microscope since the exact focus position must be determined using a series of images, unless some external measurement methods are used. As with the scanning soft X-ray microscope, times of exposure can be reduced significantly by using a high brightness source such as an undulator.

In both scanning and direct imaging microscopes, the best results have been obtained by instruments using synchrotron radiation. At this stage of instrumental development, the steady, long duration output of a synchrotron source is ideal for microscopy since focussing, alignment and instrumental adjustments can be made with the beam fully illuminating the specimen and detector. Single pulse sources such as laser or pinched gas plasma sources are not yet suitable.

Focussing with Multilayer Mirrors

In addition to the use of Fresnel zone plates, a number of efforts have been reported to construct reflecting optics to focus soft X-rays. Multilayer coatings, analogous to a quarter wave stack, are made consisting of alternating layers of different refractive index. The more heavily absorbing layer is placed at the antinodes of the standing wave which result from the interference of the incident and reflected waves. Spiller reports using material combinations such as ReW-C, W-C, ReW-B and AuPd-c [25]. The layers must be deposited so the total thickness error is less than 1/10 the wavelength over the area of the mirror. The interfaces must be smooth and stable. Reflectivities of a variety of multilayers have been measured for wavelengths as short as 5 nm. Spiller describes attempts to construct a scanning X-ray microscope at Brookhaven (NSLS, 740 MeV storage ring) [26]. Cerrina and co-workers at the University of Wisconsin 1 GeV storage ring are designing and constructing a scanning microscope using a Schwartzschild objective [27].

The instrumental development for microscopes using reflecting optics is not as far advanced as that using zone plates. Results on the scale (50-75 nm) of those reported for zone plate microscopes have not been reported for these kinds of microscopes.

Microradiography Using Soft X-rays

The best reported resolutions using soft X-rays to produce images have been made using a simple microradiographic technique (of the order of 20 nm). The method involves placing the specimen of interest in contact or close proximity to the recording medium and recording a "shadowgraph" of the specimen (Figure 6).

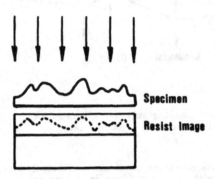

FIGURE 6:   Topographic images formed in the surface of polymer resists. The more heavily irradiated regions of the resist dissolve faster in a solvent.

The major advance recently has been the extensive use of grainless polymer films to record the image. The resolution possible with these resists, used also for lithography processes in the fabrication of microcircuits, is limited by the molecular structure of the polymer and not by the grain size of the emulsion. The positive resist is sensitive to the radiation, and higher absorbed doses result in lower molecular weight locally. The dissolution rate in the solvent developer depends on the local molecular weight, and the image appears as topographic contrast in the surface of the exposed and developed resist. The topography is examined with electron microscopy (SEM, TEM or STEM). Since the resist is itself sensitive to radiation, and significant mass loss is measured (up to 52%) [28] in the electron microscope, a metal shadowed carbon replica is taken from the resist image and the replica studied in the electron microscope.

A number of advantages of this method are immediately clear. Any source of soft X-rays, pulsed, low intensity or synchrotron radiation can be used, with the major difference among these sources being the time of exposure ranging from nanoseconds in pulsed sources to 40 hours or more in laboratory sized electron impact sources [29]. The resolution of the method is better than any other. Very large areas of specimen can be examined with high resolution. The illuminating radiation can be monochromatic, using line emission sources, monochromatized synchrotron radiation or can be broad band "white light".

The difficulties in using the method include the low sensitivity of high resolution resists used as a detector, the multiple processing steps necessary to develop the image, and the lack of direct image processing. In spite of these drawbacks, the technique has been used to produce images in synthetic polymers and in metallic thin films. An example is seen in Figure 7, in which the microstructure of an Al-Cu evaporated film is revealed.

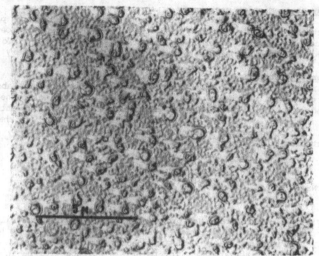

FIGURE 7:   Soft X-ray microradiograph of AlCu (evaporated film).
             PMMA resist exposed with 18A X-rays

The copper rich phases appear quite clearly, and if compared to electron micrographs of the same specimen, show small structures down to scales of 20 nm. With monochromatic radiation from synchrotron sources, the diffraction around the small particles becomes a serious problem, and the fringes obscure the small detail near the particles.

Quantitative and Analytical Microscopy

Soft X-ray imaging is still at the early stages of instrumental development when compared to electron microscopy, but a number of interesting new kinds of microscopy are being introduced.

1.   Phase contrast microscopy.
Schmahl and co-workers have recently used the information available in the phase shift component of the atomic scattering factor to produce soft X-ray phase contrast images in their direct imaging (zone plate) microscope [30].

2.   Soft X-ray holography.
The possibility of examining specimens thicker than those in transmission electron microscopy has spurred interest in obtaining 3-dimensional images using X-ray holography [31]. Diffraction patterns have been recorded as the first experimental step in obtaining a hologram.

3.   Diffraction patterns with soft X-rays.
Sayre and co-workers have succeeded in obtaining diffraction patterns with soft X-rays from non-periodic structures [32] [33]. Photons with a wavelength of 3.2 nm from a toroidal grating monochromator on the 740 MeV storage ring at Brookhaven were used. It appears that diffraction pattern imaging using undulator sources is possible. The illumination requirements (monochromaticity, divergence) are less severe than those for holography.

4. X-ray photoemission microscopy.

The focussed X-ray beam used in a scanning microscope produces photoelectrons in the specimen, and the pixel by pixel photoelectron spectra can be recorded using a cylindrical mirror analyzer. To obtain enough signal, the high brightness undulator radiation is necessary [34].

5. Absorption edge mapping.

Subtraction of images taken from one area of the specimen above and below an elemental absorption edge can be readily done using scanned X-ray images and with more difficulty using microradiography. Any elements which have an absorption edge in the soft X-ray range, including the light elements, can be so mapped. Quantitative aspects of the problem, such as the detectable limits, must yet be measured.

CONCLUSIONS

Imaging with soft X-rays has progressed rapidly, with the availability of high intensity, high brightness sources, and improved methods of accurately fabricating small structures necessary for focussing optical elements. The major applications for materials science will be in imaging intermediate thickness material at high resolution, and microchemical analysis using absorption edge mapping and photoelectron spectroscopy. The use of the diffracted information to produce images is potentially extremely important to a wide variety of synthetic materials. Much of the effort in soft X-ray imaging has been devoted to instrumental and methodology developments, and very few applications in materials science have been reported. With the increased availability of soft X-ray microscopes, this should rapidly change.

ACKNOWLEDGEMENTS

This work has been supported by NSERC (Canada), NRC, OCMR, and the Center for Chemical Physics at The University of Western Ontario. Discussions with D. Sayre, P.C. Cheng, K.H. Tan, and J. Kirz among many others, has been very helpful.

REFERENCES

1 D. Sayre, J. Kirz, R. Feder, D.M. Kim, E. Spiller, Ultramicroscopy 2, 337 (1977).

2 P.C. Cheng, D.M. Shinozaki, K.H. Tan, in X-ray Microscopy: Instrumentation and Biological Applications, edited by P.C. Cheng and G.J. Jan (Springer-Verlag, Berlin, 1987), p. 65.

3 G. Schmahl, D. Rudolph, X-ray Microscopy, (Springer-Verlag, Berlin, 1984).

4 D. Sayre, M. Howells, J. Kirz, H. Rarback, X-ray Microscopy II, (Springer-Verlag, Berlin, 1988).

5 J. Kirz, Annals N.Y. Acad. Sci 342, 273 (1980).

6 D.M. Shinozaki, R. Feder, Treatise on Materials Science and Technology 27, 111 (1988).

7 D.J. Nagel, IEEE Trans. Nucl. Sci. NS-26, 1228 (1979).

8 H.M. Epstein, P.J. Mallozzi, B.E. Campbell, Proc. SPIE 385, 141 (1983).

9 A.L Hoffman, F.F. Albrecht, E.A. Crawford, P.H. Rose, Proc. SPIE 537, 198 (1985).

10 D.J. Nagel, C. Brown, M. Peckarar, M.L. Ginter, J. Robinson, T.J. McIlrath, Appl Opt. 23, 1428 (1984).

11 A. Damerell, E. Madraszek, F. O'Neill, N. Rizvi, R. Rosser, P. Rumsby, in X-ray Microscopy II, edited by D. Sayre, M. Howells, J. Kirz, H. Rarback, (Springer-Verlag, Berlin, 1988), p. 43.

12 J. Trebes, S. Brown, E.M. Campbell, N.M. Ceglio, D. Eder, D. Gaines, A. Hawryluk, C. Keane, R. London, B. McGowan, D. Mathews, S. Maxon, D. Nilson, M. Rosen, D. Stearns, G. Stone, D. Whelan, ibid., p. 30.

13 J.S. Pearlman, J.C. Riordan, J. Vac. Sci. Technol. 19, 1190 (1981).

14 W. Neff, J. Eberle, R. Holz, R. Richter, R. Lebert in X-ray Microscopy II, edited by D. Sayre, M. Howells, J. Kirz, H. Rarback (Springer- Verlag, Berlin, 1988), p. 22.

15 H. Rarback, D. Shu, Su Cheng Feng, H. Ade, C. Jacobsen, J. Kirz, I. McNulty, Y. Vladimirsky, D. Kern, P. Chang, in X-ray Microscopy II, edited by D. Sayre, M. Howells, J. Kirz, H. Rarback (Springer- Verlag, Berlin, 1988), p. 194.

16 G. R. Morrison, M.T. Browne, C.J. Buckley, R.E. Burge, R.C. Cave, P. Charalambous, P.J. Duke, A.R. Hare, C.P.B. Hills, J.M. Kenney, A.G. Michette, K. Ogawa, A.M. Rogoyske, T. Taguchi, ibid., p. 201.

17 B. Niemann, P. Guttmann, R. Hilkenbach, J. Thieme, W. Meyer-Ilse, ibid., p. 209.

18 H. Rarback, D. Shu, S.C. Feng, H. Ade, J. Kirz, I. McNulty, D.P. Kern, T.H.P. Chang, Y. Vladimirsky, N. Iskander, D. Attwood, K. McQuaid, S. Rothman, Rev. Sci. Instrum. 59 (1), 52 (1988).

19 V. Bogli, P. Unger, H. Beneking, B. Greinke, P. Guttmann, B. Niemann, D. Rudolph, G. Schmahl, in X-ray Microscopy II, edited by D. Sayre, M. Howells, J. Kirz, H. Rarback (Springer-Verlag, Berlin, 1988), p. 80.

20 C.J. Buckley, M.T. Browne, R.E. Burge, P. Charalambous, K. Ogawa, T. Takeyoshi, ibid., p. 88.

21 G. Schmahl, D. Rudolph, P. Guttmann, O. Christ in X-ray Microscopy, edited by G. Schmahl, D. Rudolph (Springer-Verlag, Berlin, 1984), p. 63.

22 A.G. Michette, Optical Systems for Soft X-rays (Plenum Press, New York, 1986).

23 B. Niemann, D. Rudolph, G. Schmahl Proc. SPIE 368, 2 (1982).

24 B. Niemann, D. Rudolph, G. Schmahl Nucl. Instru. Meth. in Phys. Res. A208, 367 (1983).

25 E. Spiller in Handbook on Synchrotron Radiation 18, edited by E.E. Koch (North Holland, New York, 1983), p.1091.

26 E. Spiller in X-ray Microscopy, edited by G. Schmahl, D. Rudolph (Springer-Verlag, Berlin, 1984), p. 226.

27 F. Cerrina, J. Imaging Science 30 (2), 80 (1986).

28 D.M. Shinozaki, B.W. Robertson in X-ray Microscopy: Instrumentation and Biological Applications, edited by P.C. Cheng, G.J. Jan (Springer-Verlag, Berlin, 1987) p. 105.

29 D.M. Shinozaki, R. Feder, Treatise on Materials Sc. and Technol. 27 (1988), 111.

30 G. Schmahl, D. Rudolph, P. Guttman in X-ray Microscopy II, edited by D. Sayre, M. Howells, J. Kirz, H. Rarback (Springer-Verlag, Berlin, 1988), p. 228.

31 M. Howells, ibid., p. 263.

32 D. Sayre, W.B. Yun, J. Kirz, ibid., p. 272.

33 W.B. Yun, J. Kirz, Acta Crysta. A43 (1987), p. 133.

34 H. Ade, J. Kirz, H. Rarback, S. Hulbert, E. Johnson, D. Kern, P. Chang, V. Vladimirsky, in X-ray Microscopy II, edited by D. Sayre, M. Howells, J. Kirz, H. Rarback (Springer-Verlag, Berlin, 1988), p. 280.

INVESTIGATION OF GRAIN BOUNDARY MIGRATION In Situ
BY SYNCHROTRON X-RAY TOPOGRAPHY

C. L. BAUER
Carnegie Mellon University, Pittsburgh, PA 15213 USA
and
J. GASTALDI, C. JOURDAN and G. GRANGE
Centre de Recherche sur les Mecanismes de la Croissance Cristalline, 13288 Marseille, France

ABSTRACT

Grain boundary migration has been investigated in prestrained monocrystalline specimens of aluminum in situ, continuously and at temperatures ranging from 415 to 610°C by synchrotron (polychromatic) x-ray topography (SXRT). In general, new (recrystallized) grains nucleate at prepositioned surface indentations and expand into the prestrained matrix, revealing complex evolution of crystallographic facets and occasional generation of (screw) dislocations in the wake of the moving boundaries. Analysis of corresponding migration rates for several faceted grain boundaries yields activation energies ranging from 56 to 125 kCal/mole, depending on grain boundary character. It is concluded that grain boundary mobility is a sensitive function of grain boundary inclination, resulting in ultimate survival of low-mobility (faceted) inclinations as a natural consequence of growth selection. Advantages and disadvantages associated with measurement of grain boundary migration by SXRT are enumerated and corresponding results are interpreted in terms of fundamental relationships between grain boundary structure and corresponding migration kinetics.

1. INTRODUCTION

Although many kinetic phenomena, such as recrystallization and various forms of grain growth, are controlled by the rate of grain boundary migration, details of underlying mechanism(s) remain largely obscure, mainly because displacement of a single, well characterized grain boundary under known conditions of driving force, impurity and defect concentration and temperature cannot be determined easily [1-3]. In fact, most experimental results are obtained through repetitive heating and cooling cycles, thereby introducing additional complicating factors related to creation, annihilation and redistribution of chemical impurities and lattice defects. In this particular investigation, grain boundary migration has been measured in predeformed monocrystalline specimens of aluminum by nucleating (recrystallized) grains at prepositioned surface indentations and recording displacement of (faceted) grain boundaries by Synchrotron (polychromatic) X-Ray Topography (SXRT) in situ, continuously and over a range of elevated temperatures, thereby eliminating complications associated with repetitive heating and cooling cycles. The purpose of this article is to demonstrate how grain boundary migration can be investigated by SXRT in order to achieve improved appreciation of relationships between grain boundary structure and corresponding migration kinetics. More complete descriptions of this research are published elsewhere [4,5].

The remainder of this article is divided into several sections: First, experimental details are outlined in Sec. 2, then measurement of grain boundary migration by SXRT is described in Sec. 3, experimental results are presented in Sec. 4 and analyzed in Sec. 5, and, last, important conclusions stemming from this investigation are summarized in Sec. 6.

2. EXPERIMENTAL DETAILS

Monocrystalline specimens, measuring approximately 2.5 x 1.5 x 0.7 mm were prepared from zone refined (99.999 wt. %) and refined (99.99 wt. %) aluminum, prestrained about 5% in tension, indented on a flat face in order to promote localized nucleation and subsequent growth of a few (recrystallized) grains and then inserted into a small furnace situated within a portable vacuum chamber, capable of achieving pressures of about 0.1 μPa. This chamber was positioned at the exit beam of collimated polychromatic x-rays at the 1.05 GeV synchrotron at the Laboratoire pour l'Utilisation du Rayonnement Electromagnetique (LURE) in Orsay, France in order to obtain sequential topographs of the nucleation and growth process in situ, continuously and at temperatures ranging from 415 to 610°C, either by high-resolution photography or by continuous video recording. Further details concerning specimen preparation, experimental apparatus, and applications of SXRT are reported elsewhere [6-8].

3. MEASUREMENT OF GRAIN BOUNDARY MIGRATION BY SXRT

Measurement of grain boundary migration by SXRT involves analysis of various diffracted images, obtained either intermittently by high-resolution photography or continuously by video recording. In contrast to conventional x-ray topography, however, polychromatic SXRT is enhanced by several features of the synchrotron beam, such as high intensity, excellent collimation, small polychromatism and large cross section, thereby allowing determination of grain misorientation θ, grain boundary inclination ϕ and grain boundary displacement Δx, as well as identification of species and density of localized defects, *in situ*, continuously and over a range of elevated temperatures [9].

Details associated with measurement of grain boundary migration by polychromatic SXRT are provided schematically in Fig. 1, wherein images emanating from diffraction of the incident x-ray beam from various (hkl) planes of one grain defining an arbitrary grain boundary, as well as image of the transmitted beam, are projected on a plane (film) aligned normal to the axis of the incident beam. Since extremities of the images correspond to traces of the grain boundary on external surfaces of the specimen, θ can be determined by comparison of diffracted images emanating from adjacent grains, ϕ can be determined by analysis of dimensions of diffracted images with respect to one grain or the other and Δx can be determined from successive positions of these images. Accordingly, grain misorientation, grain boundary inclination and grain boundary displacement can be monitored and recorded *in situ*, continuously and over a range of elevated temperatures for appropriately prepared specimens containing one or more grain boundaries. Ultimate resolution is limited by sensitivity of the recording method, which approaches $\pm 1°$ for θ and ϕ, and ± 5 μm for Δx.

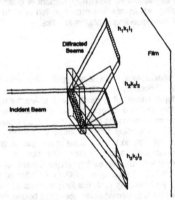

Figure 1. Schematic illustration of (three) images (hatched parallelograms) emanating from diffraction of the incident x-ray beam from various (hkl) planes of one grain defining an arbitrary grain boundary (cross-hatched parallelogram), as well as image of the transmitted beam, projected on a plane (film) aligned normal to the axis of the incident beam. Grain misorientation, grain boundary inclination and grain boundary displacement can be determined by analysis of the diffracted images.

4. EXPERIMENTAL RESULTS

Typical results for the aforementioned nucleation and growth process at an indentation in prestrained 99.99 wt. % aluminum are presented in Fig. 2, wherein x-ray topographs of the (002) reflection after (a) nucleation of a single, well defined grain at 560°C, (b) further annealing for 46.7 min at 560°C, (c) further annealing for 59.3 min at 570°C and (d) further annealing for 22.3 min at 595°C are pictured. A new (recrystallized) grain has formed at the indentation and subsequently grown (expanded) into the prestrained matrix, thereby progressively transforming from a random to a faceted configuration. In general, grains resulting from this type of nucleation and growth process are characterized by a distribution of facets expanding at unequal rates into the deformed matrix. Moreover, a dislocation network, characterized by densities ranging between 10^3 and 10^4 cm/cm^3, is clearly visible within the expanding (recrystallized) grain. These dislocations are generated in the wake of the moving grain boundary with a predominant screw component [10].

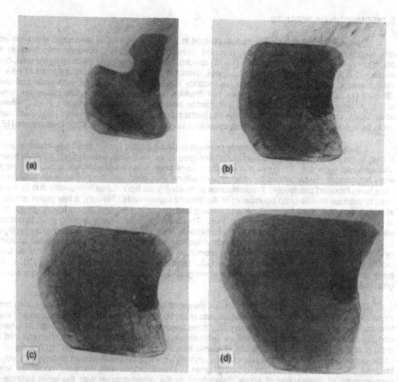

Figure 2. Growth of a (recrystallized) grain in prestrained 99.99 wt. % aluminum, wherein x-ray topographs of (002) images after (a) development of a single, well defined grain at 560°C, (b) further annealing for 46.7 min at 560°C, (c) further annealing for 59.3 min at 570°C and (d) further annealing for 22.3 min at 595°C are pictured.

Evolution of faceted grain boundary traces, similar to those depicted in Fig. 2, have been recorded for several specimens and over a range of annealing conditions. In general, displacement is a linear function of time, with corresponding slope equal to grain boundary velocity v. Grain boundary velocity for various elevated temperatures, grain boundary facets, grain misorientations, and impurity levels are then analyzed by plotting logarithm of v as a function of reciprocal temperature. In general, a broad range of curves are obtained, characterized by activation energies ranging from about 56 to 125 kCal/mole, although systematic variation of slope and/or intercept for various grain boundary inclinations (facets) and specimens characterized by different impurity levels (99.999 and 99.99 wt. % Al) cannot be discerned. Typical results are summarized in Table I, wherein grain boundary velocity v and corresponding activation energy Q are reported for selected grain boundaries characterized by approximate rotation of 38° ($\Sigma = 7$) about a common <111> axis and inclination ϕ relative to the growing grain in refined (99.99 wt. %) aluminum. These results, and others, are discussed in the following section of this article.

Table I. Velocity v and concomitant activation energy Q for selected grain boundaries characterized by approximate rotation of 38° ($\Sigma = 7$) about a common <111> axis and inclination ϕ relative to the growing grain in refined (99.99 wt. %) aluminum.

Character	ϕ {hkl}	v (560°) (10^{-5} cm/s)	Q (kCal/mole)
Mixed	{100}	3.9	100
Tilt	{110}	5.5	86
Mixed Tilt	{111}	1.4	125

5. DISCUSSION OF RESULTS

Typical experimental results, as summarized in Table I, indicate that measured activation energies are substantially larger than those reported previously for grain boundary migration in similar polycrystalline [11,12] and bicrystalline [13-15] specimens of aluminum. (At lower temperatures, however, measured activation energies tend to be lower, even for equivalent grain boundary inclinations.) These values extend from the general range for vacancy migration (12 kCal/mole) and grain boundary self diffusion (15 kCal/mole) to that for volume self diffusion (35 kcal/mole). In certain cases, an increase in activation energy near certain selected coincidence orientations has been reported, [14,16] whereas, in other cases, variations in activation energy at special boundaries were not observed [13]. Such variations could be due to the complex facets which develop during grain boundary migration in aluminum [15,17].

A likely explanation for the disparity between results reported herein and those reported previously is that, in the present investigation, activation energies for faceted grain boundaries; i.e., grain boundaries characterized by a specific angle of inclination, have been measured, whereas activation energies for grain boundaries characterized by random inclinations (but often with controlled grain misorientations) have been reported previously. Predominance of faceted grain boundaries is probably due to the method used to produce mobile grain boundaries in the present experiments. Namely, a few grains are nucleated at prepositioned surface indentations and subsequently grow in a manner which favors growth selection of the *least* mobile grain boundary inclinations. Therefore, the most rapidly moving grain boundaries are eliminated from the overall distribution of inclinations, resulting in a (closed) faceted grain configuration. In this regard, the present experimental method facilitates investigation of grain boundaries characterized by high mobility with respect to grain misorientation but low mobility with respect to grain boundary inclination. Apparently, grain boundary mobility (and corresponding activation energy) is a sensitive function of grain boundary inclination, even when grain boundary mobility does not vary appreciably with grain misorientation.

Results stemming from this investigation demonstrate certain relationships between grain boundary structure, as defined by grain misorientation and grain boundary inclination, and corresponding migration kinetics. Namely, special grain boundary structures associated with near-coincident grain misorientations and low-index grain boundary inclinations (with respect to the growing grain) are characterized by low grain boundary mobility, due to difficulty of atomic transfer from the shrinking to the highly ordered planes of the growing grain. Indeed, the rate of grain (crystal) growth is probably related to density of ledges and terraces at the extremity of the growing grain. These ledges and terraces occur as a natural consequence of intersection of screw dislocations in the growing grain with the grain boundary, thus providing an unexhaustible supply of favorable transfer sites. Observations that dislocations with a predominant screw component are generated in the wake of moving grain boundaries support this particular growth mechanism.

It is useful to compare advantages and disadvantages of SXRT for measurement of grain boundary migration. The principal advantage of SXRT is that grain misorientation θ, grain boundary inclination ϕ and grain boundary displacement can be determined *in situ*, continuously and over a range of elevated temperatures, either by high-resolution photography or video recording, thereby eliminating complications associated with usual repetitive heating and cooling cycles. Unfortunately, presumed presence of ledges and terraces, as well as other sub-micrometer defect configurations, cannot be verified by SXRT because of the relatively poor spatial resolution and, therefore, underlying mechanisms for grain boundary migration cannot easily be identified. Moreover, current experimental design does not allow preselection of θ, ϕ and concomitant driving force, as in several previous investigations. Finally, high-mobility grain boundaries with respect to θ but low mobility with respect to ϕ are measured by SXRT, whereas grain boundary mobility is usually determined with respect to θ only in conventional methods involving repetitive heating and cooling cycles. It may be concluded that different grain boundary mobilities are measured by SXRT and conventional methods and, therefore, it is not yet possible to compare critically corresponding results.

6. SUMMARY

Grain boundary migration has been investigated in prestrained monocrystalline specimens of aluminum *in situ*, continuously and at temperatures ranging from 415 to 610°C by SXRT. In general, new (recrystallized) grains nucleate at prepositioned surface indentations and expand into the prestrained matrix, revealing complex evolution of crystallographic facets and occasional generation of (screw) dislocations in the wake of the moving boundaries. Analysis of corresponding migration rates for several faceted grain boundaries yields activation energies ranging from 56 to 125 kCal/mole, depending on grain boundary character. It is concluded that grain boundary mobility is a sensitive function of grain boundary structure, as defined by *both* grain misorientation and grain boundary inclination, resulting in ultimate survival of low-mobility (faceted) inclinations as a natural consequence of growth selection.

Acknowledgements

Support by the Centre National de Recherche Scientifique and the National Science Foundation under Grant INT-8413893 is gratefully acknowledged.

References

1. *Recrystallization of Metallic Materials*, F. Haessner, Ed., Riederer Verlag GmbH, Stuttgart (1978).

2. A. H. King and D. A. Smith in *Grain Boundary Structure and Kinetics*, R. W. Balluffi, Ed., American Society for Metals (1980).

3. J. Gastaldi and C. Jourdan, *J. Crystal Growth* 52, 361 (1981).

4. J. Gastaldi, C. Jourdan, G. Grange and C. L. Bauer, *Proceedings of the 1988 Spring Meeting of the Materials Research Society* (in press).

5. J. Gastaldi, C. Jourdan, G. Grange and C. L. Bauer, *phy. stat. sol.* (in press).

6. J. Gastaldi and C. Jourdan, *phy. stat. sol. (a)* 49, 529 (1978).

7. J. Gastaldi, C. Jourdan, P. Marzo, C. Allasia and J. N. Jullien, *J. Appl. Cryst.* 18, 77 (1982).

8. J. Gastaldi and C. Jourdan, *phy. stat. sol. (a)* 97, 361 (1986).

9. J. Gastaldi and C. Jourdan in *Applications of X-Ray Topographic Methods to Materials Science*, S. Weissman and J. F. Petroff, Eds., Plenum Press (1984).

10. J. Gastaldi, C. Jourdan and G. Grange, *Phil. Mag. A*, 971 (1988).

11. P. Gordon and R. A. Vandermeer, *Trans. AIME* 224, 917 (1962).

12. C. Frois and O. Dimitrov, *Mem. Sci. Rev. Met.* 59, 643 (1962).

13. B. B. Rath and H. Hu, *Trans. AIME* 236, 1193 (1966).

14. D. W. Demianczuk and K. T. Aust, *Acta Met.* 23, 1149 (1975).

15. M. S. Masteller and C. L. Bauer, *Acta Met.* 27, 483 (1979).

16. E. M. Fridman, Ch. V. Kopetskii and L. S. Shvindlerman, *Soviet Phys. Solid State* 16, 1152 (1974).

17. J. Gastaldi and C. Jourdan, *Phil. Mag. A* 50, 309 (1984).

WHITE BEAM SYNCHROTRON TOPOGRAPHIC STUDIES OF THE EFFECTS OF LOCALIZED STRESS FIELDS ON THE KINETICS OF SINGLE CRYSTAL SOLID STATE REACTIONS.

MICHAEL DUDLEY
Dept. of Materials Science and Engineering, SUNY at Stony Brook, Stony Brook, NY 11794.

ABSTRACT.

White Beam Synchrotron Topography has been used to determine the role of localized stress fields in the solid state polymerization of single crystals of the diacetylene PTS. Results indicate that the stress fields due to grown in dislocations can accelerate local reaction kinetics in thermally induced polymerization reactions, although no such effects were previously observable in photolytically or radiolytically induced reactions. Results are analyzed in an analogous fashion to the treatment of the nucleation of solid state phase transformations at dislocations. Good agreement was found between approximate theoretical treatments and experimental observation. The response of the monomer crystal to the inhomogeneous stresses generated as a result of inhomogeneous reaction and the implications regarding local reaction kinetics are discussed in detail.

INTRODUCTION.

The influence and role of stress fields on reactivity is important from several different points of view. The most basic influences derive from the fact that solid state reactions are generally accompanied by a change in molecular volume and/or shape. As a consequence, when reaction occurs, even homogeneously (i.e. randomly), inside a macroscopic sample such as a large single crystal, stresses are induced in much the same way as they are in solid state phase transformations. The magnitude of these stresses can, in the worst case, prevent reaction from taking place, or, in less extreme cases influence the choice of reaction pathway or the reaction mechanism involved. Reaction induced stresses can also be one of the driving forces giving rise to heterogeneous reaction occurring at, for example, dislocations. The inhomogeneous strain fields which are generated as a result of this heterogeneous reaction can jeopardize the micromechanical integrity of the crystal as a whole. Clearly, then, in order to be able to exert some control over the solid state reaction process, understanding the influence of stress on reactivity is of paramount importance. Further, such understanding is central for promoting the potential for the development of novel materials generated by new solid state reactions.

Some insight into the influence of stress on single crystal reactivity has been recently afforded by the elegant work of McBride et al [1],[2] on the UV induced decomposition of diacyl peroxides, and their derivatives. Generally these decompositions, which are presumed to occur homogeneously throughout the crystals, involve the fragmentation of the peroxide molecules into a radical pair and two CO_2 molecules. Using the asymmetric stretch mode of CO_2 as a stress gauge,

McBride et al translated measured FTIR peak shifts into pressures to which the CO_2 molecules were subjected during reaction. New features observed in the spectra at higher conversions (around 0.1%) were interpreted to be indicative of the eventual overlap and interaction between the reaction induced stress fields of adjacent reaction sites, dictating new reaction pathways which yielded new products with different FTIR spectra. Unfortunately, although such studies provide vital information, FTIR spectroscopy cannot directly reveal the spatial distribution of reaction induced stresses, and details regarding the long range influences of these stresses can therefore only be inferred from this kind of study.

It has previously been shown (for example see [3],[4]) that filtered white beam synchrotron topography is a technique which is ideally suited to monitor the spatial distribution of strain in large single crystals undergoing solid state polymerization reactions. The large field of view and high strain sensitivity of the technique is ideally suited to both reveal the nature and monitor the development of inhomogeneous strain fields which result whenever reaction occurs preferentially at defects such as dislocations. Such preferential reactivity can, in itself, be a manifestation of the influence of stress on reactivity, although other factors might also potentially be involved, depending on the particular reaction, such as the production of favorable stereochemical arrangements, or the trapping of excitation energy at dislocation sites. In cases where the significant influence can be identified as reaction induced stress, monitoring the response of the crystal to the stresses which result from the preferential reaction can potentially afford insight into the general nature of the influence of stress on reactivity.

EXPERIMENTAL.

Single crystals of PTS (2,4-hexadiyne diol bis(p-toluene sulphonate)) were grown at $0^{\circ}C$ by self nucleation in saturated acetone solutions. Detailed discussion of the methodology utilized in the topographic studies can be found in earlier publications (for example see [3],[4]). Basically, filtered white beam transmission Laue patterns were recorded on large area, relatively fast (to minimize exposures) X-ray film from crystals which were being subjected to various reaction inducing stimuli. These stimuli included the X-ray beam itself, UV light and heat provided by a small hot stage. The diffraction geometry was chosen so as to maximize the number of useful reflections recorded in a single exposure, to optimize the potential for strain field analysis. Using this methodology, non-destructive, time lapse images could be recorded revealing the defect structure and general strain distribution of crystals undergoing reaction.

RESULTS.

Previous results obtained at the SRS Daresbury (UK) [5], have indicated that during X-ray induced reaction, no generation of localized strain fields attributable to preferential reactivity occurring at defects, or generation of defects, become evident. Similar results were obtained for the

UV induced reaction although in this case, reaction occurs in a thin subsurface layer which then prevents further reaction. As pointed out by Braun et al [6], in general it is important to maximize the uniformity of reaction rate as a function of depth to minimize unwanted stresses which may influence the progress of reaction. While this is possible for the X-ray and thermally induced reactions, in the case of UV induced reaction such inhomogeneity could not be avoided due to the nature of the interaction. As a consequence only the X-ray and thermally induced reactions can be usefully discussed.

Figure 1 shows a series of filtered white beam topographs recorded on relatively fast film from a PTS crystal undergoing thermally induced reaction. Note the formation of lobes of contrast centered on the central dislocation group. The size of these lobes decreases as a function of time, as the crystal is held at the same temperature in the hot stage. Eventually, as reaction approaches completion, the strain contrast is replaced by simple orientation contrast, produced by a microcrack which has propagated across the crystal in both directions from the central region of crystal.

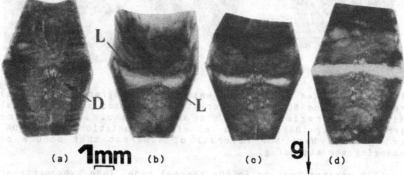

(a) **1mm** (b)   (c)   **g** (d)

Figure 1. Series of filtered white beam topographs (g=$\overline{1}$04, λ =0.8Å) recorded at SRS Daresbury from a PTS crystal undergoing thermally induced polymerization. D denotes growth dislocation images which appear almost end on in (a) which is recorded from the monomer crystal. L denotes lobe contrast which develops centered on this dislocation group in (b). Note the slight reduction in size of this contrast feature in (c). Note the orientation contrast in (d), recorded after reaction had progressed thoughout the whole crystal.

It is postulated that the lobe type contrast observed is due to the strain field generated by more rapid reaction kinetics occurring locally in the vicinity of the central dislocation group. Eventually, this strain field diminishes as a result of relaxation of the stress field via three possible processes, (1) the formation of the microcrack, observed optically after reaction was complete, and via orientation contrast on the topographs, (2) through a gradual diminution of the strain gradient between reacted and unreacted regions, as thermally induced reaction in the bulk "catches up" with that near the dislocations, or (3) through a similar diminution of the strain gradients, but this time caused by reaction being induced by the stress field associated with the inhomogeneous reaction. The latter influence is not however expected to be

significant since rather than aiding reaction, the stress field
due to the inhomogeneous reaction is likely to hinder reaction
since it is likely to put the surrounding crystal in a state of
tension, rather than the required compression.

The hypothesis that the lobe type contrast was due to
preferential reaction occurring near the central dislocation
group was tested as follows. A region of enhanced reaction was
induced in a similar region of a similar crystal by synchrotron
irradiation through a small pinhole. The resultant strain field
was characterized under conditions similar to those used in
the thermal case. The results can be seen in figure 2.

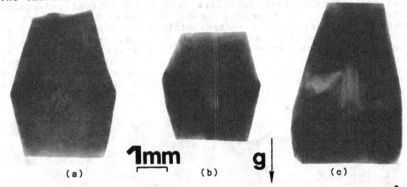

Figure 2. Series of filtered white beam topographs (λ =0.8 Å)
recorded at the NSLS from a crystal irradiated in the central
dislocated region through a small pinhole. (a) before
irradiation (g=$\overline{1}$04), (b) and (c) after irradiation ((b) g=$\overline{1}$04,
(c) g=10$\overline{4}$). Note the reversal of contrast that occurs on
changing the sign of g.

It appears that as in the thermal case, lobe type contrast
is formed centered on the reacted region, verifying that the
contrast observed during thermally induced reaction is due to a
strain field arising from preferential reaction occurring near
the central dislocation group. In the pinhole case, contrast
reversal is observed when g is reversed. This type of behavior
is to be expected from dynamical type images formed through tie
point migration in regions of uniform misorientation gradient
[7]. Due to the similarity between the contrast observed here
with that observed during thermally induced reaction, some
insight into the nature of the strain field induced during
thermally induced reaction is afforded. It is interesting to
note that although in this case, as in the thermal case,
microcrack formation is observed centered on the reacted
volume, no significant diminution of the observed strain field,
indicative of a significant stress relaxation is observed. This
could be an indication that in the thermal case, the diminition
of the lobe contrast is due to process (2) rather than (1).

DISCUSSION.

Preferential thermally induced reaction occurring at
dislocation sites in PTS has previously been observed by
Schermann et al [8], using optical microscopy. In their

discussion it was concluded that, since no energy trapping is expected in PTS, the energy of the dislocation could be significant in promoting preferential reaction.

In solid state reactions, the strain energy associated with reaction induced stress is a necessary contributor to any reaction activation barriers. Following the theory developed for solid state phase transformations (for review see [9]), dislocation distortion fields can become significant in promoting preferential reactivity if they allow some relaxation of reaction induced stress, thus effectively reducing the reaction activation barrier, and thereby potentially influencing reaction kinetics. Generally speaking, the polymerization reaction in PTS can be broken down into two basic steps, chain initiation (formation of a biradical dimer), and chain propagation. In the thermally induced polymerization of PTS, it has been shown that measured reaction activation energies are primarily associated with chain initiation events, whereas in the photo-induced reaction the much smaller measured activation energies are primarily associated with chain propagation events (for review see [10],[11]). This difference was explained on the basis that in the photo-induced case, the photon energies (UV or X-ray) are much larger than the initiation barrier, so that upon irradiation at room temperature, a single excited monomer molecule readily reacts with an adjacent molecule to form a biradical dimer. The biradical dimer can then undergo fission back to two monomer molecules or initiate a chain and propagate, with the latter being readily achieved at room temperature and higher. Thus, reductions in strain energy are expected to be insignificant in the photo-induced case since the effective reaction barrier is relatively small and is expected to be easily overcome at room temperature [10]. In situ X-ray polymerization studies conducted at 70°K [12] also revealed homogeneous reaction, suggesting the validity of similar conclusions at low temperature. On the other hand in the thermally induced reaction, since there remains an effective reaction barrier, that to chain initiation, any reduction in strain energy can potentially influence local reaction kinetics. Following Eshelby [13], the interaction energy between the elastic field of the dislocation with that due to the volume misfit associated with reaction can be written,

$$E = - \int_V S^d_{ij} \, e^T_{ij} \, dV \tag{1}$$

Where e^T_{ij} is the "stress free" reaction strain, which would correspond to polymer forming without the constraints of the monomer matrix, and S^d_{ij} is the dislocation stress field. The integration is performed over the volume of the "inclusion." For PTS, calculation yields the following stress free reaction strain (accompanied by a small rotation) corresponding to the conversion of the monomer unit cell to that of the polymer,

$$e^T_{ij} = \begin{pmatrix} -0.00511 & 0 & 0.00285 \\ 0 & -0.04660 & 0 \\ 0.00285 & 0 & -0.00533 \end{pmatrix} \tag{2}$$

Insufficient elastic constant data prevents exact determination of the stress field of the dislocations of interest (mixed, with Burgers vector b=[010]), although one can assume that finite values are expected from all terms in the stress tensor except S_{13}, so that equation (1) will reduce to,

$$E = - \int_V (-0.00511 \, S_{11} - 0.04660 \, S_{22} - 0.00533 \, S_{33}) \, dV \tag{3}$$

For preferential reaction to be promoted, E must be negative. Since e^T_{22} is by far the largest strain term, the term e^T_{22} S_{22}, is likely to dominate. S_{22} corresponds to regions of compressive and tensile stress in the direction of the Burgers vector, and negative values of S_{22} (i.e. compressive stress) are expected to lead to negative values of the interaction term, which can, in the thermal polymerization case, influence local reaction kinetics. The stresses from several closely spaced dislocations are expected to combine to accentuate this effect. Intuitive considerations would predict S_{22} to be important since this stress component provides shrinkage along the chain axis, which is necessary for reaction to occur.

CONCLUSIONS.

1.   The relaxation of reaction induced stress in the combined distortion fields of a small group of dislocations is the primary influence causing preferential reaction in PTS. Approximate strain energy calculations show that regions of compression in the dislocation stress field are likely to be the most important in promoting such preferential reaction.
2.   The strain field generated as a result of preferential reaction can be observed out to distances of the order of millimeters from the reaction site. Simple energetic considerations lead to the conclusion that the associated stress field is likely to hinder rather than aid further reaction in surrounding volumes.

References.
1.   J.M. McBride, B.E. Segmuller, M.D. Hollingsworth, D.E. Mills, and B.E. Weber, Science, 234, 830, (1986).
2.   M.D. Hollingsworth and J.M. McBride, Mol. Cryst. Liq. Cryst., 161, 25, (1988).
3.   M. Dudley, J.N. Sherwood, D. Bloor and D. Ando, Mol. Cryst. Liq. Cryst., 93, 223, (1983).
4.   M. Dudley, J.N. Sherwood, D. Bloor and D. Ando, in "Polydiacetylenes", D. Bloor and R.R. Chance (eds.), NATO ASI Series E (Applied Sciences), No. 102, p. 87, Martinus Nijhoff, Dordrecht, (1985).
5.   M. Dudley, J.N. Sherwood, D. Bloor and D. Ando, J. Mater. Sci. Lett., 1, 479, (1982).
6.   H.-G. Braun and G. Wegner, Mol. Cryst. Liq. Cryst., 96, 121, (1983).
7.   Y. Ando and N. Kato, J. Appl. Cryst., 3, 74, (1970).
8.   W. Schermann, G. Wegner, J.O. Williams, and J.M. Thomas, J. Polymer Sci., Polymer Phys. Ed., 13, 753, (1975).
9.   J.W. Christian, "Theory of Phase Transformations in Metals and Alloys," Part 1, Equilibrium and General Kinetic Theory, 2nd Ed., Pergamon (1981), p. 468.
10.   H. Eckhardt, R.R. Chance, and T. Prusik, in ref. 4.
11.   H. Gross, W. Neumann and H. Sixl, Chem. Phys. Lett., 95, 584, (1983).
12.   M. Dudley and J.N. Sherwood, Unpublished work.
13.   J.D. Eshelby, Proc. Roy. Soc., A241, 376, (1957).
Acknowledgements.
Work performed in part on the topography beamline (X-19C, NSLS) which is supported by the US DOE (Grant No.DE-FG0284ER45098). Support is also acknowledged from the NSF under Grant No. DMR 8506948. Initial experiments, performed with J.N. Sherwood and D. Bloor at the SRS Daresbury, were supported by the SERC (UK).

AUTOMATIC ORIENTATION MAPPING WITH
SYNCHROTRON RADIATION

G. GOTTSTEIN

Department of Metallurgy, Mechanics & Materials Science
Michigan State University
East Lansing, MI 48824

ABSTRACT

Synchrotron radiation can be utilized to obtain Laue back reflection patterns of microscopic areas with very short exposure time. A method is presented to fully automatically evaluate Laue patterns. The experimental setup and procedure are outlined to determine the orientations of a large number of contiguous grains, and consequently to enable orientation mapping. Preliminary experimental results are presented.

1. INTRODUCTION

The distribution of orientations in a polycrystalline solid is referred to as its crystallographic texture. The texture of a material is of interest, since during material processing almost always a nonrandom orientation distribution is introduced, for instance by cold rolling and subsequent annealing, the traditional forming process for metals. A texture can be detrimental(e.g. earing behavior during deep drawing of steel) or beneficial(e.g. grain oriented transformer steel). Conventional texture determination is usually carried out by pole figure measurements which provide the intensity contours of a set of crystallographic axes in the stereographic projection (Fig.1a). While the pole figure gives only the distribution of crystallographic axes, the orientation distribution function (ODF) can be computed from several pole figures (Fig.1b).

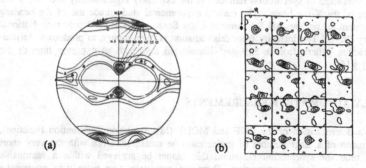

Fig. 1 {111} pole figure (a) and ODF (b) of pure Cu, rolled 95 % [1].

If only the volume average orientation distribution is of interest, then the ODF and its analysis provide adequate information. However, the ODF does not render information on how grains with particular orientations are related topologically. For many purposes it is important to know how the grains are oriented individually and how their orientations correlate with that of the neighboring grains. This is in particular true for determining the misorientation across grain boundaries and thus, the frequency of occurrence of misorientations of contiguous grains, for instance with regard to the orientation relationship of recrystallizing grains to their nucleation and growth environment. Inevitably, this requires the orientation determination of a large number of individual grains. For a common commercial material with a grain size of 30 μm and an illuminated area of 3×10 mm^2, the intensity at each point of the pole figure is subsumed from the contribution of about 50,000 grains. To achieve a comparable statistical significance with single orientation measurements, on the order of 10,000 grain orientations must be determined. There are a variety of techniques to measure the orientation of individual grains, the most common being X-ray Laue back reflection. While this technique is simple and appropriate for large grains (grain size $\phi > 1$ mm) or single crystals, a collimation of the X-ray beam on the order of $\phi \approx 10$ μm, as necessary to illuminate individual grains, which leads to a drastic loss of intensity and correspondingly, to substantially increased exposure times. Even when high intensity X-rays are employed (e.g. rotating anode) exposure times on the order of 10h are necessary to obtain a satisfactory Laue pattern. Assuming 10h exposure time, this would require a total of 100,000h or 11y exposure time. In conclusion, this would be an unrealistic task.

There are basically three different techniques to measure the orientation of small grains, namely selected area diffraction (SAD) by TEM [2], SAD channeling patterns [3] or backscatter Kikuchi patterns [4] by SEM, and Laue back reflection with synchrotron radiation [5]. The disadvantage of SAD is the limited field of view, so that only few grains can be evaluated in each sample, and a large number of samples have to be prepared, a time consuming, costly, and tedious task. Channeling or backscatter Kikuchi patterns have proven to be a very powerful technique for this purpose, and they have the advantage over X-rays, that the area of interest can be imaged and evaluated with regard to orientation. The major disadvantage of channeling patterns is their sensitivity to crystal perfection. With increasing deformation, the pattern quality degrades, and 5% strain is usually sufficient to essentially eliminate a useful channeling pattern. This is particularly important for the evaluation of deformed, partially or dynamically recrystallized microstructures, where always an essential volume fraction will be deformed.

The advantage of synchrotron radiation is the extremely high intensity and very small divergence of the X-ray beam. The current experimental study made use of the bending magnet system at the National Synchrotron Light Source (NSLS) of Brookhaven National Laboratory (BNL). The intensity from this radiation is high enough to produce a satisfactory diffraction pattern from a 5×5 μm^2 illuminated area with an exposure time of the order of 1 s [5].

2. FULLY AUTOMATIC MEASUREMENTS

For a determination of the ODF and MODF (Misorientation Distribution Function), the orientation of a large number of grains must be measured. Even with the very short exposure time for synchrotron radiation, this cannot be achieved within a reasonable timeframe without further automation. There are essentially two more time consuming steps involved in the procedure, namely processing of the recording medium, currently a high speed X-ray film, and evaluation of the diffraction pattern with regard to orientation and orientation relationships.

The recording of the diffraction pattern on X-ray films does not provide a solution to rapid large scale orientation measurements, since the film processing, including film mounting, dismounting, developing and digitizing requires at least in the order of 15 min for each exposure, not to mention the cost involved. A solution to this problem is a position sensitive area detector, which has been developed recently. Several such detectors are now commercially available [6].

A fast evaluation of the diffraction pattern requires a fully automatic computer processing of the digital detector output. A fast and usually unambiguous method is the identification of zones in a diffraction pattern, since the visible zones almost always belong to low index zone axes. The only necessary input besides the beam-specimen geometry is the identification of points which belong to the same zone. While the zones are easily discerned by viewing an exposed film, it is not trivial for a computer to identify the points of a common zone. This can be achieved by a Hough transformation, as recently introduced in pattern recognition algorithms [7,8]. The general problem consists of finding a curve of known shape (e.g. straight line, parabola etc.) through a subset of points in a point pattern. Since one can confine the consideration to straight lines for zone recognition a rapid evaluation is possible. We have developed a fast computer code based on pattern recognition algorithms and zone identification to fully and automatically evaluate the orientation from digitized Laue patterns, or equivalently from the digital output of a position sensitive detector.

An experiment will begin with a metallographic investigation of the specimen with regard to grain size and gain morphology, to determine the minimum beam translation between consecutive exposures. This is necessary, because the illuminated area cannot be imaged during or between exposures, so that the sample has to be scanned in predetermined steps to avoid omission of grains for a small grain size or aquisition of unnecessarily many data for large grain sizes. Then the specimen is mounted on an x-y positioning table and the specimen-detector distance is very accurately determined (crucial!). A reference point and direction is defined to allow a correlation of orientation and microstructure. The specimen is then translated to its starting position, and a control computer organizes exposure, data acquisition from the area detector, primary data processing and storage, specimen translation to the next prescribed point, exposure, etc., until the desired area is completely scanned. All secondary evaluation, e.g. computation of ODF and MODF, can be carried out after the measurement, if desirable, on a different computer.

3. FIRST RESULTS

A first experiment, however using a high speed X-ray film rather than a position sensitive detector, was carried out at the NSLS at BNL, beamline 13A. The beam size was $5 \times 5 \ \mu m^2$. A high speed X-ray film served to record the Laue back reflection pattern.

Copper and nickel specimens were investigated. Since the results for all specimens were essentially similar, we confine our report here to the results on a nickel specimen. With a beam size of $5 \times 5 \ \mu m^2$ an exposure time of the order of 1s was sufficient to reveal a defined Laue pattern. Owing to the small beam size the diffraction spots were very small and thus, allowed very accurate determination of their position and hence, of the crystal orientation.

The diffraction patterns were digitized and automatically evaluated by a computer program with respect to crystal orientation determination. An example is given in Fig. 2.

The orientation is denoted in terms of the axis (hkl) of the direction normal to the crystal surface and the direction [uvw] of the y-axis parallel to the crystal surface, and the corresponding {111} pole figure. The orientation is also given in terms of the closest Miller indices (<10) with the deviation indicated by the respective DEV in Fig. 2.

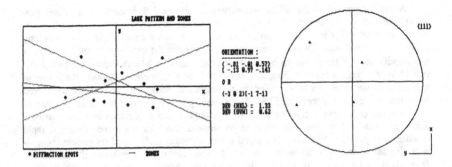

Fig. 2 Diffraction pattern and zones (a), and evaluated orientation (b) of a grain in a Ni Polycrystal.

The scanned path is indicated by a broken line in Fig. 3. Between measurements the sample was moved 20 μm in the plane perpendicular to the stationary beam. The points identify some locations of measurement. Four grains were intersected by this particular path, and the respective diffraction patterns and orientations are indicated in Fig. 3, as related by the corresponding numbers. Grains 2 and 4 are seen to have the same orientation and the evaluation reveals that grain 1 is related to grains 2 and 4 by an approximately 40°<111> rotation. The relatively straight boundaries between grains 1 and 2,4 are the {111} twist boundaries, which are usually found to be predominantly straight [9]. Grain 3 has an orientation entirely different from grains 1,2 and 4.

4. CONCLUDING REMARKS

It has been demonstrated that it is possible to fully automatically, and thus in a reasonable time frame, determine the orientation of a large number of contiguous grains and consequently obtain an orientation and misorientation mapping. There are two disadvantages in comparison of this method with SEM channeling or Kikuchi patterns; namely concurrent imaging and accessibility of high energy research facilities. The concurrent imaging is not believed to be a major problem, since for an automatic evaluation imaging is unnecessarily time consuming and also not indispensable, since a correlation of microstructure and orientation mapping can be obtained subsequent to the measurement. The accessibility problem can be essentially mitigated by instituting a dedicated beam line for texture research in solids at one of the national synchrotron light sources. The advantages over SEM channeling patterns is the considerably lower sensitivity to crystal imperfections and the immediate applicability also for nonconductive materials, like ceramics or composites.

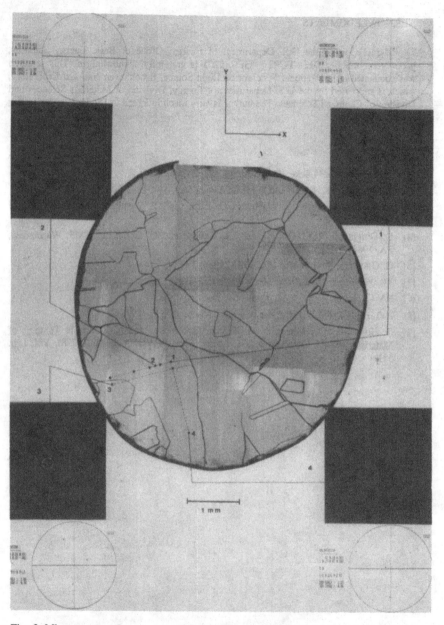

Fig. 3 Microstructure, Laue patterns and {111} pole figures of respective grain orientations in a Ni polycrystal. The pursued path is indicated by a broken line. The dots mark some position - among may more - where Laue patterns were taken. (Note that most small diffraction spots are lost during reproduction)

ACKNOWLEDGMENTS

The support of the U.S. Department of Energy, Office of Basic Energy Sciences, under grant number DE - FG02 - 85ER45205 is gratefully acknowledged. This research was conducted at the National Synchrotron Light Source, Brookhaven National Laboratory, which is supported by the U.S. Department of Energy, Division of Materials Sciences and Division of Chemical Sciences. The author is very much indebted to Dr. D. Cox for setting up the beam line.

REFERENCES

[1] J. Hirsch and K. Lücke; Acta Met. 36 , 2863 (1988).

[2] H.J. Perlwitz, K. Lücke and W. Pitsch; Acta Met. 17 , 1183 (1969).

[3] S. Hanada, T. Ogura, S. Watanabe, O. Izumi, and T. Masumoto; Acta Met. 34 , 13 (1986).

[4] D. Dingley; in ICOTOM 8 , ed. J.S. Kallend and G. Gottstein; (TMS, Warrendale, PA, 1988), p. 189.

[5] G. Gottstein; Scripta Met. 20 , 1791 (1986).

[6] J.R. Helliwell; Nucl. Instr. Meth. 201 , 153 (1982).

[7] P.V.C. Hough; U.S. Patent 3,069,654, Dec. 18, 1962.

[8] R.O. Duda and P.E. Hart; Ass. Comput. Mach. 15 , 11 (1972).

[9] G. Gottstein, H.C. Murmann, G. Renner, C. Simpson and K. Lücke; in Textures of Materials , ed. G. Gottstein and K. Lücke (Springer Verlag, Berlin, 1978), Vol. I, p. 511.

REFLECTIVITY OF SOFT X-RAYS BY POLYMER MIXTURES

T. P. Russell*, W. Jark†, G. Comelli‡ and J. Stöhr*

*IBM Research Division Almaden Research Center 650 Harry Road San Jose, California 95120-6099
†Present address: HASYLAB-DESY F41, 2000 Hamburg, Federal Republic of Germany.
‡Present address: Sincrotrone Trieste, Padriciano 99, 34012 Trieste, Italy.

ABSTRACT

The specular reflectivty of soft x-rays has been used to investigate the surface behavior of mixtures of poly(vinylidene fluoride), PVF_2, and poly(methyl methacrylate), PMMA. Mixtures where the concentration of PMMA is 0.8 where investigated as a function of the preparation and annealing conditions. It was found that mixtures prepared by rapidly casting from dimethylformamide, DMF, exhibited a gradient in the concentration of the two components. The gradient extended over large distances and was such that PMMA, the higher surface energy component was located preferentially at the surface. With annealing at 165°C, the width of the gradient diminished but the concentration of PMMA remained high at the surface. Films of the mixtures prepared under a slower solvent evaporation procedure produced a reflectivity profile characteristic of a phase separated mixture. The extent of phase separation was reduced with annealing. In all cases the roughness of the surface of the films was found to be 5.0Å or less.

INTRODUCTION

Evaluating the dependence of the composition as a function of depth for a polymer mixture has been limited by the experimental techniques available. For example, X-ray photoelectron spectroscopy, XPS, can be used to determine the surface and near surface composition in mixtures. XPS has excellent depth resolution but is limited in that only the first 50Å can be investigated[1]. Techniques such as forward recoil spectrometry have greater depth penetration but are limited in that the depth resolution is on the hundreds of angstroms scale [2]. The specular reflectivity of x-rays and neutrons [2] circumvent these shortcomings. Reflectivity of radiation, where the wavelength of the radiation is short, has the advantages that the depth resolution is excellent, ca. 10Å, yet the penetration depth is on the thousands of angstroms scale. Thus, x-ray and neutron reflectivity provide unique and sensitive means of determining the depth dependence of the atomic constituents and, consequently, the composition as a function of distance from an interface. The only limitation is that the contrast, i.e. electron density for x-rays and neutron scattering length density for neutrons, be sufficient to resolve the profile. In this report, we present an investigation of the concentration profile and surface characteristics of thin films of polymer mixtures cast from solution. These non-equilibrium specimens are then investigated as a function of annealing at temperatures above the glass transition temperature.

Stern [4] and Heavens [5] have developed exact formalisms describing the reflectivity of a homogeneous thin film taking into account all multiple reflections and absorption in the film. This formalism is readily extended to a multilayer system or a system containing a concentration gradient which can be considered as a histogram comprised of consecutive thin layers with slight concentration changes. Defining θ as the grazing angle of incidence between the incident x-ray beam and the surface plane, the magnitude of the scattering vector k_j in the jth layer is defined as

$$\tilde{k}_j = \frac{2\pi}{\lambda} \sqrt{\tilde{\varepsilon}_j - \tilde{\varepsilon}_0 \cos^2\theta} \tag{1}$$

where λ is the wavelength, $\tilde{\varepsilon}_j$ is the complex dielectric constant of the jth layer and $\tilde{\varepsilon}_0$ is the complex refractive index of the surrounding medium which for vacuum is 1. In this and following equations, the tilde denotes a complex number. The reflection coefficient between the jth and $(j + 1)$th layer is denoted as $\tilde{r}_{j,j+1}$ and is defined as

$$\tilde{r}_{j,j+1} = \frac{\tilde{k}_j - \tilde{k}_{j+1}}{\tilde{k}_j + \tilde{k}_{j+1}} \tag{2}$$

If the substrate is considered as the $(j + 1)^{th}$ layer in a specimen comprised of j layers, then the reflection coefficient of the jth layer closest to the substrate with thickness d_j is given by

$$\tilde{r}_{j-1,j} = \frac{\tilde{r}_{j-1,j} + \tilde{r}_{j,j+1}e^{2id_jk_j}}{1 + \tilde{r}_{j-1,j}\tilde{r}_{j,j+1}e^{2id_jk_j}} \tag{3}$$

This recursion relation is then used to calculate the reflectivities of each of the remaining $j - 1$ layers up to $\tilde{r}_{0,1}$. The reflectivity is then given by

$$R = \tilde{r}_{0,1}\tilde{r}_{0,1}^* \exp^{-((4\pi/\lambda)\sin\theta)^2 <z^2>} \tag{4}$$

where the asterisk denotes the complex conjugate. The exponential term included in Eq. (4) accounts for surface roughness where $<z^2>$ is the mean square value of the surface roughness.[6] The dielectric constant $\tilde{\varepsilon}_j$ can be calculated from the atomic scattering factors $f = f_1 + if_2$ which for the soft x-ray case have been tabulated by Henke et al.[7] according to

$$\tilde{\varepsilon}_j = 1 - \frac{r_0\lambda^2}{\pi} \sum_p n_p(f_{1,p} - if_{2,p}) \tag{5}$$

where r_0 is the classical electron radius and n_p is the number of atoms of element p per unit volume with atomic scattering factors $f_{1,p}$ and $f_{2,p}$.

EXPERIMENTAL

A dry mixture of 80% by weight of poly(methylmethacrylate) PMMA having a weight average molecular weight, M_w, of 500,000 and a ratio of M_w to the number average molecular weight, M_n, of 1.05 with 20% by weight of poly(vinylidenefluoride) PVF_2 (M_w = 120,000; $M_w M_n$ = 1.6) was dissolved in dimethylformamide, DMF. The first specimen was prepared by spin coating the solution onto a 2.5 cm diameter quartz wafer until dry. The specimen was placed in vacuum to remove the residual solvent. In a second case, the solution was spin coated onto a quartz wafer to coat the wafer with the solution. The specimen was transferred to a hot plate at 80°C under a flow of nitrogen to evaporate the solvent slowly and then placed under vacuum to remove residual solvent.

Measurements were performed using a vacuum compatible, triple axis diffractometer which has been described in detail elsewhere.[8,9]. As shown in Figure 1, specular reflectivity measurements were performed by rotating the specimen at an angle θ and the detector by an angle 2θ. In this manner the diffraction vector is maintained normal to the film surface. Off-specular measurements were also performed by setting the incident beam at a fixed angle with respect to the specimen and rotating the detector through an angle Ω. At the specular condition $\Omega = 2\theta$. This type of scan rotates the diffraction vector by and angle Ω off the surface normal, thereby, placing a component of the diffraction vector in the plane of the specimen. It should be noted that the finite size of the specimen prohibited the measurement of the reflectivity below an angle of $\theta = 0.8°$ without substantial geometric corrections. All experiments were performed on beamline III-4 at the Stanford Synchrotron Radiation Laboratory using x-rays of wavelength 0.795 nm with $\Delta\lambda/\lambda = 0.01$ at a chamber pressure less than 5×10^{-5} Torr. All reflectivity profiles were obtained over time intervals of 300 seconds during which time the incident x-ray flux did not change significantly.

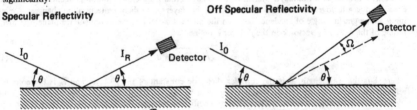

Figure 1. Reflectivity geometries defining specular and off-specular conditions.

RESULTS AND DISCUSSION

In Figure 2 is shown the specular reflectivty of the spin coated 80/20 PMMA/PVF$_2$ along with the off-specular measurements at integral values of the angle of incidence. Focusing on the specular profile initially, it is seen that the reflected intensity decreases monotonically as a function of the angle of incidence with a high frequency oscillation characteristic of the total thickness of the specimen. In order to amplify this oscillation, the specularly reflected profile was divided by a five point smoothed profile. The log of this ratio is shown in Figure 3 as a function of the angle of incidence. Fourier transformation of the data in Figure 3 shows a sharp maximum corresponding to a total film thickness of 2.5x10^{3}Å. Consequently, the radiation is penetrating through the entire thickness of the specimen and the specular profile

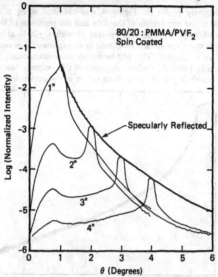

Figure 2. Specular and off-specular reflectivity profiles of an 80/20 PMMA/PVF$_2$ mixture rapidly cast from DMF.

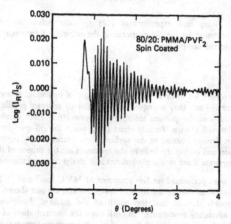

Figure 3. Log of the ratio of the measured specular reflectivity to a five point smoothed profile for a rapidly cast 80/20 PMMA/PVF$_2$ mixture.

is characterizing the composition variation through the specimen. Using Equations (1)-(5), it was possible to fit the specularly reflected profile with a linear concentration gradient that extended over 1.5×10^3 Å where the gradient ranged from a PMMA concentration at the vacuum/film surface of 1.0 to a bulk concentration of 0.8 (in volume fraction). A comparison of the measured and calculated reflectivity profiles is shown in Figure 4. The preponderance of PMMA at the surface is clearly evident by the fact that the critical angle occurs at a value of θ less than $0.8°$. The critical angle of PMMA is $0.75°$, whereas that for PVF_2 is $1.02°$. This results is somewhat surprising since it indicates that PMMA, the higher surface energy component ($\gamma_{PMMA} = 39.0$ dyne/cm [10] and $\gamma_{PVF_2} = 25.0$ dyne/cm [11]) is preferentially located at the surface.

The calculated reflectivity profile in Figure 4 was obtained using a roughness parameter of $<z[2]>^{1/2}$ of 4.0Å. Within the constraints of the data and the precision of the angular motion of the diffractometer, this value of $<z[2]>^{1/2}$ can be reported to within $\pm 20\%$ at most. In fact, since the specular reflectivity depends upon the surface roughness in an exponential manner, the measurements are quite sensitive to the surface roughness. This will be shown more dramatically in subsequent data. It is important to emphasize the fact that the surface roughness of this polymer film and most non-crystalline polymer films is relatively small despite the long chain nature of the molecules.

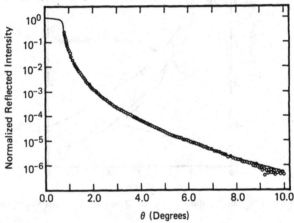

Figure 4. Calculated (line) and experimental (O) specular reflectivity profiles for an 80/20 PMMA/PVF$_2$ mixture prepared by rapidly evaporating the solvent. Experimental data below $\theta = 0.8°$ have been removed due to the sample size.

The off specular profiles for integral values of the angle of incidence, shown in Figure 2, display some characteristic features. First, they are equal to the specularly reflected profile when the θ-2θ condition is met. All of the off-specular profiles are strongly asymmetric. Each exhibits a maximum at the critical angle which is the well-known Yoneda effect [12]. Since the off-specular scattering arises from variations in the electron density parallel to the surface, then lateral density correlations in the electron density at both the surface and within the bulk of the specimen are the origins of the scattering. It is not possible, at present, to separate these two contributions in a straightforward manner [13].

Annealing studies were performed on this specimen at 165°C for 3 and 18 hours. The specularly reflected profiles look similar to that of the original specimen and are not shown here. Fitting the data via the histogram approach showed that the width of the gradient diminished from 1.0×10^3Å to $2.5 \times 10[2]$Å for the two annealing treatments. In both cases the concentration of PMMA at the surface remained high. At 165°C the mixture is well above the glass transition temperature such that molecular motions should be fairly rapid. However, it is clear that the films have not attained equilibrium in the time allotted, since the higher surface energy component is still preferentially located at the surface. This result suggests that the molecular motions of the polymer molecules near a boundary are slower than those found in the bulk. Due to limited beam time measurements could not be performed to pursue this point further. As with the initial specimen, the surface roughness values were small, 3.7Å and 3.0Å, respectively. The difference between these is not felt to be significant.

The specular reflectivity and off-specular reflectivity profiles for the mixture cast via a slow solvent evaporation process are shown in Figure 5. The specular profile shows a shoulder occurring at an incidence angle near $\theta = 1.0°$ but, otherwise, monotonically deceasing. As with the data for the spin cast specimen, though not evident in the data as drawn, there is, also, a high frequency oscillation corresponding to a total film thickness of 2.0×10^3Å. The unusual feature of the specular profile is that the shoulder is observed with no further oscillations of comparable magnitude. This feature alone eliminates the possibility of there being an overlayer of either PVF_2 or PMMA, since an oscillation in the reflectivity profile characteristic of the thickness of this layer would have been easily observed experimentally over the angular range investigated. In fact, assuming any reasonable gradient in the concentration of the constituents did not produce a suitable fit to the reflectivity profile. The best fit to the experimental data was obtained by using a wieghted sum of the reflectivities of pure PMMA and PVF_2. Shown in Figure 6 is a comparison between the measured reflectivity profile and the sum of 72% of the reflectivity of PMMA and 28% that of PVF_2 with $<z[2]>^{1/2} = 3.5$Å. As can be seen the agreement between the calculations and the experiment are quite good. The calculations were insensitive to any gradients in the concentration and were dominated by the reflectivities of the pure components. These results strongly suggest that the mixture has undergone phase separation near the film surface during the casting process. This result is surprising for this polymer pair, since the two are known to be miscible. However, Nishi and Wang

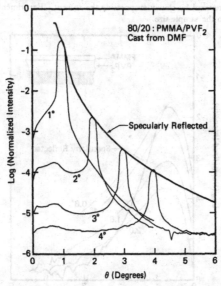

Figure 5. Specular and off-specular reflectivity profiles of an 80/20 PMMA/PVF_2 mixture slowly cast from DMF.

[14] observed a similar behavior on the bulk mixtures where they clearly observed that the PVF_2 separated from the mixture and, in their case, crystallized.

The off-specular reflectivity profiles show characteristics that are similar to that of the spin coated specimen. However, it is evident that the off-specular scattering is weaker. This result can be described in part by the smaller surface roughness. Another contributing factor must be associated with the reduction in the concentration fluctuation scattering caused by the phase separation, provided the domains near the surface are not spatially correlated and are larger than several thousand angstroms.

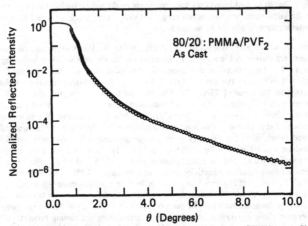

Figure 6. Calculated (line) and experimental (O) specular reflectivity profiles for an 80/20 PMMA/PVF$_2$ mixture prepared by slowly evaporating the solvent. Experimental data below $\theta = 0.8°$ have been removed due to the sample size.

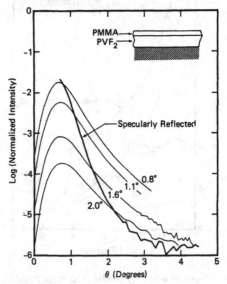

Figure 7. Specular and off-specular reflectivity profiles for a bilayered specimen comprised of a layer of PMMA on top of a layer of PVF$_2$ on a silicon substrate. The experiment is designed to measure the interdiffusion of the two polymers.

Annealing this specimen for 3 and 18 hrs. at 165°C produced reflectivity profiles that were similar to that of the initial specimen. The major difference was that the extent of phase separation appeared to change with time. After 3 hours the reflectivity profile could be fit with a sum of 80% of the reflectivity of PMMA and 20% that of PVF$_2$ and after 18 hours 85% and 15%, respectively. Consequently, as in the case of the spin cast specimen, the approach to equilibrium was quite slow, much slower than that in the bulk.

The sensitivity of the reflectivity technique to the surface roughness can be easily demonstrated by the following experiment that was actually prevented due to a surface roughness. Given the electron density difference of PVF$_2$ and PMMA, an attempt was made to monitor the interdiffusion of the two components by preparing a bilayered specimen. Here, as shown in the inset of Figure 7, a film of PVF$_2$ was spin coated onto a silicon wafer where the thickness of the film was ca. 3×10^3Å. On a separate substrate, a 300Å layer of PMMA was prepared. This was floated off the substrate onto water and the PVF$_2$ coated substrate was used to pick up the floating film, thereby, forming a bilayered specimen. This preparation technique has been shown to produce a specimen from which reflectivity measurements can easily be made [3]. The specular and off-specular reflectivity profiles for the PVF$_2$/PMMA bilayer are shown in Figure 7. It is clear from these results that there is essentially no specularly reflected radiation and only off-specular scattering is evident. By $\theta = 2.5°$ what was measured as the specular reflectivity has dropped to near the resolution limits of the detector and the reflectivity does not approach unity. at angles near the critical angle. Contrary to the other data shown, the off-specular scattering shows no evidence of a maximum in the vicinity of the θ-2θ condition. It is evident that the surface of the specimen is very rough and has caused a total loss of the specular scattering. In fact, it is not possible to determine a roughness parameter. The origin of this large surface roughness is the crystallization of the PVF$_2$. PVF$_2$ crystallizes into lamellae that are approximately 60Å in size which also form near the surface causing the surface to be rough. This results points to the fact that measurement of the specular reflectivity of crystalline polymers, in general, will be hampered, if not prevented, by the formation of lamellae which cause a very uneven surface.

ACKNOWLEDGMENTS

These experiments were performed at the Stanford Synchrotron Radiation Laboratory which is supported by the Department of Energy, Division of Basic Energy Sciences and the National Institute of Health, Biotechnology Resource Program, Division of Research Resources.

REFERENCES

1. A. Dilks, *Electron Spectroscopy: Theory, Techniques and Applications*, Vol. 4, C.R.Brundle and A. D. Bukin, eds. (Academic Press, London, 1981), p. 277.

2. P. J. Mills, P. F. Green, C. J. Palmstrom, J. W. Mayer and E. J. Kramer, *Appl. Phys. Lett.* **45**, 9 (1984).

3. T. P. Russell, A. Karim, A. Mansour and G. P. Felcher, *Macromolecules* **21**, 1890 (1988).

4. F. Stern in *Solid State Physics*, Vol. 15, F. Seitz and D. Turnbull (Academic Press, New York, 1963).

5. O. S. Heavens, *Optical Properties of Thin Solid Films* (Dover Publications Inc., 1965).

6. P. Beckman and A. Spizzichino, The Scattering of Electromagnetic Waves From Rough Surfaces (Pergamon Press, Ltd., New York, 1963).

7. B. L. Henke, P. Lee, T. J. Tanaka, R. L. Shimbabukuro and B. K. Fujikawa, *At. Dat. Nucl. Dat. Tables* **27**, 1 (1982).

8. W. Jark and J. Stöhr, *Nucl. Instr. and Meth.* **A266**, 654 (1988).

9. W. Jark, G. Comelli, T. P. Russell and J. Stohr, *Thin Sol. Films*, accepted.

10. D. A. Olsen and A. J. Osternas, *J. Appl. Polym. Sci.* **13**, 1523 (1969).

11. A. H. Ellison and W. A. Zisman, *J. Phys. Chem.* **58**, 260 (1954).

12. Y. Yoneda, *Phys. Rev.* **131**, 2010 (1963).

13. S. K. Sinha, E. B. Sirota, S. Garoff and H. B. Stanley, *Phys. Rev. B*, submitted.

14. T. Nishi and T. T. Wang, *Macromolecules* **8**, 909 (1975).

SYNCHROTRON MICROTOMOGRAPHY OF COMPOSITES

S. R. STOCK*, J. H. KINNEY**, T. M. BREUNIG*, U. BONSE***, S.
D. ANTOLOVICH*, Q. C. JOHNSON**, and M. C. NICHOLS****
*Mechanical Properties Research Laboratory and School of
Materials Engineering, Georgia Institute of Technology, Atlanta,
GA 30332-0245, USA
**Chemistry and Materials Science Department, Lawrence
Livermore National Laboratory, Livermore, CA 94550, USA
***Department of Physics, Dortmund University, FRG
****Exploratory Chemistry Division, Sandia National Laboratory,
Livermore, CA, USA

ABSTRACT

X-ray computed tomography (CT) uses absorption profiles
from many different viewing directions to reconstruct the two-
dimensional distribution of x-ray absorptivity within a slice
of the sample. The tunability, high brightness and parallelism
of synchrotron radiation are critical to high resolution
(0.001mm), high contrast (1%) CT or microtomography. In situ
study of samples multiple times during the course of an
experiment is exciting to consider.

Continuous fiber SiC/Al composites were deformed under
three-point bending, and the resulting damage and fiber
arrangement were revealed with synchrotron microtomography.
Several hundred slices of 0.012 mm thickness were recorded
simultaneously using 25 keV radiation and a phosphor
screen/charge coupled device (CCD) detector. Reconstruction was
with the filtered back projection method. Low density regions
were observed in the matrix in regions of highest stress where
cracking is expected.

INTRODUCTION

Advanced composites have received recently a tremendous
amount of attention, and metal matrix composites (MMC) are
filling important niches as structural materials. The
operational conditions for spacecraft and space structures, for
example, demand high specific stiffness and strength, high
thermal and electrical conductivity and dimensional stability
over wide temperature ranges [1]. Continuous-fiber MMC possess
these properties and minimize the effects of thermal deformation
[2].

The utilization of MMC for primary structural applications
requires adequate design safety margins and lifetimes under all
anticipated loading conditions and environments. This goal
demands sound understanding of fracture, creep (if applicable)
and fatigue behavior of the composites which in turn
necessitates a broad experimental data base [3]. Failure modes
(damage) such as matrix cracking, fiber-matrix debonding, and
fiber fracture act in concert and are responsible for the

Mat. Res. Soc. Symp. Proc. Vol. 143. ©1989 Materials Research Society

overall property degradation. The inhomogeneous microstructure of MMC is accompanied by large sample-to-sample variability in macroscopic properties. These variations make it imperative that the state of damage be monitored throughout the test of each MMC specimen.

Characterization, therefore, must be nondestructive if incidental specimen differences are not to obscure fundamental damage accumulation mechanisms. Synchrotron x-ray microtomography offers the necessary attributes for damage characterization in composites. Spatially-broad, well-collimated and very intense beams of synchrotron x-radiation are the key to rapid characterization of volumes greater than 1x1x1 mm^3. Synchrotron radiation's broad spectrum allows selection of the most appropriate wavelength(s) for a given specimen; contrast can be greatly enhanced by comparing images recorded with wavelengths on either side of the absorption edge of the element of interest [4-6].

In computed tomography (CT), the spatial variation of x-ray absorption in a thin slice or cross-section of the sample is recorded for various projection directions or views, the number of which is an important factor in determining the spatial resolution in the reconstructed image. Figure 1 illustrates how the projected absorption profile P_θ can vary as a function of viewing direction: in the view shown to the left, the uniformly absorbing "phase" allows a narrower, but more intense, transmitted beam to pass through the sample than it does in the view on the right. The different views are combined via a reconstruction algorithm, and the two-dimensional map of x-ray absorption within the slice is obtained. The filtered back projection method is the most popular reconstruction method (see example [7]), and it is the method used below.

MICROTOMOGRAPHY

Spatial resolution distinguishes industrial or medical CT from microtomography. Most nondestructive inspection units have resolution no better than 0.1 mm, although a specialized unit with 0.050 mm resolution has been constructed [8], and current

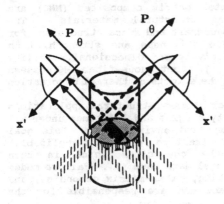

Figure 1. Illustration of the variation of projected absorption profile P_θ as a function of viewing direction.

projections forecast a limitation of 0.025 mm [9]. In microtomography, special detectors and/or collimators are employed to obtain spatial resolution approaching 0.005 mm. Elliott and coworkers use pinhole collimation and a single conventional detector to achieve high resolution: each absorption profile is produced by translating the narrow beam across the sample [10-12]. This scheme is simple but slow and has been used with both synchrotron and microfocus radiation. Parallel data collection using two-dimensional detectors has been used with synchrotron radiation to record projections of many slices simultaneously [13-15].

The apparatus used to record the data presented below is based on a two-dimensional CCD (charge-coupled-device) coupled through a lens to a fine-grained phosphor screen or single crystal [13]. X-rays pass through the specimen and are converted to visible light, allowing great flexibility in changing magnification. A rotation stage positions the sample so that different viewing directions (angles) can be recorded, and stage movements, data acquisition and reconstructions are controlled with a Microvax II computer.

The continuous fiber SiC/Al composite is a 1.5 mm thick panel and consists of eight layers of unidirectional fibers in an aluminum matrix (Figure 2). The fiber volume fraction is about 0.40, each fiber has a 0.04 mm carbon core and a total diameter of 0.14 mm (AVCO SCS-8) and the fibers lie along the length of the $1.5 \times 1.5 \times 20$ mm^3 microtomograpy samples. Three-point bending, about an axis perpendicular to the fibers, is the mode of deformation, and audible popping is an indication of fiber cracking during bending of the panel and of the smaller microtomography specimens.

The data for the reconstructions is from monochromatized bending magnet radiation (25 keV energy) at HASYLAB during less than optimal beam conditions. The unstable beam has introduced ring artifacts into some of the slices despite normalization for detector and beam nonuniformity between each view. Filtering of the power spectrum, presently underway, should ameliorate this effect. Sixty views are available for each sample, a considerable undersampling necessitated by the short beam

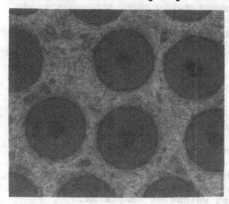

Figure 2. Opitcal micrographs of the polished surface normal to the fibers of the SiC/Al composite.

0.1mm

lifetimes, and the pixel size in the reconstructions is about 0.012 mm. A total of 150 slices of 0.012 mm thickness were obtained from each of the two samples.

Figure 3 shows slices of both samples. One slice is from relatively undeformed material of B4, and another is from the most heavily distorted portion of that sample. The distortion of the shape of sample B4 clearly indicates the most heavily deformed volume. The carbon cores are visible in the reconstructions (gray levels similar to the background outside of the sample), and comparison of the core positions in the two slices reveals the displacement of the intact fibers from their positions in the undamaged composite. Breaks in fibers also are seen with the matrix filling the volume between the ends of the fibers. Figure 3c shows slices of sample B3 which was less deformed than B4. The ring artifacts are less pronounced, but undersampling still causes the grainy image. Figure 4a shows a three-dimensional view of B3 with different "sectioning" planes revealing the straight carbon cores of the fibers, and Figure 4b shows only the fiber cores. The SiC portions of the fibers cannot be distinguished from the surrounding Al matrix because the gray levels have been quantitized for equal increments of absorption (i.e. to $1/255 \pm 0.1\%$). As more sophisticated data representation becomes available, small differences such as these may become more apparent.

PROSPECTIVE

Increased spatial resolution, higher contrast and the ability to study larger samples are important goals. Contrast better than 5% with spatial resolution of 0.001 mm can be achieved, but considerable instrumentation development is necessary. Damage accumulation and cracking during mechanical testing are of considerable interest, and in situ deformation studies are straight-forward to implement. Nondestructive identification of the path of macroscopic cracks and the morphology of microcracks (e.g. matrix cracks, interface failure

 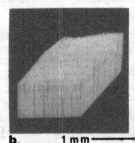 

a.　　1 mm ——　　**b.**　　1 mm ——　　**c.**　　1 mm ——

Figure 3. Reconstructed slices of the SiC/Al composite with the fibers' carbon cores appearing darker than the matrix.
a) Lightly deformed volume of sample B4.
b) Heavily deformed volume of sample B4.
c) Typical volume of sample B3.

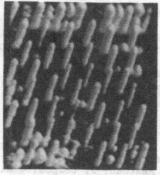

Figure 4. Three-dimensional views of sample B3. a) Different sectioning planes. b) Image showing only the fibers' carbon cores.

or fiber breakage) is one area in which microtomography would be invaluable. Knowledge of crack initiation and propagation stages, along with a measure of quantity of damage present, would allow prediction of the remaining life of a structure via continuum mechanics and finite element methodologies.

Damage is very difficult to detect experimentally because of crack closure when externally applied loads are removed. Current methods for assessing the damage state are based on stiffness loss [16-18] and do not adequately locate and quantify the damage present. If somewhat larger specimens were studied with resolution better than 0.005 mm and with in situ loading, the effect of closure on crack propagation rates could be directly investigated. The physical crack opening under different loads could be compared with the indirect methods of inferring crack closure. These measurements would help elucidate crack growth mechanisms.

Achievement of 0.001 mm spatial resolution in microtomography over millimeter-wide dimensions will fill the gap between electron microscopy and statistical sampling techniques such as small angle scattering on the one hand and macroscopic measuring techniques on the other. The previously inaccessible size range is where microscopic processes link with macroscopic behavior (and associated continuum models), and, as such, its study is critical for physically based modeling.

Results presented above on SiC/Al MMC are very encouraging. Better sampling and statistics and optimization of contrast through selection of beam energy are important next steps. Recent experiments at HASYLAB using both an asymmetric single crystal magnifier and a single crystal scintillator have demonstrated a 25-fold increase in spatial resolution: this means, at least for small samples, spatial resolution <0.005 mm has been achieved [19]. Newly completed microfocus tomography of the same SiC/Al MMC has successfully detected cracks in the unloaded state [20]. The next synchrotron data obtained from damaged ceramics should be very interesting indeed.

ACKNOWLEDGEMENTS

This research was partially supported by NSF grant ENG/MSM-8614493, by DOE W-7495-ENG-48 and by BMFT, West Germany. Gratitude is also due to the staff of HASYLAB (Hamburg, West Germany) where this work was performed and to Lockheed Aeronautical Systems which provided the material.

REFERENCES

1. H. H. Armstrong, "Satellite Applications of Metal-Matrix Composites," SAMPE National Symposium, May 1979.
2. M. F. Amateau, J. Composite Matls. 10, (1976).
3. M. H. Kwal and B. K. Min, J. Composite Matls. 18, 619 (1984).
4. R. Glocher and W. Frohnmayer, Ann. Phys. Leipzig 76, 369 (1925).
5. L. Grodzins, Nucl. Instrum. Methods 206, 541 (1983); 206, 547 (1983).
6. J. H. Kinney, Q. C. Johnson, M. C. Nichols, U. Bonse and R. Nusshardt, Appl. Optics 25, 4583 (1986).
7. A. C. Kak and M. Slaney, Principles of Computerized Tomographic Imaging, (IEEE Press, New York, 1987).
8. F. H. Sequin, P. Burstein, P. J. Bjorkholm, F. Homburger and R. A. Adams, Appl. Optics 24, 4117 (1985).
9. R. A. Armistead, Advanced Matls. and Processes/Metals Programs, March 1988, p. 42.
10. J. C. Elliott and S. D. Dover, J. Microscopy 138, 329 (1985).
11. D. K. Bowen, J. C. Elliott, S. R. Stock and S. D. Dover, in X-ray Imaging II edited by L. V. Knight and D. K. Bowen (SPIE Bellingham, WA, 1986) p. 94.
12. S. R. Stock, A. Guvenilir, T. L. Starr, J. C. Elliott, P. Anderson, S. D. Dover and D. K. Bowen, in Advanced Characterization Techniques in Ceramics (in press).
13. J. H. Kinney, Q. C. Johnson, U. Bonse, M. C. Nichols, R. A. Saroyan, R. Nusshardt, R. Pahl and J. M. Brase, MRS Bull. XIII, 13 (1988).
14. T. Hirano, K. Usani, K. Sakamoto and Y. Suzuki, in Photon Factory Activity Report 1987 (KEK Report 87-2, Tsukuba, Japan) p. 187.
15. B. P. Flannery, H. Deckman, W. Roberge and K. D'Amico, Science 237, 1439 (1987).
16. W. Hwang and K. S. Han, J. Composite Matls. 20, 125 (1986).
17. W. S. Johnson, NASA Tech. Mem. 89116, March 1987.
18. J. Aboudi, Composites Sci. and Tech. 28, 103 (1987)
19. R. Nusshardt and U. Bonse, private communication (1988).
20. S. R. Stock, J. C. Elliott, T. Breunig, A. Guvenilir, S. D. Antolovich and P. Anderson, unpublished data (1988).

SOFT X-RAY ABSORPTION MICROSCOPY OF SURFACES WITH SYNCHROTRON RADIATION

G. R. HARP and B. P. TONNER, Dept. of Physics and Laboratory for Surface Studies, University of Wisconsin - Milwaukee, 1900 E. Kenwood Blvd., Milwaukee, WI 53211

ABSTRACT

As a spectroscopic technique, x-ray absorption near-edge structure (XANES) of core-levels using synchrotron radiation is in wide-spread use for the determination of the molecular composition of solid surfaces. A common detection method measures the yield of secondary electrons, which is proportional to the x-ray absorption coefficient for sufficiently high photon energy. In the experiments reported here, we show how the secondary electrons emitted as a result of photoabsorption can be used to generate a magnified image of the sample surface, with fundamental spatial limits determined by the deBroglie wavelength of the emitted electrons. Contrast in the secondary electron spatial distribution contains both topographical and chemical information about the sample surface. The use of tunable synchrotron radiation enables us to separate these contributions to the microscopic image, and to spatially resolve surface chemical composition as reflected in micro-XANES spectra.

1.0 SPATIALLY RESOLVED X-RAY ABSORPTION FROM SURFACES

One of the important areas of development in microscopy is the coupling of spatial resolution with simultaneous spectroscopy of a fundamental electronic property of the sample. For example, in transmission electron microscopy, spatially resolved electron-energy-loss spectroscopy (TEM-EELS) can be measured by energy filtering the scattered electron image [1].

Additional challenges face the microscopist interested in determining composition of surface layers and thin films, since depth resolution becomes as important as the lateral resolving power of the microscope. The most common surface analytical probe for core-level micro-spectroscopy is scanning Auger microscopy (SAM), in which the surface sensitivity arises from the short mean-free-path of the Auger electron. The SAM technique relies on a focussed incident electron beam for its spatial resolution, which has the advantage of being easily packaged into a laboratory instrument. However, there are some compelling reasons which motivate the replacement of the incident electron beam with an incident photon beam from a synchrotron.

The most obvious difference between photoemission and electron excited emission is the quantum mechanical one: a photon is discretely absorbed. This results in a large increase in the signal/background ratio of photoemission spectroscopy as compared to Auger spectroscopy. Another advantage of core-level photon spectroscopies is a reduction in sample damage, since soft x-ray photons create fewer valence excitations per core-level aborption event than do high energy electron beams. Combining the S/B ratio and core-level vs. valence excitation quantum yield, it has been estimated that photon spectroscopy has a two-order of magnitude advantage overall in sensitivity (or minimum areal resolution) for the same sample damage threshold as compared to electron probe techniques [2,3].

Although photon spectroscopies have long been recognized to have such potential advantages in microanalysis, the difficulty of generating intense monochromatic beams has precluded significant microscopic applications in the past. We have recently demonstrated the successful combination of photoemission microscopy with soft x-ray absorption spectroscopy using synchrotron radiation [4]. The philosophy of the technique is to use the monochromatic synchrotron beam as the spectroscopic probe, and derive spatial information from an image generated using conventional electron optics. The spectroscopic energy resolution is determined by the line-widths of core-level absorption features, which can be as narrow as a few tenths of an eV. The spatial resolution is ultimately limited by the electron deBroglie wavelength, although limitations due to available photon flux and optical aberrations determine the resolution in practice.

2.0 MICRO X-RAY ABSORPTION SPECTROSCOPY WITH A PHOTOEMISSION MICROSCOPE

A schematic diagram of a simple apparatus used to acquire the micro-XANES spectra reported here is shown in Fig. 1. A monochromatic beam of soft x-rays illuminates a region of interest on the sample surface. Secondary electrons, produced by various photoabsorption processes, are accelerated and focussed by a two-element electrostatic immersion lens, equipped with an aperture in the focal plane. In an immersion objective lens, the sample is placed in a large electric field. This field accelerates the emitted secondary electrons, and matches the large angles of emission to the small acceptance angles defined by the focal plane aperture. A microchannelplate image intensifier transfers the magnified electron image to a phosphor screen. This light image is then digitized using video techniques.

Fixed wavelength photoemission microscopes (PEM) are currently in use in a few laboratories [5,6]. These instruments have shown a capability for high resolution (100Å), ultra-high vacuum compatible surface structure studies of a wide variety of samples, from metal-semiconductor interfaces to DNA [5]. The light source in these instruments is typically a Hg discharge lamp, so that the photon energy barely exceeds the sample work function. Under these conditions, the image in the PEM contains contrast arising primarily from two sources: surface topography which alters the local electric field, and variations in local chemical composition as reflected in a change in work function.

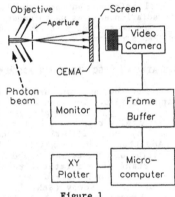

Figure 1

By incorporating synchrotron radiation in the PEM, we have been able to separate the topological and chemical contributions to the contrast of the secondary electron image. As the incident photon energy is varied, changes in the image contrast are observed which reflect the spatially-resolved photoyield from the sample surface. We have shown that the incident photon energy can be chosen to provide high contrast between metal-metal and metal-semiconductor lateral interfaces, which is a crude form of chemical analysis [4]. The important new development is that this technique can be made quantitative, and used for surface micro-x-ray absorption spectroscopy (μXAS).

For photon energies high enough to ionize core-levels, the total secondary electron yield has been shown [7] to be proportional to the optical absorption coefficient,

$$Y(\hbar\omega) \sim \hbar\omega \, \mu(\hbar\omega) \, D_{eff} \qquad (1)$$

an expression which has been empirically verified for photon energies spanning the range from 100 eV to several kV [8]. The quantum yield of the secondaries is high, since there can be more than one secondary electron emitted into the vacuum for each photon absorbed within D_{eff}.

A complete description of the secondary electron yield due to photoabsorption includes the kinetic energy distribution, angular distribution, and effective sampling depth of the yield as a function of kinetic energy (K). The kinetic energy distribution of secondary electrons for a wide range of materials can be described by the function

$$dn(K) = 6W^2 \, \frac{K}{(K + W)^4} \, dK \qquad (2)$$

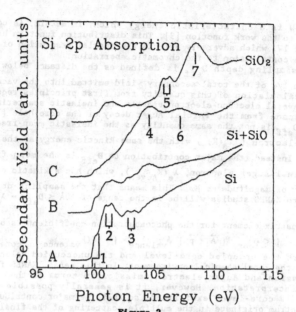

Figure 2.
Micro-XANES spectra from a Si(111)/SiO₂ desorption experiment. All spectra are from a 5μ diameter sample area. The spin-orbit splitting is indicated for peaks 1-3 and 5. Curve A at the bottom is from a region of clean Si(111), curve D at the top is from the native oxide.

The top curve in Fig. 2D shows the micro-XANES spectrum from a Si(111) surface covered with ≈15Å of native oxide. The Si features from the substrate can be seen in the photon energy range from 99.5 - 103.5 eV. The new features (5) and (7) are identified with SiO_2 by comparison to previous bulk absorption spectra. Our data resolves the spin-orbit splitting in feature (5), which appears as a single peak in previous area-averaged experiments [14].

Annealing the Si(111)/SiO₂ film to 1000C produces a surface with regions that yield spectra similar to 2B and 2D. Previous scanning Auger microscopy studies of Si(100)/SiO₂ have shown that desorption occurs by a reaction $SiO_2 + Si \rightarrow 2SiO$ which takes place at the edges of voids [15]. Our micro-XANES spectrum 2B from Si(111)/SiO₂ shows the presence of an additional spectral feature at 104 eV, which has been previously identified as being from Si(II)O in bulk studies. The relative intensity of this feature, in comparison the clean Si edge at 99.5 eV, suggests a sub-monolayer coverage of SiO on the exposed areas from which SiO_2 has been depleted. No direct evidence for the presence of SiO has been presented in previous electron microscopy experiments, so we must be cautious in interpreting these preliminary results. It is possible, however, that electron beam decomposition of SiO takes place during SAM measurements in this system.

The spectrum shown as 2C occurs in small regions near the end-point of the desorption experiment, when the majority of the sample surface is clean and produces spectrum 2A. The new spectral features (4) and (6) have not been conclusively identified.

in which W is a characteristic energy which, in metals, is comparable in magnitude to the work function [9]. This distribution function has an energy width of 1.1W, which adversely affects the spatial resolution of micro-XANES by a large contribution to the chromatic aberration.

The sampling depth D_{eff} is defined as the distance below the surface from which 1/e of the total secondary yield emitted into the vacuum originates. A calculation of this quantity from first principles requires attention to several electron-electron elastic and inelastic scattering processes which cascade from the initial Auger decay of the core-hole [10]. In particular, D_{eff} is not the same quantity as the inelastic mean-free-path of a primary electron ($\lambda_e(K)$) with the same kinetic energy as the secondary electron. Instead, the major contribution to D_{eff} is the mean-free-path of the original Auger electron, $\lambda_e(K_{Auger})$, which has a kinetic energy of the same order of magnitude as $\hbar\omega$. This means that the sampling depth in soft x-ray micro-XANES studies will be in the range of $10\text{Å} \leq D_{eff} \sim \lambda_e(\hbar\omega) \leq 100\text{Å}$.

The matrix element for the photoabsorption coefficient is of the form

$$\mu(\hbar\omega) \sim | < \psi_{core} | \vec{A} \cdot \vec{p} | \psi_{valence} > |^2 \ \delta(E_{valence} - E_{core} - \hbar\omega) \qquad (3)$$

where both the occupied core-level and the unoccupied valence level may depend on the local chemical environment. In this sense, XANES falls between photoemission and Auger electron emission in terms of the complexity of spectral interpretation. However, it is generally possible to select a 'simple' 1s core-level excitation, so that the major contributions to the XANES spectrum originate in the multiple scattering of the final state [11]. When polarized synchrotron radiation is used as the source, the dipole selection rules inherent in (3) can be used to determine molecular orientations [12].

3.0 EXAMPLES OF THE MICRO-XANES TECHNIQUE IN SURFACE STUDIES

The simple microscope shown in Fig. 1 has been used to evaluate the potential of the micro-XANES technique, and related photoemission microscopies with synchrotron radiation. The spatial resolution of this instrument was limited to approximately 2μ, in part by incomplete AC magnetic shielding and vibration isolation. The topographical images of a variety of surfaces, taken with photon energies from 10 - 200 eV can be found in recent references [4]. In this report we present the results of micro-XANES measurements made during a survey of a number of heterogeneous surfaces.

All of the micro-XANES spectra shown were accumulated under similar conditions. Synchrotron radiation from 3 and 6m TGM monochromators on bending-magnet beamlines at the Aladdin and NSLS synchrotron facilities was used. The μXANES spectra were acquired from sample regions of $5\mu \times 5\mu$ in size, signal averaging for 0.4 s per data point.

Results from a kinetic study of the desorption of SiO_2 from silicon single crystal surfaces is shown in Fig. 2. All spectra are displayed in as-acquired form without further processing, other than to convert the horizontal axis from a wavelength to an energy scale. The bottom curve (A) is from a 5μ diameter region of the Si(111) surface at the endpoint of a desorption experiment. This spectrum is identical to that originally measured by Brown using bulk films [13]. This spectrum establishes the reliability of the micro-XANES detection technique. Each of the principle structures in the clean Si micro-XANES shows a characteristic L_{23} spin-orbit splitting of 0.58±0.05 eV (see features labelled 1, 2 and 3 in Fig. 2). The 2p edges are evident (feature 1), as well as the core-exciton (2) and band-structure peaks (3).

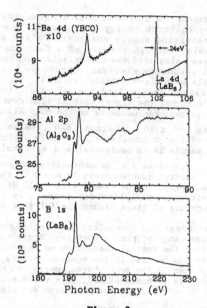

Figure 3.

Example Micro-XANES spectra from sample regions of 5μ diameter. These are raw data, except for the B 1s spectrum which has a polynomial subtracted.

The micro-XANES technique is applicable to a wide variety of materials, as shown in Fig. 3. Again, these are from 5μ diameter regions of the indicated samples. The narrow intrinsic line-width of core-exciton XANES is illustrated in the 4d spectra of Ba (from $YBa_2Cu_3O_{7-\delta}$) and La (from LaB_6).

The total linewidth of 0.24 eV shown for the La 4d line includes an instrumental broadening of 0.13 eV. The Al 2p edge structure is shown for a small particle (about 20μ diameter) of amorphous alumina which inadvertently migrated onto a silicon substrate. The bottom spectrum of Fig. 3 shows the rich structure in the B 1s XANES. The sample is a thin film of sputter deposited LaB_6. As this film is annealed, the relative intensity of the narrow lines from 190-195 eV photon energy are altered. We believe the intensity variation is due to polarization dependences which become apparent as the film crystallizes. The change in the intensity of these B 1s features is sufficient to distinguish individual crystallites in the secondary electron image by an appropriate choice of incident photon energy.

4.0 FUTURE DIRECTIONS

Although the instrumentation used in the feasibility studies is somewhat crude, it has enabled us to quantify the expected performance of an optimized photoemission microscope for spatially resolved spectroscopy. The μXANES technique derives its spectral resolution from the photon energy bandpass and the intrinsic line-width of the photoabsorption matrix element (which can be as small as 0.1 eV). The spatial resolution is determined by electron optics, and the primary factor affecting the ultimate resolution is currently the magnitude of the electron yield. Because of this, the practical spatial resolution of μXANES is much smaller than what can be achieved in micro- photoelectron spectroscopy, simply because the yield of secondary electrons is much higher than that of primary photoelectrons.

In order to overcome the chromatic aberration of the electron optics, the broad secondary electron energy distribution must be filtered by an appropriate image-preserving energy analyser. Polack and Lowenthal have addressed this problem through the use of a Castaing-Henry prism-mirror filter [16]. They have calculated the transmission and spatial resolution of an energy-filtered secondary electron microscope, and find that ≈500Å resolution is possible with a transmission of 1.5% .

In order to reach this level of spatial resolution and still acquire spectra in reasonable times, the higher spatial-flux density of an undulator light source is required. Measurements on the NSLS X13 undulator show that this device produces $1\text{-}4 \times 10^{10}$ photons/$(s\text{-}0.1\%BW\text{-}\mu m^2)$ referred to the source. It is expected that a monochromator with 1% efficiency can demagnify the source onto the sample 2X in horizontal and vertical directions. This results in $\geq 10^8$ photons/$(s\text{-}\mu m^2)$ of monochromatic photons at the sample. If the total secondary electron quantum yield is $\eta \sim 0.1$, then the count rate from a 100Å diameter region on the sample will be 10 Hz, which is adequate for pulse-counting micro-channelplate detectors. The newly installed X1 undulator should produce ~4 times higher spatial flux-density than the X13 insertion device.

One of the advantages of the micro-XAS techniques is that they are applicable over a wide range of photon energies. Since the photon optics are not required to define the spatial resolution on the sample, ordinary grazing incidence mirrors can be used in the monochromators, which allows the use of continuously tunable incident photon energy. An obvious extension of the μXANES experiments is to micro-SEXAFS. This would permit the measurement of molecular structure with Å-scale accuracy from sub-micron size domains on the sample.

REFERENCES

1. A. J. F. Metherell, Adv. in Optical and Electron Microscopy $\underline{4}$, 263 (1971).
2. J. Cazaux, Appl. of Surface Sci. $\underline{20}$, 457 (1985); Ultramicroscopy $\underline{12}$, 321 (1984).
3. J. Kirschner, in Springer Series in Optical Sciences 43, eds. G. Schmahl and D. Rudolph (Springer, Berlin, 1984), p. 308.
4. B. P. Tonner and G. R. Harp, Rev. Sci. Instrum. $\underline{59}$, 853 (1988); J. Vac. Sci. Technol. A, to be published Feb. 1988.
5. O. H. Griffith and G. F. Rempfer, Adv. in Optical and Electron Microscopy $\underline{10}$, 269 (1987).
6. W. Telieps and E. Bauer, Surface Sci. $\underline{162}$, 163 (1985).
7. W. Gudat and C. Kunz, Phys. Rev. Lett. $\underline{29}$, 169 (1972).
8. A. Erbil, G. S. Cargill III, R. Frahm and R. F. Boehme, Phys. Rev. B. $\underline{37}$, 2450 (1988).
9. B. Henke, J. A. Smith and D. T. Atwood, J. Appl. Phys. $\underline{48}$, 1852 (1977).
10. O. Hachenberg and W. Brauer, Advances in Electronics and Electron Physics (Academic Press, New York, 1959), Vol. 11, p. 413.
11. P. Durham, Comp. Phys. Comm. $\underline{25}$, 193 (1982); Sol. State Commun. $\underline{38}$, 159 (1981).
12. J. Stohr, J. L. Gland, W. Eberhardt, D. Outka, R. J. Madix, F. Sette, R. J. Koestner and U. Dobler, Phys. Rev. Lett. $\underline{51}$, 2414 (1983).
13. F. C. Brown and O. P. Rustgi, Phys. Rev. Lett. $\underline{28}$, 497 (1972).
14. A. Bianconi, Surface Sci. $\underline{89}$, 41 (19790; A. Bianconi and R. S. Bauer, Surface Sci. $\underline{99}$, 76 (1980).
15. G. W. Rubloff, K. Hofmann, M. Liehr and D. R. Young, Phys. Rev. Lett. $\underline{58}$, 2379 (1987); R. Tromp, G. W. Rubloff, P. Balk, F. K. LeGoues, and E. J. Van Loenen, Phys. Rev. Lett. $\underline{55}$, 2332 (1985).
16. F. Polack and S. Lowenthal, Springer Series in Optical Sciences (Springer-Verlag, Berlin, 1984), vol. 43, edited by G. Schmahl and D. Rudolph, p. 251.

X-RAY FLUORESCENCE MICROPROBE IMAGING
WITH UNDULATOR RADIATION

MARK L. RIVERS[1,2], STEPHEN R. SUTTON[1,2] AND BARRY M. GORDON[2]
[1] Department of the Geophysical Sciences, The University of Chicago,
5734 S. Ellis Avenue, Chicago, IL 60637
[2] Department of Applied Science, Building 815, Brookhaven National Laboratory, Upton,
NY 11973

ABSTRACT

Synchrotron x-ray fluorescence experiments were performed using a prototype undulator for the Advanced Photon Source installed on the CESR storage ring at Cornell
University during a run in May, 1988. Fluorescence spectra were collected from a number
of standard references and unknowns. Thick target minimum detectable limits (MDL)
were about a factor of two higher than those obtained using white bending magnet radiation at the NSLS. The higher MDLs could be due to lower polarization and/or imperfect
alignment of the Si(Li) detector. Thin target MDLs were about 10 times lower than the
NSLS since the undulator produced a usable spot size which was also 10 times smaller.
Several one dimensional multi-elemental scans and two dimensional images were made
with 10 μm resolution and 30 ppm MDL. These experiments demonstrate that undulators on the proposed Advanced Photon Source will be ideal for a trace element x-ray
fluorescence microprobe with excellent elemental sensitivity and spatial resolution.

INTRODUCTION

Undulator radiation from high energy storage rings should make an ideal source
for a synchrotron x-ray fluorescence (XRF) microprobe [1]. The radiation is extremely
intense, highly collimated in both the horizontal and vertical directions and is monochromatic with a bandwidth of about 1%. High energy storage rings are required to produce
hard x-rays (>10 keV) from an undulator because the relatively low magnetic field of
an undulator produces lower energy x-rays at a given electron or positron energy than
a bending magnet or wiggler. Suitable storage rings include the CESR ring at Cornell
University, PEP at Stanford University, the proposed Advanced Photon Source (APS) at
Argonne National Laboratory and the proposed European Synchrotron Radiation Facility
(ESRF) in Grenoble. The high intensity and collimation of undulator radiation should
permit one to construct a useful microprobe using only pinholes, and/or small and easily
fabricated optical elements to produce beam diameters below 1 μm. The spectral characteristics of the undulator source are ideal for synchrotron XRF. The radiation consists of
a fundamental peak and a number of harmonics. The width of these peaks depends upon
the emittance of the storage ring and the physical details of the undulator, but in general
the peaks have widths of a few percent in $\Delta E/E$. Figure 1 shows the predicted brightness
(photons/sec/mrad^{-2}/100 mA/0.1of an undulator with 61, 3.3 cm periods installed on
CESR and on the proposed APS storage ring. Also plotted is the brightness of a bending
magnet on the x-ray ring at the National Synchrotron Light Source (NSLS). For collimated XRF microprobe experiments two characteristics of the spectrum are critical. The
first is clearly the brightness, since higher brightness permits one to use smaller beam
diameters. The ratio of the beam brightness above an element's absorption edge to the

brightness at its fluorescent energy is also important, since photons above the edge are required to excite the fluorescence, while those near the fluorescent peak form the background by elastic and inelastic scatter into the detector. In the bending magnet spectrum, even with absorbers in the beam, the spectrum is quite flat and this ratio near unity. The undulator spectra are much better, since they have above/below edge ratios as high as 10^4. By varying the gap on the undulators, the energies of the peaks and valleys can be changed to optimize the sensitivity for particular elements.

EXPERIMENTS

Experiments were conducted at the Cornell High Energy Synchrotron Source (CHESS) during a dedicated undulator running period at the end of May, 1988.

Storage ring

The CESR storage ring was run in a low emittance mode to optimize the undulator output. The electron beam size and divergence in the undulator straight section were estimated to be σ_x=1000 μm, σ_y=60 μm, σ'_x=50 μrad, σ'_y=11 μrad. The electron energy was 5.37 GeV and the current during the XRF experiments was only 2-5 mA. The low currents were sufficient to produce optimum count rates on the Si(Li) detector. Higher ring currents resulted in unsafe radiation levels outside the hutch, probably due to bremsstrahlung radiation produced in the undulator straight section and scattered off apparatus in the hutch.

Undulator

The undulator used was a prototype built for the APS. It consists of 61 periods, with a 33 mm period. The preliminary design at the APS calls for this device to be tunable through a K range of 0.4 - 2.5. The vacuum chamber at CHESS was larger, and the available K range was about 0.6 to 1.35, resulting in fundamental energies of 4.4 - 7.1 keV.

Fluorescence apparatus

The experiments were performed in the A3 hutch, with the sample approximately 21.5 m from the undulator source. The experimental apparatus used to collect the fluorescence spectra was similar to that described previously for the collimated x-ray microprobe on beamline X-26 at the NSLS [2]. Samples were mounted on an X-Y-Z-θ stepper motor stage with 1 μm precision. The target was observed with a Nikon Optiphot microscope with a 5X objective, reflected light illumination and a TV camera for viewing outside the hutch. A Si(Li) detector was mounted at 90 degrees to the incident beam, in the horizontal plane of the storage ring. This geometry minimizes the scattered background by taking advantage of the polarization of the synchrotron radiation.

A set of primary slits 18.5 m from the source, 3 m upstream from the sample, was set to 100 μm width in the horizontal and vertical directions. Assuming that the 2σ source dimensions are 2000 μm and 120 μm in the horizontal and vertical directions respectively, these slits will produce a beam spot at the sample with dimensions approximately 450 μm in the horizontal and 140 μm in the vertical. This was confirmed by measuring the image of the beam on x-ray sensitive film.

Several approaches were tried to further collimate the beam with the goal of produc-

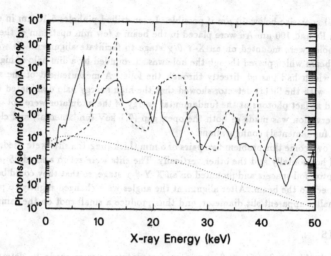

Figure 1. Solid curve: predicted central brightness of an undulator with 61 periods, 33 mm period, at K=0.6, installed on the CESR storage ring running at 5.37 GeV. Dashed curve: predicted central brightness of a similar undulator with 158 periods, 33 mm period at K=0.4 installed on the APS running at 7 GeV. Dotted curve: predicted central brightness of a bending magnet on the x-ray ring at the NSLS running at 2.5 GeV.

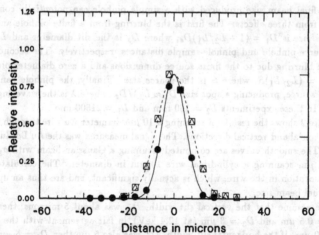

Figure 2. Circles: measured Au Kα fluorescence intensity scanning a 10 μm diameter Au wire vertically through the beam. Solid curve: computed intensity of a Gaussian beam with σ=4.5 μm passing over the wire. Open squares: measured intensity of horizontal scan of the wire. Dashed curve: predicted intensity of a Gaussian beam with σ=7 μm passing over the wire.

ing a final spot size below 10 μm if possible. Laser drilled pinholes 5-10 μm in diameter in 25 μm Pt and 100 μm Au were placed in the beam a few mm upstream of the sample. The pinholes were mounted on an X-Y-θ-χ stage to facilitate alignment. It was found that the beam which passed though the hole was surrounded by a dim halo of high energy photons which had passed directly through the foils. A measurement of the scattered spectrum with the Si(Li) detector showed that the high energy halo dominated the spectrum, and in fact photons at the fundamental energy of the undulator were not observed. This observation was made in both the open gap (7.1 keV fundamental) and closed gap (4.4 keV fundamental) configurations.

To overcome this problem two pairs of 6 mm thick tungsten slits were used, one pair mounted horizontally and the other vertically. The slits were set to a fixed 25 μm width with Kapton foil spacers and mounted on an X-Y-θ-χ stage, so that they could be aligned with respect to the beam. After alignment the angles were changed slightly to produce a much smaller apparent slit diameter, and thus produce a small spot on the sample.

RESULTS

With this configuration fluorescence measurements were made to determine the beam size. In addition a number of reference materials and unknown samples of geochemical and biological interest were measured, and two scanning fluorescence images were made.

Beam size

The final beam size produced with a simple pinhole can potentially contain contributions from three effects. The first is the blurring from a finite pinhole and a point source. Its size is $D_1 = (1 + L_2/L_1)D_s$, where D_s is the slit diameter and L_1 and L_2 are the source-pinhole and pinhole-sample distances, respectively. The second effect is penumbral blurring due to the finite source dimensions and a zero diameter pinhole. Its size is $D_2 = (L_2/L_1)S$, where S is the source size. Finally, the pinhole can act as a diffraction source, producing a spot size $D_3 \approx L_2\lambda/D_s$, where λ is the wavelength of the radiation. In these experiments $L_2 = 160$ mm and $L_1 = 21500$ mm.

Figure 2 shows the results of scanning a 10 μm diameter Au wire through the beam in the horizontal and vertical directions. The signal measured was the Au Lα fluorescence intensity. The smooth curves are computed assuming a Gaussian beam with σ_y=4.5 μm and σ_x=7 μm scanning a cylindrical wire 10 μm in diameter. The calculated curves neglect absorption in the wire, which is actually significant, and are thus an upper limit on the actual beam size.

If we assume that the vertical slit width, D_s, was about 5 microns, then $D_1 = 5$ μm, $D_2 = 0.5$ μm and $D_3 = 3$ μm (at 12.4 keV), in fair agreement with the measured fwhm of 10 μm. If the horizontal slit width, D_s, was also 5 μm then $D_1 = 5$ μm, $D_2 = 7$ μm and $D_3 = 3$ μm, again in agreement with the measured fwhm of 16 μm. These beam sizes can be reduced by decreasing L_2 and D_s, perhaps producing a spot as small as 1 μm.

Fluorescence experiments

Measurements on a number of standard materials, including NBS glasses, synthetic sulfides and thin foil standards resulted in thick target detection limits of a few ppm, and

289

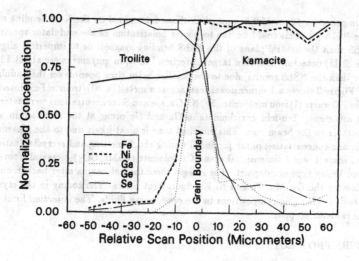

Figure 3. One dimensional scan across an interface between troilite (FeS) and kamacite (Fe-Ni metal) in the Four Corners (I) iron meteorite. Ten-fold enrichments in Ga and Ge occur at the grain boundary in a band comparable to the beam size of 10 μm.

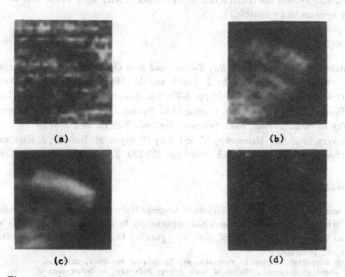

Figure 4. 31 × 31 scanning x-ray fluorescence image of a zoned dolomite crystal $(CaMg(CO_3)_2)$ from a Mississippi Valley type ore deposit. Pixel size = 10 μm × 10 μm, 2 second collection time per pixel. (a) Ca K α, (b) Mn K α, (c) Fe K α, (d) Ni K α.

were in general a factor of 2-3 worse than those determined on the X-26 bending magnet at the NSLS [3]. This could be due to poorer polarization of the undulator spectrum at CHESS than the central plane of the NSLS bending magnet, or to imperfect alignment of the Si(Li) detector. The thin target detection limits, in gm/cm^2, were about 10 times better than the NSLS results, due to the smaller beam sizes possible on the undulator.

Figure 3 shows a 1-dimensional scan across a metal(Fe-Ni)/troilite(FeS) interface in the Four Corners (I) iron meteorite. Fe, Ni, Ga, Ge and Se concentrations were determined in 10 μm steps. Ten-fold enrichments in Ga and Ge occur at the interface in a band comparable to the beam size. This enriched zone is most likely due to the presence of a siderophile-concentrating phase produced during shock melting and recrystallization.

Figure 4 is a 2-dimensional scan of a dolomite $(CaMg(CO_3)_2)$ crystal from a Mississippi Valley-type ore deposit. The images show the net counts after background subtraction for the Ca, Mn, Fe and Ni Kα fluorescent peaks. The zoning in the crystals is the result of changing compositions in the ore-forming fluids. The detection limit in this image is about 30 ppm.

FUTURE PROSPECTS

These experiments demonstrate that the undulator allows a tenfold reduction in beam area, with 100 times less ring current, than the NSLS bending magnet. The proposed APS at Argonne will be an even better source (Figure 1). The higher machine energy will permit the undulator fundamental to be tuned to much higher energy, $\approx$13 keV, and thus permit one to excite the K or L lines of all elements with the fundamental or second harmonic. The lower source emittance will also produce a much cleaner spectrum, with large peak/valley ratios and higher brightness. With the APS and a relatively simple focussing system the construction of a sub-micron XRF microprobe with sub-ppm sensitivity appears to be feasible.

REFERENCES

1. M.L. Rivers, in *Synchrotron X-Ray Sources and New Opportunities in the Earth Sciences: Workshop Report*, edited by J. Smith and M. Manghnani, (Argonne National Laboratory Report ANL/APS-TM-3) pp. 5-22; S.R. Sutton *et. al.*, *ibid.*, pp. 93-112.

2. A.L. Hanson, K.W. Jones, B.M. Gordon, J.G. Pounds, W. Kwiatek, M.L. Rivers, G. Schidlovsky and S.R. Sutton, Nucl. Instrum. Methods B24-25, 400, 1987.

3. M.L. Rivers, in *X-Ray Microscopy II*, edited by D. Sayre, M. Howells, J. Kirz, and H. Rarback (Springer-Verlag, New York, 1988) pp. 233-239; S.R. Sutton, *ibid.* pp. 438-431.

ACKNOWLEDGMENTS

We would like to thank P. J. Vicarro of Argonne National Laboratory and D. Bilderback of CHESS for providing us with the opportunity to use the undulator. We would also like to thank all of the CHESS staff who gave up their Memorial Day weekend to make this run possible.

*Work supported in part by Processes and Techniques Branches, Division of Chemical Sciences, Office of Basic Energy Sciences, US Department of Energy, under Contract No. DE-AC02-76CH00016; Director's Office of Energy Research, Office of Health and Environmental Research, US Department of Energy Contract No. DE-AC03-76SF00098; applications to biomedical problems by the National Institutes of Health as a Biotechnology Research Resource under Grant No. P41RR01838; applications in geochemistry by National Science Foundation Grant No. EAR-8618346; and applications in cosmochemistry by NASA Grant No. NAG 9-106.

PART VI

Abstracts of Unpublished Papers

Abstracts of Unpublished Papers

X-RAY SCATTERING STUDIES OF THE MELTING OF LEAD

SURFACES. P. H. Fuoss , L. J. Norton, AT&T Bell Laboratories, Holmdel, NJ 07733; S. Brennan, Stanford Synchrotron Radiation Laboratory, SLAC Bin 69, P. O. Box 4349, Stanford, CA 94309.

Of fundamental importance in understanding melting is the question of premelting (i.e. do surfaces melt at a lower temperature than the bulk). We report the results of an x-ray scattering experiment that has directly measured liquid correlations in Pb(110) and Pb(111) surfaces at temperatures substantially below the bulk melting point of Pb. Surprisingly, significant liquid scattering is observed at $\approx 230^\circ K$ or $< 1/2 T_m$ on both the (110) and the (111) surfaces. On the (110) surface, analysis of the scattered intensity yields 2-3 monolayers of liquid from $575^\circ K$ to $599^\circ K$ while substantially less than 1 monolayer of liquid is present on the (111) surface at $599^\circ K$.

METALS IN BETA-RHOMBOHEDRAL BORON.* Joe Wong, Lawrence
Livermore National Laboratory, P.O. Box 808, Livermore, CA
94550; Glen A. Slack, General Electric Corporate Research
& Development, P.O. Box 8, Schenectady, N.Y. 12301.

The bonding and local atomic structure of a series of 3d
metal-beta boron solid solution are investigated using a
combination of x-ray absorption near-edge structure
(XANES) and extended fine structure (EXAFS) technique
utilizing intense synchrotron radiation as a light
source. The corresponding metal diborides MB_2, (M = Sc,
Ti, V, Cr) were also measured and used to model the coordination environment of these metal sites in beta-boron.

As the d-states are progressively filled up in going
across the 3d series from Ti to Cu, the pre-edge feature
at 1-2 eV becomes less well-defined as is clear from the
XANES of Mn, Fe, and almost disappears in the spectra of
Co, Ni and Cu. The transition at 20 eV is strong and
remains quite similar across the 3d series, as expected,
since the final states are 4p in character, which are
empty in all the 3d elements and is dipole allowed. It is
noted that Cu exhibits a relatively narrower 4p transition
at lower energy (17 eV).

The XANES spectral variation are also studied as a
function of dopant concentration for the case of Ti, V,
Cr, Fe, Co, Ni and Cu. Finally, doubly doped (V/Cu)
system is studied and compared with the singly-doped V and
Cu materials.

*Work performed under the auspices of the U.S. Department
of Energy by the Lawrence Livermore National Laboratory
under contract number W-7405-ENG-48.

EXAFS OF NEAR MONOLAYER AND SUBMONOLAYER COVERAGE FILMS.

Troy W. Barbee, Jr., Joe Wong, Lawrence Livermore National
Laboratory, Livermore, CA 94550.

The high intensities available using synchrotron source
insertion devices potentially extend the range of adatom
surface coverage on well prepared surfaces that may be
experimentally studied to very low levels. In this paper
we report the results of a fluoresence EXAFS study of a
0.5 nm thick hafnium film in the vicinity of its L_{III}
absorption edge (9561 eV). This film was deposited on a
platinum-carbon multilayer having a period of 31.7Å.
These results indicate that EXAFS studies of species at
surface coverages of 0.01 monolayer are easily experimentally accessible.

Work performed under the auspices of the U.S. Department
of Energy by Lawrence Livermore National Laboratory under
contract #W-7405-Eng-48.

IN SITU STUDY OF THE BONDING OF IMPURITIES IN METAL-IMPURITY-METAL LAMINATE COMPOSITES.

E. V. Barrera, Department of Materials Science & Engineering, University of Pennsylvania, Philadelphia, PA 19104; S. M. Heald, Materials Science Department, Brookhaven National Laboratory, Upton, Long Island, NY 11973; H. L. Marcus, Center for Materials Science & Engineering, The University of Texas at Austin, Austin, TX 78712.

The local structure of interface or near interface impurity elements has been observed by means of extended x-ray absorption fine structure spectroscopy (EXAFS). Multilayered samples were prepared using reactive ion sputtering, electron beam evaporation , and molecular beam deposition methods with vacuum conditions ranging from 10^{-5} to 10^{-9} Pa. The interface layers of titanium, gallium, and arsenic were deposited in partial to multiple atom layers alternating with 5 to 10 nm thick matrix layers of nickel, cobalt, or aluminum. Oxide phases, atoms in solution, and local ordering were identified. Ti oxides and silicon arsenide were used as standards for the EXAFS analysis. Auger electron spectroscopy complemented the EXAFS analysis.

Support was provided by the Office of Naval Research under contract no. N00014-83-K-0143 and EXAFS experimets were conducted at Cornell High Energy Synchrotron Source at Cornell University and the National Synchrotron Light Sourse at Brookhaven National Laboratory. Support for NSLS beam line X-11 was provided by Department of Energy under contracts DE-AS05-80-ER10742 and DE-AC02-76CH00016.

X-RAY SCATTERING ABOVE 100KeV.* J. B. Hastings,
National Synchrotron Light Source, Brookhaven National Laboratory, Upton NY 11973

Conventional x-ray scattering studies have been
limited to photon energies (wavelengths) in the 5
to 20 KeV (approx. 2A to 0.5A) regime. With these
energies absorption lengths limit the volume of
illumination to the first tens of microns of
samples. If it were possible to use x-rays of
very high energies, true bulk (tens of millimeters) samples could be studied. The availability, intensity and resolution possible with
high energies will be discussed and their role in
the expanding field of x-ray scattering presented.
Preliminary studies at the Cornell High Energy
Synchrotron Source (CHESS) will form the basis of
these discussions.

*This work was performed under the auspices of the
U.S. Dept. of Energy, under contract no. DE-AC02-
76CH00016.

EPITAXIAL LAYERS OF FCC FE ON CU(001)

R.F.Willis,R.Morra,F.d'Almeida,G.Mankey,M.Kief
Dept.Physics,Penn State University,Pa. 16802

Metastable films of fcc Fe epitaxed to Cu(001)
show unusual ferromagnetic properties(1). Changes
in the lattice volume occur with film thickness in
the 1 to 5 monolayer range, reflecting changes in
the magnitude of the spin moments and the magnetic
exchange coupling. The associated uniaxial strain
is manifest in strong magnetic anisotropy, layer
by layer lattice expansion, phonon softening, and
surface layer atomic reconstructions. We present
LEED results of a detailed structural investigation,
neutron reflectance measurements of the spin
moments, surface magneto-optic Kerr effect studies
of the magnetic anisotropy, and angle resolved
photoemission of the electronic states.
The observed effects are due to the strain-related
kinetic energy associated with the formation of
the atomic spin moments, predicted by spin-density
-functional calculations of bulk fcc Fe(2).

(1) R.F.Willis,J.A.C.Bland,W.Schwarzacher, J.Appl.
    Phys. 63,4051 (1988)
(2) V.L.Moruzzi, Phys. Rev. Letters 57,2211 (1986)

MULTILAYER MONOCHROMATORS FOR THE SOFT X-RAY AND EXTREME
ULTRAVIOLET. <u>Troy W. Barbee, Jr</u>., Lawrence Livermore
National Laboratory, Livermore, CA 94550; Piero Pianetta,
Stanford Synchrotron Radiation Laboratory, Stanford, CA
94309.

Simple multilayer structures and multilayer diffraction
gratings are now of sufficient quality to be used as
optical elements in synchrotron radiation source instrumentation. In this paper results obtained with a multilayer two element monochromator will be presented. Three
specific types of results will be discussed. First, transmission measurements of the absorption cross-sections of
elemental thin films in the energy range 50 to 2000 eV
will be presented and used to demonstrate the performance
of the monochromator. Second, application of this monochromator in x-ray lithography research will be described
and the advantages of the broad bandpass of multilayer
optics demonstrated. Third, use of this monochromator in
scattering studies of long period structures will be
discussed. The potential for the use of multilayer
diffraction gratings in high resolution monochromator
applications will also be considered.

Work performed under the auspices of the U.S. Department
of Energy by Lawrence Livermore National Laboratory under
contract #W-7405-Eng-48.

Author Index

302

Subject Index

Printed in the United States
By Bookmasters